Integrating Agriculture and Aquaculture

Integrating Agriculture and Aquaculture

Editor: Geoffrey Gilbert

www.callistoreference.com

Callisto Reference,
118-35 Queens Blvd., Suite 400,
Forest Hills, NY 11375, USA

Visit us on the World Wide Web at:
www.callistoreference.com

ISBN: 978-1-63239-877-2 (Hardback)

The publisher's policy is to use permanent paper from mills that operate a sustainable forestry policy. Furthermore, the publisher ensures that the text paper and cover boards used have met acceptable environmental accreditation standards.

Trademark Notice: Registered trademark of products or corporate names are used only for explanation and identification without intent to infringe.

Printed in the United States of America.

Cataloging-in-Publication Data

Integrating agriculture and aquaculture / edited by Geoffrey Gilbert.
    p. cm.
Includes bibliographical references and index.
ISBN 978-1-63239-877-2
1. Integrated agricultural systems. 2. Integrated aquaculture. 3. Agriculture. 4. Aquaculture.
5. Agricultural systems. 6. Agricultural engineering. I. Gilbert, Geoffrey.
S494.5.S95 I58 2017
630--dc23

Table of Contents

Preface

The purpose of the book is to provide a glimpse into the dynamics and to present opinions and studies of some of the scientists engaged in the development of new ideas in the field from very different standpoints. This book will prove useful to students and researchers owing to its high content quality.

Aquaculture is the cultivation and harvesting of marine organisms under controlled conditions for commercial purposes. This book on integrated agriculture and aquaculture talks about landscape and cultivation practices that serve to improve aquaculture and marine harvesting. Biological pest control mechanisms are another area of great importance for maintaining healthy yield. Farming of aquatic fish is sometimes complemented with other organisms such as crustaceans, molluscs and algae. It picks up individual branches and explains their need and contribution in the context of a growing economy. This book brings forth some of the most innovative concepts and elucidates the unexplored aspects of aquaculture. It will serve as a valuable source of reference for graduate and post graduate students.

At the end, I would like to appreciate all the efforts made by the authors in completing their chapters professionally. I express my deepest gratitude to all of them for contributing to this book by sharing their valuable works. A special thanks to my family and friends for their constant support in this journey.

Editor

Analysis of effluent discharge in to natural forest in Bangladesh

Awal, Mohd Abdul

Environmental Scientist, Ministry of Environment and Forest, Health & Pollution Research Farm, Long Island City, New York, USA

Email address:

abdul-awal2004@yahoo.com

Abstract: Natural forest resources like Sundarbans mangroves in Asia including Bangladesh, India, and East Africa previously contained a much fuller range of species (Seidensticker, and Hai, 1983; Khan, 1997). In the Southeast Asian region, species diversity of mangroves was previously much higher, where approximately two-thirds of all species and 70% of the major vegetation types with 15% of terrestrial species in the Bangladesh-India-Malayan realm have already been destroyed (Ellison, 1998, 2000). Despite this designation, this natural forest resources (Sundarbans) in Bangladesh has been facing tremendous problems, including that of dieback (top-dying), shrimp farming, human destructions, deforestations, illicit fellings, miss-management of the main tree species (*Heritiera fomes*) which is affecting millions of trees (Awal, 2007). The cause of this dieback is still not well understood unknown. The present work has investigated one of the possible factors that might be causing this top-dying, namely the concentrations of various chemical elements present in the sediments, particularly heavy metals, though other chemical parameters such as the pH, salinity, moisture content of the sediment and nutrient status were also assessed. A questionnaire survey was conducted among different groups of people inside and outside of Sundarbans to explore local perceptions as to the possible causes of top dying. This confirmed the increase in top-dying prevalence (Awal, 2007). Despite various hypotheses as to the causes of this top-dying, the underlying causes are still not well understood. The present work has explored some of the possible factors involved, focusing particularly on the relationship between the amount of top-dying in different places and the concentrations of a number of chemical elements present in the soil and water, in order to test the hypothesis that chemical pollution might be responsible. Other factors such as the pH, salinity and nutrient status were also assessed. The vegetation structure was assessed in terms of tree height, bole diameter, species present, and regeneration status; and the intensity of top-dying within the plots was recorded on a rank scale. Most of the elements studied had no significant correlation with the top dying of *Heritiera fomes*. However, Sn, Exchangeable K, and soil pH were significantly related, and three elements, namely Pb, Zn, Ni, were also close to significance. Sn concentration is negatively associated with top dying. Soil pH varied significantly in the different plots. Exchangeable K was positively associated with the tree diameter whether the top dying was severe or mild (Awal, 2007).

Keywords: Shrimp Farming, Chemical Contamination, Abnormal Elemental Concentration, Chemical Contamination, Health Problems, Causal Factors, Heavy Metal Concentrations, Pollution, *Natural Resources Degradations,* Sundarbans, Top-Dying

1. Introduction

Bangladesh is literally a treasure-trove of rich and variegated natural beauty interspersed with enchanting landscape, mighty meandering rivers, exotic flora and fauna, picturesque resorts, long sunny beaches, tropical natural mangrove forests, fascinating art and architecture, ancient relics and archaeological sites and colorful tribal life. As a vacation land, Bangladesh has many facets like Sundarbans natural forest resources in Bangladesh Mosaic of Bangladesh; 2006; External Publicity Wing; Ministry of Foreign Affairs; Government of the People's Republic of Bangladesh; p: 1-145). However, the Sundarbans is the largest single mangrove forest in the world (*figure 1.0*), occupying about 6,029 km^2 in Bangladesh and the rest in India (Iftekhar & Islam, 2004). The Sundarbans supports a diverse fauna and flora (e.g. Prain, 1903; Siddiqi *et al.*, 1993, Iftekhar, 2006), approximately one million people of Bangladesh and India depend on it directly for their livelihood (Iftekhar & Islam,

2004), and also it provides a critical natural habitat which helps protect the low lying country and its population from natural catastrophes such as cyclones (e.g. Blasco *et al.*, 1992; Iftekhar, 2008).

It is a land of enormous economic potentials, inhabited by diligent and hard-working people who have a love for heritage. All of these together make Sundarbans a colorful mosaic of nature's splendor and bounty (Awal, 2007, 2009, 2014).

1.1. Natural Forest Resources in Bangladesh

History records that till the advent of the British in the eighteenth century; Bangladesh had an enviable position in the entire region and was known as the legendary land of affluence and prosperity. The country has almost achieved green revolution. Sub-sectors like fishery, livestock rearing, and forestry are growing by leaps and bounds. The Bay of Bengal is literally a treasure- trove of sea fish and other wealth. Bangladesh's major natural forest resource is Sundarbans. There is also bright prospect of striking oil and gas. Huge deposits of coal, limestone, peat, bitumen, hard rock, lignite's, white clay etc. have already been indentified and projects are being implemented for their harnessing for productive use(Mosaic of Bangladesh; 2006; External Publicity Wing; Ministry of Foreign Affairs; Government of the People's Republic of Bangladesh; p:1-145).

1.2. Destruction of Natural Forest

Coastal lands cover 6% of the world's land surface (Tiner, 1984). Coastal and wetlands everywhere are under threat from agricultural intensification, pollution, major engineering schemes and urban development, (UN-ESCAP 1987; 1988). Mangroves in Asia including Bangladesh, India, and East Africa previously contained a much fuller range of species (Seidensticker, and Hai, 1983; Khan, 1997). In the Southeast Asian region, species diversity of mangroves was previously much higher, where approximately two-thirds of all species and 70% of the major vegetation types with 15% of terrestrial species in the Bangladesh-India-Malayan realm have already been destroyed (Ellison, 1998, 2000). The Indo-Pacific region is known for its luxuriant mangroves. The mangrove zone of Bangladesh is about 710 km long including several tiny islands (Rahman, *et al.,* 2003). In the present day the Indo-Malayan mangroves are confined to Sundarban reserved forests, mainly in Bangladesh (figure 1.0). According to Miller *et al.* (1985, 1981), this forest had been affected by direct human destructions, human settlement and agricultural activities during and under both the Bengal Sultanate (1204-1575) and the Mughal Empire (1575-1765). At the arrival of British rule in 1765, the Sundarbans forests were double their present size and significant exhaustion of the growing stock led to dwindling by 40% - 45% between 1959 and 1983 (Chaffey *et al.,* 1985).

1.3. Shrimp Farming

Sundri *(Heritiera fomes)* and other important floral species

in surrounding areas of Sundarbans have probably been adversely affected by the establishment of shrimp farms for shrimp cultivation (Personal observation, 1993-97; Currie, 1984). Massive extraction of wild post larvae shrimp affects the stocks of natural resources. Satellite images show the expansion of the shrimp farming industry within the wetlands of the Gulf including Sundarbans recently, which changes the mangroves, lagoons and estuaries, and results in a net loss of habitat for native and migratory birds, fishes, mollusks, crustaceans, mammals, biological diversity and aquatic resources severely (Personal observation, 1994-1998). Along with habitat changes, biodiversity is strongly impacted by the practice of catching post larval shrimp along with any other accompanying fauna, and separating out the shrimp while exterminating the other fauna by allowing the rest of the catch to die on the ground, or applying chemicals that do not harm shrimp but kill other species (Personal observation, 1993-98). Post larvae shrimp usually make up approximately 10% of the catch; leaving 90% to be exterminated (reviewed by FAO, 1982). Shrimp culture might be associated with deteriorating mangrove conditions in the western part of the Sundarbans (Chaffey, Miller and Sandom, 1985).

According to Phillips, 1994, tropical shrimp farming has a long history, dating back at least 400 years (e.g. the 'tambaks' of Indonesia, 'bheris' and 'gera' of Bangal and tidal ponds in Ecuador) the expansion of the industry over the last 15 years has been extremely rapid and its environmental impact is now the subject of grave concern (Phillips, 1994). In the Khulna District of Bangladesh, artificial shrimp farms (locally called ger) are flooded with brackish water in the dry season months for shrimp culture (Nuruzzaman, 1990). For economic reasons associated with the high price of shrimp, such partial or complete switches from rice farming to aquaculture are putting further pressure on the remaining mangroves. Thailand has lost a total of 203,000 ha, 52% of the mangrove resource, due to shrimp-culture since 1961 (Anon, 1975; Briggs, 1991). Similar events are taking place in many other areas of the world; in Indonesia, most of the 300,000 ha of land being used to culture shrimp was ex-mangrove forest and the government of that country is planning to raise this figure to more than 1 million ha. By 1985, Java had lost 70% of its mangroves, Sulawesi 49% and Sumatra 36% (Anon, 1975). A similar scenario exists in the Philippines where mangrove areas have shrunk from 448,000 ha in 1968 to 110,000 ha in 1991 and such destruction has had a devastating effect on coastal fisheries and has led to the marginalization of subsistence fisherman and the erosion of shorelines (Singh, 1988).

It is estimated that 100 organisms (Personal observation in Sundarbans as Head of East Wildlife Sanctuary from 1993 to 1998) are destroyed for every shrimp-fry collected to supply extensive shrimp ponds ('gher') in Bangladesh; as many as 80% of the people in some coastal areas of the country are engaged in aquaculture seed collection (Personal observation, 1995-97). There is growing evidence that the environmental impacts of shrimp farming play a significant role in disease outbreaks and subsequent crop loss, as a result of the overloading of the carrying capacity of the environment (Personal observation, 1995-97; Phillips *et al.,* 1994). The result of this is an accumulation of wastes in the surrounding ecosystems which may lead to severe and sometimes irreversible problems (Personal observation, 1993-98). In the Khulna District of Bangladesh, poldered rice fields ('gher') are flooded with brackish water in the dry season months for shrimp culture, then a rice crop is grown in the wet seasons when the field can be flushed with freshwater (e.g. Nuruzzaman,1990). In Asia large tracts of back mangrove were cleared initially for agriculture, especially rice farming (reviewed by FAO, 1981, 1982).

1.4. Effluent Discharge

Effluent discharge from intensive shrimp farming has been put at 1.29 billion cubic meters of effluent per year in Thailand (FAO Report, 1997), although it has not been assessed in Bangladesh. However, effluent discharge is a problem of Sundarbans (Personal communication Mongla Port Authority, Khulna, 1996).

A large number of industries are discharging untreated effluents directly into the river at Khulna which is carried down to the Sundarbans forests. The polluting industries are Khulna Newsprint Mill (KNM), Hardboard Mills, and some match factories, fish processing units, Goalpara power station, some Jute mills and Khulna shipyard. KNM alone continuously discharge nearly 4500 m^3/ ha of waste water containing high levels of suspended solids (300-500mg/l) and sulphur compounds (UN-ESCAP, 1987). Moreover, resuspension of dredging material for port development is a potential threat to the Sundarbans due to long lasting toxicological effects from heavy metal pollution.

1.5. Indirect Human Influence

Sundarbans plays a vital role for human survivability from cradle to grave including tangible and intangible benefits. Coastal lands include some of the most productive ecosystems and have a wide range of natural functions. Wetlands are also one of the most threatened habitats because of their vulnerability and attractiveness for 'development'. The first global conservation convention (Ramsar Convention) focused solely on coastal lands and wetlands like Sundarbans, and this Convention has recently been strengthened and elaborated with regard to the wise use of all coastal forests such as Sundarbans.

Sundarbans protects people and resources from strong tidal surges, hurricanes, tides and from waves. Progressive reclamation of the Sundarbans over the last 150 years has resulted in the loss of substantial masses of mangrove forests.

Bangladesh is a poor country, the size of Wisconsin, bursting with a population nearly half (of the total population) of the United States. On top of rampant illiteracy, poverty, and disease, the country suffers year after year from devastating natural disasters (Emilie,R; and Sandhya, S, 2006).

The major portion of the land is low with a maximum height of 10 m above mean sea level (Alam, 1990). Every year, tornados strike at peoples, buildings, plants and wildlife in the coastal belt at the beginning and end of autumn season. Although less obvious than habitat loss, the indirect effects of agriculture on mangroves, though the diversion of freshwater by agriculture irrigation schemes, or run–off of agricultural chemical residues into mangroves, have also been significant factors associated with deteriorating mangrove conditions in the Indus Delta and in the western part of the Sundarbans (Chaffey, Miller and Sandom, 1985). Although very little information is available, there is great concern in Asia regarding environmental impacts from agricultural pesticides, some of which are known to be highly toxic to shrimp (Phillips, 1994), as well as concern about the consumption of aquaculture products which can expose consumes to high levels of contaminants (Pullin, 1993). Bangladesh experiences many kinds of pollution (Emilie,R; and Sandhya, S, 2006), and this is likely to affect ecosystems as well as human health. Bangladesh is the one of the most densely populated countries (World Fact book, July 2006). Bangladesh is facing many problems including environmental pollution, high population density and poverty, these problems being interlinked with each other. These unique coastal tropical forests are among the most threatened habitats in the world. They may be disappearing more quickly than inland tropical rainforests, and so far, with little public notice. The Sundarbans provide critical habitat for a diverse marine and terrestrial flora and fauna, and 3.5 million people depend on Sundarbans forests and waterways for their survival (Anon, 1986; Chaffey *et al*, 1985). Enrichment and illicit removal of timber and firewood from the forests are the major forest conservation problems in Sundarbans. Approximately 2.5 million people live in small villages surrounding the Sundarbans, while number of people within 20 km of the Sundarban boundary was 3.14 million (Islam, 1993). The annual average destruction of forest land in the country was 8000 ha in 1980 and subsequently it increased to 38000 ha in 1981-90 according to FAO (1993). But probably the rate of destruction of forest is more severe than the official statistics as it is very difficult to estimate the real picture (Awal, 2007}. Deforestation affects one eighth of the country's land areas (Awal, 2007). Approximately 100,000 to 200,000 people work inside the Sundarbans for at least 6 months, while the number of people entering the forest in a year could be as high as 3,000,000 (Hussain and Karim, 1994). Of these, about 25,000 people work in fish drying and 60,000 – 90,000 people in shrimp post-larvae collection inside the Sundarbans (FAO, 1994). About one million people are engaged in shrimp larval collection in the rivers and creeks around the outside of the Sundarbans (Chantarasri, 1994).

The Sundarbans, like other tidal forests, is tolerant of natural disturbances such as the cyclones and tidal waves of the Bay of Bangle, but it is highly vulnerable to human disturbances (Seidensticker, 1983). Most of the abuses found in professional forestry management elsewhere have been observed here as well, such as excessive cutting of stocks in the auctioned area and connivance between purchasers and forestry staff to cut wider areas than sanctioned (Bari, 1993). 157 major oil spills in tropical seas between 1974 and 1990 (Burns *et al.,* 1993). Deep mud coastal habitats may take 20 years or more to recover from the toxic effects of such oil spills (Phillips, 1994).

2. Methodology

In this section the various field and laboratory methods used in this study will be discussed as below:

2.1. Field Sampling Methods

The Sundarbans Reserved Forest is located at the south west corner of the Ganges River Delta close to the Bay of Bengal, mainly on the sea-shore line, river banks, channels, and small creeks *(figure 1.0)*. The location of the Sundarbans within Bangladesh has been shown in *Figure 3.0.*

Figure 3.0. Photograph of part of the Sundarbans coastal area, near compartment number 26, where the trees have been cut down to be replaced by a fishing village.

2.1.1. Site Selection and Location of the Study Area

General reconnaissance of possible sites was made by visiting all the possible regional areas before categorizing and selecting plots for sampling. It was decided to sample from the Chandpai area which is the mostly human accessible and ecologically polluted area (in *Figure 2.0)*. Three compartments from this regional area (range), namely numbers 26, 28 and 31, were selected because they were believed to represent a range of severity of top-dying disease,

based on relevant maps, documents, literature, consultations with forest professionals, and surrounding peoples. The location of these compartments within the Chandpai area, and the location of this area in the wider Sundarbans is shown in *Figure 2.0.* Among the three compartments, compartment number 26 was selected as an area highly affected by top-dying, where most of the trees were affected severely. Compartment 26 had pronounced human activities, and also in places is undergoing rapid housing development involving extensive construction activities due to the presence nearby of the Range HQ office in Chandpai (in *figure 2.0)*. Compartment number 28 was selected as a moderately affected area. This compartment has various human activities including boat making grounds, football-playing grounds, and cattle-grazing fields, all types of major soil erosion, a moderate amount of construction activities and the presence of communities of fishermen *(figure 3.0)*. Compartment number 31 was chosen as being relatively little affected by top-dying disease.

Of the three chosen compartments, the nearest compartment to Mongla port is compartment 31, with comparatively modest human activities, but which nonetheless involve clear-cutting of natural vegetation, replanting with other species rather than mangrove or other native species, all types of soil erosion, and construction activities present. Within each of the three compartments, detailed observations of the regeneration and sampling of soil and water took place within three 20 m x 20 m plots, chosen to reflect a range of top-dying intensities (High, medium and Low for that area). The sampling was conducted in a randomized block design, in that a plot was sited within a particular top-dying intensity block, but the precise location of that plot was randomized so as not to bias the detailed data collection. Thus in total nine plots were sampled, representing a range of top-dying intensities.

Intensive field data collection was made among these nine selected plots (in *Figure 2.0)*. Observations were performed from observation towers during low and high tides, also traversing the forest floor and vegetation on foot, as well as using a speed boat, trawlers, country-boats, and a launch as required to gain access.

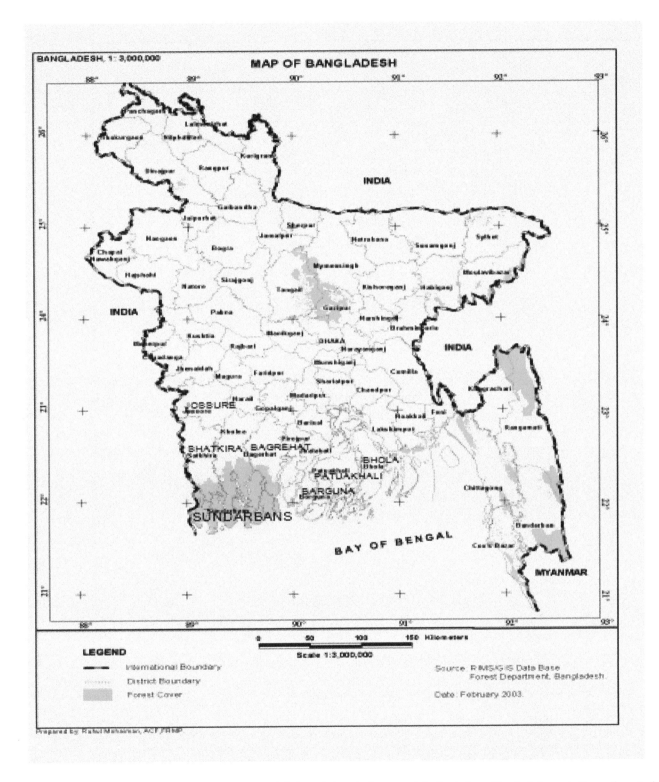

Figure 1.0. *Map showing the administrative districts of Bangladesh, including the location of the Sundarbans (the shaded area in the south-west of the country)*

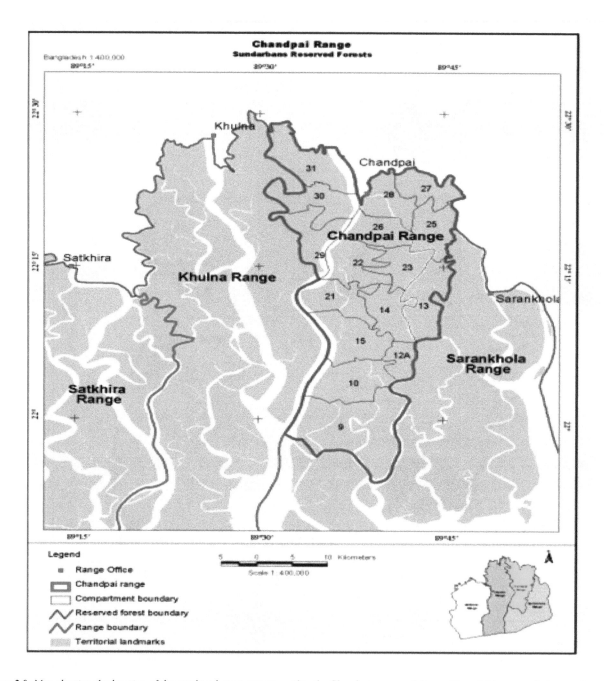

Figure 2.0. *Map showing the location of the numbered compartments within the Chandpai area, and the position of this area relative to other parts of the Sundarbans in Bangladesh (darker shaded area).*

Table 1.0. *Mean (±1 S.E.) and extreme heavy-metal elemental concentrations (ppb) in Sundarbans, together with comparisons with values from other published sources. An asterisk denotes a value below the limits of detection. Comparable data could not be found for all elements.*

| Element | Values from this study | | | | Values reported elsewhere |
	Minimum	Mean	S.E.	Maximum	(Data refer to sediments unless otherwise stated; number within brackets indicates source in footnote)
Al	0.89	16332.44	854.17	37570.00	420 – 585 (soil,[1]); 8089000 – 46100000 ([1]); 500 (spring and well water,[2])
As	*	4.56	0.24	10.06	3150 – 6830 ([7])
B	0.55	19.20	2.14	103.80	2600 (spring and well water,[2])
Ba	0.59	52.41	2.37	141.80	300 (spring and well water,[2]); 141 (coastal soils,[5])
Bi	*	0.40	0.02	0.74	
Cd	0.15	0.55	0.03	1.62	0.52 – 0.92 (soil,[1]); 300 – 13520 ([1]); 43 – 147 ([4]); 0.8 (coastal soils,[5]); 11 – 65 ([7])
Co	5.93	31.31	5.65	143.60	0 – 7.9 (ocean water,[3]); 3800 – 26000 ([1]); 10.6 (coastal soils,[5]); 5540 – 15500 ([7])
Cr	3.11	15.72	3.39	114.90	7 (spring and well water,[2]); 1480 – 8560 ([4]); 41.2 (coastal soils,[5]); 12.8

| Element | Values from this study | | | | Values reported elsewhere |
	Minimum	Mean	S.E.	Maximum	(Data refer to sediments unless otherwise stated; number within brackets indicates source in footnote)
Cu	1.85	10.52	1.71	43.76	(water,[6]); 33200 ([6]); 19500 – 46100 ([7]) 12.2 – 16.6 (soil,[1]); 12940 – 85600 ([1]); 22 (spring and well water,[2]); 22 – 37.2 (ocean water,[3]); 2270 – 14730 ([4]); 23.1 (coastal soils,[5]); 3.8 (water,[6]); 18200 ([6]); 6950 – 31600 ([7])
Fe	25.82	173891.10	9883.85	248200.00	634 – 820 (soil,[1]); 8080000 – 52000000 ([1]); 63 (spring and well water,[2]); 6.2 – 131.5 (ocean water,[3]); 38.5 (water,[6]); 7110000 ([6])
Hg	*	6.41	1.47	83.30	66 – 180 ([4]); 1.8 (water,[6]); 6320 ([6])
Mn	0.70	436.80	14.69	697.00	4980 – 438000 ([1]); 25 (spring and well water,[2]); 1.8 – 40.8 (ocean water,[3]); 3738 (coastal soils,[5]); 7.4 (water,[6]); 412000 ([6])
Mo	0.20	1.62	0.46	26.15	24 (spring and well water,[2])
Ni	7.58	76.08	18.84	1127.00	10800 – 37400 ([1]); 3 (spring and well water,[2]); 0 – 12.1 (ocean water,[3]); 24.5 (coastal soils,[5]); 15900 – 44600 ([7])
Pb	0.32	19.30	0.98	34.19	1.0 – 1.76 (soil,[1]); 1460 – 10400 ([1]); 2 (spring and well water,[2]); 3440 – 15590 ([4]); 74.0 (coastal soils,[5]); 2.3 (water,[6]); 12800 ([6]); 8046 – 15700 ([7])
Rb	0.18	36.37	1.65	76.94	
Sb	*	0.09	0.05	2.93	30 – 94 ([7])
Sc	*	6.05	0.37	8.98	
Se	*	0.17	0.05	1.43	
Sn	*	0.61	0.16	9.68	219 – 654 ([7])
Sr	0.18	27.77	0.89	44.17	2200 (spring and well water,[2])
Ti	4.61	475.39	26.26	1350.00	72 – 341 ([7])
V	0.09	32.93	1.14	51.65	13 (spring and well water,[2]); 18500 – 46900 ([7])
Y	0.03	6.60	0.34	16.69	
Zn	2.30	73.60	2.23	112.50	35.0 – 56.2 (soil,[1]); 120 - 62200 ([1]); 2.4 – 20 (ocean water,[3]); 72.5 (water,[6]); 43200 ([6]); 24300 – 76000 ([7])

[1] Balasubramanian, 1999. [2] Bond, R G & Straub, C P (eds), 1973 [3] Braganca & Sanzgiri, 1980. [4] IUCN Reports 1987. [5] McGrath & Loveland (1992). [6] Sarkar, S.K. et al. 2003 (Premonsoon data from the mouth of the Ganga estuary near Gangasagar used). [7] Zöckler, C & Bunting, G 2006.

Fieldwork was performed in October, 2003 to March, 2004. Locations of sampling points were determined using a Global Positioning System with a precision of 5-10 m. For one typical plot, in compartment 31, the altitude was recorded as 4.4 m above sea level.

2.1.2. Vegetation Recording Methods in the Field

Within each of the nine 20mx20m plots, each adult tree was assessed for three parameters. The diameter at 1 m height was recorded (in cm) by using a tree diameter-measuring tape or slide calipers depending on girth. The tree height to the top of the crown was determined mainly by ocular estimation but some heights were checked by using Clinometers at a set distance of 20 m to test the accuracy of such ocular estimations.

Finally, the status of the tree in respect of the amount of top-dying was assessed by using a four point qualitative scale of intensity, namely; not affected, little affected, moderately affected or highly affected by top-dying. This was later expressed as a semi-quantitative or rank scale of 0 to 3 respectively, so that a median rank value could be calculated and used as an index of top-dying intensity in that plot. After that, the total number of seedlings (individuals of the tree species <1 m tall), and saplings (young trees >1 m tall with a diameter of trunk of < 10 cm), were counted within the plots. Care was taken to ensure that trees, saplings and seedlings were not counted more than once or missed in the counting process. After recording, adult trees were marked with white chalk to segregate those marked trees from other trees, seedlings and saplings; red paints were applied to all seedlings and saplings as they were recorded.

2.1.3. Soil and Water Sampling Methods

As stated above, from the three selected compartments, a total of nine plots of 20m x 20m were selected. From each of these plots, seven soil samples were collected; one from the centre of the plot, four (one each) from all the corners, and two from the middle sides of the plot. Therefore a total of 63 soil samples were taken. Also nine water samples were collected from nearby rivers, creeks or channels, one from the area of each of the sampled plots. Soil samples were collected from 0-30 cm soil depth by using a stainless steel spatula and steel cylinder (d=5.25 cm), and all soil samples were kept in sealed plastic bags. Water samples were collected directly in pre-cleaned plastic-containers. Marking and labelling was performed with a detailed description of the selected sampling site on both the soil-containing plastic bags and water containers, and preserved in portable coolers until arrival at the laboratory at Dhaka University for initial chemical analysis. This field sampling method followed the W.H.O, U.K, and E.P.A systems of standard laboratory and field sampling principles, rules and regulations. Rainfall for the area during sample collection was not notably different from the respective monthly averages for the Sundarbans of recent years (shown in *Table 2.0* for reference); there was no heavy intensity of rainfall within one month before sampling.

Table 2.0. Showing previous monthly average rainfall data.

Year	1996	1997	1998	1999	2000	2002
Location	Mongla (mm)	Mongla (mm)	Mongla (mm)	Mongla (mm)	Mongla (mm)	Mongla (mm)

January	2	19	29	1	24	13
February	25	35	65	0	9	5
March	24	102	149	0	15	32
April	89	94	91	25	134	74
May	119	241	234	202	288	206
June	453	204	229	262	309	952
July	385	486	304	435	356	389
August	357	422	471	466	209	441
September	133	334	553	568	327	492
October	274	40	110	321	224	62
November	1	4	207	12	5	89
December	25	13	0	0	0	0

Any evidence of changes was recorded, sometimes obtained through asking local people and forestry staff, or from personal observations. In particular, any soil erosion and diversion of the river's position or of new channels and creeks observed during the data collection period were recorded, as were signs of siltation changes.

2.1.4. Questionnaire Survey of Local People

In order to establish the views of local people about the incidence and causes of top-dying, a questionnaire was prepared for asking peoples either individually or in groups. This survey was done among people living or working in the 17 Sub-Districts of Sundarbans, making a distinction between those living within and outside of Sundarbans. They were asked whether they had seen the top-dying disease of *Heritiera fomes* (Sundri) in Sudarbans for a long time, either through living within the Sundarbans or through visiting Sundarbans for their daily work, for their professional work such as forestry officials, for fishing or for collecting wood as wood cutters, for seasonal honey collection, or other purposes. Groups were made up among targeted people in all locations and from all categories mentioned above, based on age, profession, and also for their sharp memory. In this way, 50 questionnaires were filled up through interview, mostly of groups and sometimes of individual people. The justification of selection of people for the questionnaire survey was that the targeted people were familiar with the top-dying problem in Sundarbans, and are related through their professions with Sundarbans directly and indirectly. The questionnaire started by establishing that the respondents were familiar with top-dying, and went on to seek their views and information on what changes they had observed and whether they had noticed possible causes. This was possible because, most of the interviewees are living within the Sundarbans for their daily activities. So, this survey was performed to receive their indigenous response and knowledge towards top dying and its present conditions, and their ideas about what leads to top-dying, as well as questions about tree regeneration and human health in Sundarbans (Awal, 2014).

2.1.5. Statistical Analysis

Initial statistical analysis of quantitative data, particularly of the elemental concentrations, consisted of calculation of arithmetic means, standard deviations and standard error values for each variable separately. Data on the severity of top-dying for each tree in a plot, which had been recorded as ' not affected', 'mildly affected', 'moderately affected', and 'highly affected', were converted into a four-point scale (0-3), so that they could be summed and an average (median) could be determined for each plot, thus producing an index based on ranked data. Comparisons of the strength of relationship between two variables were assessed by correlation: the Pearson's product-moment correlation coefficient where both variables were fully quantitative or the Spearman's rank correlation coefficient where the top-dying index was one of the variables. In the case of the Spearman's coefficient, the probability of the outcome was determined by using the approximation to a t-statistic appropriate to these tests (Sokal and Rohlf, 1981). Occasionally, a Pearson's correlation coefficient was calculated where top-dying was one of the variables, in order to check on the extent of the difference between the rank and quantitative versions for these data. Data on frequencies of seedlings or saplings in each of the plots and compartments were tested by x^2 contingency table analysis to determine whether there was an association between the selected plot type (severely, moderately or little affected by top-dying) and the three chosen compartments. A similar consideration of the different compartments as comprising one factor, and the plot type as a second, was used to test the pattern of elemental concentrations and other variables by a 2-factor analysis of variance test with replication. This allows an assessment of the significance, not only of the two factors separately but also of the interaction term linking the two factors. It should be noted that the plot type was not a strictly controlled factor, since the three categories of top-dying intensity were relative to each other within any one compartment and might not have been exactly equivalent between the three categories; interpretation of the results from these tests therefore needs to bear this in mind. MINITAB Release 14 Statistical Software has been used for windows on CD-ROM, 2004 edition for all data analysis, both statistical and graphical, except for those produced automatically by the Excel package attached to the ICPMS.

3. Results

These results are indicated as follows: if one number (all values in ppb) is given it is a mean, otherwise if a range is given they are the minimum and maximum; the number is followed by the type of material from which the data come, with no text indicating it is from sediments (the most common material reported in the literature); finally, the number in brackets indicates the numbered reference source, the sources being indicated in the legend. Besides attempting to establish whether the element concentrations are elevated or not, it is valuable to explore whether there is any marked spatial (as opposed to random) variation in the concentrations found.

4. Discussion

Shrimp cultivation is a serious problem for the heavy metal contamination in the water, soil, plant, fishes, animals etc. Coastal lands cover 6% of the world's land surface (Tiner, 1984). Coastal and wetlands everywhere are under threat from shrimp-cultivation, agricultural intensification, pollution, major engineering schemes and urban development (UN-ESCAP 1987; 1988). The Indo-Pacific region is known for its luxuriant mangroves. The distribution of mangroves in the Indo-West Pacific bio-geographical region has been outlined in Macnae (1968). However, the mangroves in half of these countries, as well as those of other regions, have since been destroyed through various pollution problems and population pressure (Peters et al., 1985). The country's food security and public health will be in danger if the water and wetlands are destroyed at the present rate (Awal,2009).Unplanned natural resources management and environmental contamination in Sundarbans are fast destroying the surface, vegetation, water, and underground fresh water sources (Awal, 2014). The primary source of fresh water for fishes, trees, human's health in Bangladesh was vital but due to heavy-metal contaminations in soil, water, faulty-vegetation management and deforestations of trees, top-dying disease and other diseases and health problems, irrigation for food production, industrialization, and public health, while industrial pollution and poor sanitation are making surface water unusable (Awal, 2007). The health of vegetation including natural flora & fauna, drinking water, soil are largely depends on uncontaminated groundwater,(Awal, 2009). Water related diseases are responsible for 80 percent of all deaths in the developing world (Awal, 2014).Data were summarized by calculating means and standard errors, and by noting minimum and maximum values, for comparison with other data reported in the literature. The spatial variability in the data was assessed by calculating two-factor analyses of variance with replication, where the factors used were the broad-scale variation between compartments, the smaller-scale variation between plots within the same compartment, and the interaction between these two sources of variation. The concentrations of the various trace metals determined by ICP-MS from our Sundarbans soils are given in Table 1.0. For completeness, the minimum and maximum values are given, as well as the mean and the standard error, in order to facilitate comparisons with other published information and to indicate the extent of variability between the different samples in the results. It is clear that due to destruction of natural resources and contaminations of soil, water for some elements the variability is considerable; for example, nickel has a maximum value many times larger than the mean, while for iron the mean and maximum are more similar but the minimum is substantially smaller.

In order to try and establish whether these values are elevated compared to other data, comparison values are included in (Table:1.0) from results published in the literature concerning the Bay of Bengal region. In order to assess this, an analysis of variance has been performed for each element separately, testing for location by using the compartment from which the samples were taken as one factor, and the plot number within each compartment as a second factor, testing also for interaction between these two sources of variation. The results of these analyses are presented in Table 1.0, showing only those elements which showed at least one F-value at or near significance. Most of the elements tested do not show any significant variation related to any of these factors, and may therefore be considered to be relatively uniform or non-consistently variable in their concentrations, at least across the compartments studied here. There were no elements that proved significant when comparing between plots, nor in the interaction term, although antimony had a close result for the plot term (p = 0.06). However, a few elements did give significant results comparing between compartments, namely bismuth, scandium, strontium and vanadium. In all four cases, the lowest recorded concentrations were from compartment number 28; in all cases except for strontium, the compartment where the highest elemental concentrations were recorded was number 26 (compartment 31 being the highest for strontium). Interestingly, a similar analysis of concentrations (not presented separately here) also showed significant differences between compartments for sodium and phosphorus, with compartment 28 again having the lowest recorded concentrations.

In particular, the metal values from sediments are likely to be much higher than those from soils, and indeed one author (Balasubramanian, reported by Swaminathan, 2000) found sediment values to be often at least one thousand times higher than values for equivalent soils. The concentration values for the various trace metals recorded in Table 1.0 are believed to be amongst the first published for soils from the Sundarbans, and as such provide baseline data for comparison with other future studies in the area. It has therefore not been possible in many instances to find appropriate comparators from the literature in order to help assess whether the data show elevated concentrations from those expected in such soils. The comparatives information from the literature presented in Table 1.0 is all from the same general region, but includes values from water (from ocean, springs or wells), from coastal soils (but not within the main mangrove areas), and particularly from mangrove sediments, as well as a few from mangrove soils. All comments based on these other sources must therefore be judged on the basis of the differences between the materials in the results likely to be obtained. A further complicating factor in interpreting these results is the high degree of variability in concentrations between different samples shown by many elements, as indicated in *Table 1.0* by the high standard error values and by the results highlighted for some elements in *Table 1.0*.

For most of the elements tabulated there are either no comparisons in the literature, or the results from our study do not appear particularly elevated compared to other results. Perhaps surprisingly, given the problems that have been identified with elevated concentrations of arsenic (As) in groundwater in Bangladesh (e.g. Nickson *et al.*, 1998;

Chowdhury *et al.*, 2000), this element was not notably high in the soils studied here. However, there were other metals which may be elevated in their concentrations. Two results appeared particularly elevated, namely those for mercury (Hg) and for nickel (Ni). In considering the result for mercury, it is recognized that the ICP-MS method of testing the soils is not the most appropriate one for obtaining an accurate determination of mercury concentration because of the potential for cross-contamination of samples from earlier ones due to retention of the element within the instrument. The problem was reduced by the use of gold-wash solution rather than nitric acid in preparation of the calibration standards. Nonetheless, the elevated concentrations of mercury in these mangrove soils can only be considered as an indication until confirmation of these values by further work involving a different analytical procedure can be completed.

There is likely to be considerable geographic variation in the extent of pollution problems in the different parts of the Sundarbans, associated both with the proximity to local polluting sources such as Mongla port and with the extent to which the area is influenced by the Ganges river, which is strongly polluted (Sarkar *et al.*, 2003). This was indeed found *(Table 1.0)* as there were significant differences between different areas in the Sundarbans with regard to at least some of the elements studied, and others were probably not significant only because of the large amount of variability between different samples within individual compartments. It is therefore perhaps not surprising that the values reported by Zöckler and Bunting (2006) were lower than ours, since their study was in the east of Bangladesh away from the Ganges and other main sources of pollutants. Also, the choice of sites in the present work emphasised areas likely to be polluted because they were near to human activity and hence more accessible. Even so, and allowing for the fact that sediment data is the only comparator medium, the data from the literature suggest that the Sundarbans is not yet as polluted as some other mangroves from the region, such as in Pakistan (e.g. IUCN, 1987).

Clearly, further work is required to confirm and extend the results reported here. The indications of potentially elevated heavy metal concentrations is a matter of concern, and a higher general pollution load is likely to contribute to the increase in top-dying observed in the Sundarbans (Rahman, 2003; Chaffey, *et al.*,1985; Chowdhuy,1984; Gibson; 1975). A likely mechanism of influence might be that greater concentrations of the trace-metals weaken the resistance of the tree to attack by pathogenic fungi. The relationship between individual trace metals and the amount of top-dying will be explored further in a separate article. It is also worth noting that local residents and those who work in the Sundarbans quite frequently reported health problems, of which problems of the skin were the most common (data from a questionnaire, included in Awal, 2007). It is possible that the high concentration of nickel (Awal, 2007), which can cause skin conditions, is leading to such complaints. Such health issues are therefore also a cause of concern and need further confirmation and elaboration.

5. Conclusion

Shrimp culture is the major problem of ecological problems in Sundarbans (Awal, 2007, 2009, 2014).Duet to illicit cutting of trees, deforestations, chemical fertilizer and all types of artificial activities within natural forest like Sundarbans is the interrelation of chemical pollution in Sundarbans (Awal, 2007).The overall conclusions from the results presented in this section are that the selection of sites has not produced clear statistical differences in the amount of top-dying evident; probably because of the way the data were collected. However, it is believed that there is notable variation between plots and compartments, and certainly this seems to be reflected in the ability of the trees to regenerate. However, the link between top-dying and the size of the trees is not clear, with tree height and diameter not being directly related consistently to amount of top-dying, although moisture content of soil was inversely related. Since the great majority of trees present in all plots is the species *Heritiera fomes*, this means that the comments above are essentially referring to the response of this species rather than that of any others. So that, the Sundri, by contrast, prefers largely fresh water in which it resembles the mesophytes, but the species is adapted to the wet swampy condition of the Sundarbans by virtue of its leaves having partly xerophytic adaptations and plentiful pneumatophores which help cope with the saline swamps of the Sundarbans.The vegetations need sound ecological balanced to survive but due to deforestations, illicit fellings, hunman destruction are responsible for the heavy metal contaminations in soil, water and vegetation(Awal, 2007, 2009, 2014).

Comparing figures in the *table 1.0*, it would suggest that about two thirds of the elements have concentrations which are elevated compare to other reference sources in the Sundarbans. This would be consistent with the evidence that heavy metals were having an influence on top-dying intensity (Awal, 2007, 2009, 2014). The elements Pb, Sn, and Zn were highlighted earlier in this discussion, and although not all of them quite reached statistical significance (Awal, 2007), the positive trend linking two of them to top-dying suggests a likely mechanism of influence, namely that greater concentration of the heavy-metal weaken the resistance of the tree to attack by the pathogenic fungi (Awal, 2007, 2009, 20014). This might well be a process that other elements contribute to as well (Awal, 2007, 2009, 2014), but has not been picked out by the analysis as showing a link because of the variability between samples inherent in the data (Awal, 2007). In this respect, the anomaly of the negative relationship indicated for Sn is harder to explain (Awal, 2007), but a possible process might be an antagonistic response of Sn and another element (Awal, 2007), so that when Sn is less abundant the other element can have a stronger (deleterious) effect on the trees (Awal, 2007), thus allowing more top-dying to occur (Awal, 2007). A further point is that variations in soil pH from site to site (shown to be significant) will also have a marked effect on the bio-availability of some of these heavy metals (Awal, 2007), and thus perhaps influence top-dying (Awal, 2007).We should protect natural resources such as Sundarbans in Bangladesh (Awal, 2009), because the future of Sundarbans is as directly dependent on the health of her wetlands, as is the future flora and fauna. Considering the limitations of the current planning process in Bangladesh, it is possible that within a few short decades, as water tables fall, rivers run dry and lakes shrivel, water–riots will become the order of the day. It may also be the case that wars on the subcontinent will more likely be fought over water than oil. And until uncontrolled development is restricted, the threat of floods and droughts due to loss of mangrove systems will continue to be present (Sahgal, 1991). Coastal lands include some of the most productive of ecosystems with a wide range of natural functions, but are also one of the most threatened habitats because of their vulnerability and attractiveness for 'development'. The first global conservation convention, the Ramsar Convention, focused solely on coastal lands and wetlands, and it has recently been strengthened and elaborated with regard to the wise use of all coastal areas such as Sundarbans.

Acknowledgements

My research work was supported financially by the Peoples' Republic of Bangladesh and the Asian Development Bank (ADB), whom I thank. Particular thanks are due to the ADB head office, Manila for their support and help. I thank sincerely Dr. W.G.H. Hale (Principal Supervisor), University of Bradford, UK, Professor Mike Ashmore (Technical Supervisor), University of York, UK and Dr. P.J. Hogarth for their advice and comments on the work; Dr. Ben Stern and the staff at the Analytical Centre, Bradford, for their help with the ICPMS analyses; Professor Sirajul Hoque, Mustafa (lab Technician)and staff at Dhaka University for providing facilities; and staff of the Forestry Service, Bangladesh Government, for field assistance. Moreover I indebted to my beloved parents Munshi Aowlad Hossain (Father: Teacher and Landlord as well as blossom friend of poor), Mrs. Ashrafunness(Mother: born Literate and Socialist, pious and friend of poor and distress), my venerated fore-fathers:Abadat Biswas (Mighty-Landlord), Golam-Rabbani-Biswas(Mighty-Landlord) my esteemed grandfathers (Munshi Bellal Hossain Biswas (Learned-Literate-Landlord), Md. Ataher (Lawyer and Powerful Jotdar and Landlord), my respected grand-mothers: Rohima Khatun (pious & friend of poor), and Alimoonnessa (social, pious and friend of poor and distress), Dr. Shajahan Kabir (Nephew), my beloved wife (Dr. Shahanaj Khatun), my beloved son (Munshi Tasneem Redowan), my beloved daughter (Marwa Ashra), my beloved brothers: Munshi Abul Azad (Officer in BAF), and Munshi Abdus Salam (Program Officer in UNDP), Munshi Abdul Rouf (Businessman), Motiar Rahman (Local Leader and social worker), and my beloved 6 sisters (Layla Anjumand Banu (Chandu), Akter Rashida Banu (Turi), Saleha Pervin (Lili), Azmeri Ferdowsi (Dolly), Jannatul Ferdowsi (Polly) and my beloved youngest sister Bedowra Ferdowsi (Jolly), Chondona, Sultana, Rono, Chapa, Loti, Urfa, Alta, Joytoon, my respected only uncle Munshi Abdur Razzak (Teacher and Jotder as well as blossom Friend of poor people),Noorjahan (unty), Kolimuddin Biswas(Aunt), my venerated maternal uncles: Anowarul Azim (Director of Family Planning), Amirul Azim (First Class-Magistrate and UNO), Aftabul Azim (Banker), Azizul Azim (Influential Leader and Govt.-Officer in Land Department), Anisul Azim (Social Leader and influential-Businessman), Aminul Azim(Dramatist), my respected Khalas (Khuki, Bulbul), as well as all family members for their inspiration and help.

References

[1]   Awal, M.A. (2007). Analysis of possible environmental factors causing top-dying in mangrove forest trees in the Sundarbans in Bangladesh. PhD thesis, University of Bradford.

[2]   Awal, M.A., Hale, W.H.G. & Stern, B. (2009). Trace element concentrations in mangrove sediments in the Sundarbans, Bangladesh. Marine Pollution Bulletin, 58(12), 1944-1948.

[3]   Awal, M.A. (2014). "Correlation between the chemical composition of the surface sediment and water in the mangrove forest of the Sundarbans, Bangladesh, and the regeneration, growth and dieback of the forest trees and people health"..Journal of Science Innovation; 2014. 2(2): pp.11-21.Science Publishing Group, USA; May 20th, 2014(2):11-21;doi: 10.11648/j/si.20140202.11.

[4]   Asian Development Bank, 1993-95. Main Plan-1993/2012. Vol.1. *Forestry Master Plan.* Asian Development Bank, Manila, Philippines.

[5]   Anonymous, 1986. *Mangroves in India*: Status Report, Government of India, Ministry of Environment and Forests, New Delhi, 150 pp.

[6]   Anonymous, 1986. *Sundri trees fast reducing.* The Bangladesh Observer, 1st. December, 1986.

[7]   Bangladesh Bureau of Statistics, 2006. *Statistical Yearbook of Bangladesh*, Statistics Division. Ministry of Planning. Dhaka, Bangladesh.

[8]   Bari, A. 1993. *Afforestation and the nutrient sink. Assistance to Fisheries Research Institute.* Mymensingh. BGD / 89 / 012, Field Document-3.

[9]   Burns, K.A; S. D. Garrity, and S.C. Levings. 1993. How many years until mangrove ecosystems recover from catastrophic spills? *Marine Pollution Bulletin* 26 (5): 239-248.

[10] Chaffey, D. R; Miller, F.R; Sandom, J. H. 1985. *A forest inventory of the Sundarbans*, Bangladesh, *Main report*, Project Report No.140, 196 pp; Overseas Development Administration, London, U.K:195-196.

[11] Chantarasri, S. 1994. *Integrated Resource Development of the Sundarban, Fisheries Resources Mangagement for the Sundarban*, UNDP / FAO, BGD / 84 / 056, Khulna, Bangladeshp: 170-172.

[12] Chowdhury, A.M. 1984. *Integrated Development of the Sundarbans, Bangladesh: Silvicultural Aspects of the Sundarbans.* FAO Report No / TCP/ BGD/ 2309 (Mf), W / R003.

[13] Chowdhury, M. I. 1984. *Morphological, hydrological and ecological aspects of the Sundarbans*. FAO report N0. FO: TCP/BGD/2309(Mf) W /R0027, 32 P.

[14] Christensen, B. 1984. *Integrated development of the Sundarbans, Bangladesh: Ecological aspects of the Sundarbans*. Reported prepared for the Government of Bangladesh. FAO report no. FO: TCP/ BGD/2309(MF) W/ R0030.

[15] Faizuddin, M. 2003. *Research on the Top Dying of Sundri in Bangladesh*: 43, Mangrove Silviculture Division, Bangladesh Forest Research Institute, Khulna, Bangldesh.

[16] Faizuddin, M. and Islam, S.A. 2003. *Generated Technology and Usable Information of the Mangrove Silviculture*. Mangrove Silviculture Division, Bangladesh Forest Research Institute, Khulna, p. 17.

[17] FAO, 1993. Forest resources assessment 1990*: Tropical countries. FAO Forestry Paper*. 112, Rome, 98-102p.

[18] FAO, 1994. Review of the state of world marine fisheries resources. *FAO Fisheries resources. FAO Fisheries Technical Approach Paper* 335:143.

[19] Gibson, I.A.S. 1975. *Reports on a visit to the People's Republic of Bangladesh,* 28 February to 1 April 1975.Unpublished Report, ODA, London, 28pp.

[20] Government of Bangladesh, (1993). *Forestry Master Plan: Executive Summary*. Asian Development Bank, UNDP/FAOBGD/88/025, Forest Department, Government of Bangladesh, Dhaka.31p.

[21] Government of Bangladesh, (2006).Mosaic of Bangladesh; 2006; External Publicity Wing; Ministry of Foreign Affairs; Government of the People's Republic of Bangladesh; p:1-145

[22] Hambrey, J. 1999. *Mangrove, Fisheries and Economic*. Aquaculture and Aquatic Resource Management Program. *Asian Institute of Technology*. Thailand: 1-4.

[23] Harris, L. D. (1984). *The Fragmented Forest*: Island Biogeographic Theory and the Preservation of Biotic Diversity. Chicago: University of Chicago Press.

[24] Hussain, Z. and Karim, A. (1994). Introduction. In: *Mangroves of the Sundarbans*. Volume 2: Bangladesh, Z. Hussain and G. Acharya (Eds.) IUCN. Bankok, Thailand. !-18 pp.

[25] Islam, M. A.1993. *Some Relevant Information about Sundarban*. Sundarbans Forest Division, Khulna, Bangladesh, p. 21.

[26] Mukharjee, A. K. 1975. The Sundarbans of India and its biota. *Journal of Bombay Natural History Society*, 72 (1):1-20.

[27] Rahman, M.A. 2003. *Genetic Approach to mitigate the top Dying Problem of Heritiera fomes in the Mangrove Forests*, Khulna University, Bangladesh, 87pp.

[28] Rahman, M.A. 2003. *Mid-term Report on Top Dying of Sundri (Heritiera fomes) and Its Management in the Sundarbans* Biodiversity Conservation Project, Khulna.109pp.

[29] Seidensticker, J. Hai; A. 1983. *The Sundarbans wildlife management plan*: conservation in the Bangladesh (cited in Chaffey et al., 1985), Bangladesh.

[30] Tiner, R. W; J R. (1984). *Wetlands of the United States*: Current Status and Recent Trends. Newton Corner, Massachusetts: U.S. Fish and Wildlife Service, Habitat Resources.

[31] UN-ESCAP, 1987. Final Report: Volume 2. *Coastal environment management plan for Bangladesh*. Bangkok, Thailand.

[32] UN-ESCAP, 1988. *Coastal environment management plan for Bangladesh*. Bangkok, Thailand: 7-34.

[33] WHO, 1981. *Resistance of Disease vectors to pesticides*. World Health Organisation, *Chronicle*, 35, 143.

Dynamics of soil fertility as influenced by different land use systems and soil depth in west Showa zone, Gindeberet district, Ethiopia

Lechisa Takele[1, *], Achalu Chimdi[1], Alemayehu Abebaw[2]

[1]Wollega University, College of Natural and Computational Science Department of Soil Resource and Watershed Management, P.O. Box 395, Nekemte, Ethiopia

[2]Ambo University, College of Natural and Computational Science Department of Chemistry, PO Box: 19, Ambo, Ethiopia

Email address:

lechisat@gmail.com (Lechisa T.), achaluchimdi@yahoo.com (Achalu C.), alemayehuabebaw@yahoo.com (Alemayehu A.)

Abstract: Land use change from natural forest to cultivated land, grazing land and subsequent changes in soil physicochemical properties was widespread in Ethiopia. Thus, assessing land use-induced changes in soil properties are essential for addressing the issues of agro-ecosystem transformation and sustainable land productivity. The aim of the study was to determine selected soil physicochemical properties of forest land, cultivated land and grazing land and make investigation among the soil properties. Standard procedures were employed for the analyses of soil parameters. One way ANOVA was employed to compare the soil parameters at particular and overall soil depth. Textural class of all land use types was clay indicating similarity in parent materials distribution of bulk density in all soil depths of cultivated land were higher compared to both forest and grazing land. Soil moisture content was significantly increasing with increasing soil depths. The highest soil pH in all soil depth was observed under forest land compared to both grazing and cultivated land. The highest soil OM contents were observed in the surface soils (0-10 cm) of forest land while least Figures were from subsurface (10-20 cm) layers of the cultivated land. TN, CEC, exchangeable (Ca, Na and Mg) of the forest land soil were improved when compared with both cultivated and grazing land soil.

Keywords: Land Use, Soil Fertility, Soil Physicochemical Properties

1. Introduction

Land use change is an important factor in global change phenomena. It is directly related to issues such as food security, water quality, soil quality and other important global life support issues [1]. Land use change and changes in soil fertility management often occur together resulting changes in soil quality including the activities of soil micro-organism [2]. As a consequence, one would expect close relationships between land use change and soil nutrient contents [3].

Land use change also affects the productivity of a soil. This manifests as changes in soil properties such as contents of available of macro and micro nutrient, organic matter and CEC [4]. Agricultural sustainability requires periodic evaluation of soil fertility status which is important in understanding factors that impose serious constraints to crop production under different land use types and for adoption of suitable land management practices [5].

Land use change affect the distribution and supply of soil nutrients by directly altering soil properties and influencing biological transformations in rooting zone. For instance, cultivation of forests diminishes the soil carbon within a few years of initial conversion and substantially lowers mineral stable of nitrogen [6]. Soil quality is a concept that integrates soil biological, chemical and physical factors into a framework for soil resource evaluation [7-9].

Soil properties such as water holding capacity, aeration, tendency to crust, and cation exchange capacity can be estimated from particle size distribution [10]. Differences in soil texture also impacts organic matter levels which broke down faster in sandy soils than in fine-textured soils this gives similar environmental conditions and fertility management because of a higher amount of oxygen available for decomposition in the light-textured sandy soils. The cation exchange capacity of the soil increases with percent clay and organic matter and the pH buffering capacity of a soil (its ability to resist pH change upon lime addition) is also largely based on clay and organic matter content [11].

Soil chemical properties are the most important among the factors that determine the nutrient supplying power of the soil to the plants and microbes. The chemical reactions that occur in the soil affects soil development and soil fertility build up. Plants are capable of absorbing and assimilating as many as forty or fifty different chemical elements. Sixteen of these chemical elements have been found to be essential to the growth of most plants. Therefore, this study was conducted with specific objective to assess and explore the status of soil physicochemical characteristics of three different land use systems along soil depth of representative area of Western Oromia Region. The result of this study

expected to add value to the up-to-date scientific documentation of the status of soil fertility and soil quality of different land uses of the study area and other similar agro-ecological environments in the country.

2. Materials and Methods

2.1. Description of the Study Area

The study is situated in Gindeberet district, West Shewa Zone of Oromiya National Regional State, Ethiopia, between astronomical grids of $9^0 21'$ to $9^0 50'$ N and $37^0 37'$ to $38^0 08'$ E. The district town, Kachisi ($9^0 32'$N and $37^0 49'$E) is geographically located approximately at the centre of the district 193 Km west of Addis Ababa [12].

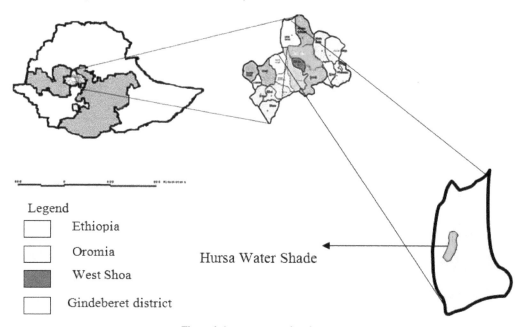

Legend
- ☐ Ethiopia
- ☐ Oromia
- ■ West Shoa
- ☐ Gindeberet district

Hursa Water Shade

Figure 1. Location map of study area.

Ten years trends of rainfall distribution showed that there was no even distribution of rainfall in each year. Rather it was highly fluctuated between the ranges of 882-2039 mm.

the same is true for temperature which was varied from 21 to 25.9 and 5.6 to 9.2°C for the maximum and minimum temperature respectively.

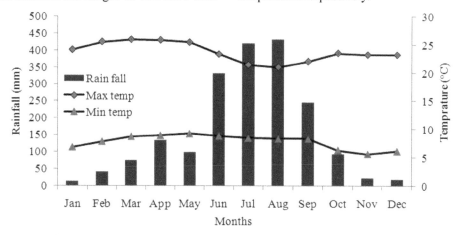

Figure 2. Mean monthly rainfall and mean maximum and minimum temperatures of the study area.

2.2. Soil Sampling and Analysis

In order to investigate soil fertility status through the analysis of some physical soil properties, Representative, intact soil samples was collected with a manual core sampler of known volume and weight, from each land-use practice unit plot in X design in five replicates from three blocks. Thus, samples were collected from four corners of a square of 20 X 20 plots, with one from the center. The samples were separated into 0–10cm, 10–20cm and 20 - 30cm horizons.

To determine the effect of different land uses on soil chemical properties, a total of 27 (3 land uses*3 soil depths*3 replicates) soil samples were taken. Then, the soil sample was taken from the pits by scuffing the wall of the soil profile for respective depth; the lowest first and the top soil at last to avoid contamination between the two layers. For each land use types and a soil samples, about 1kg of soil samples was taken. Then, the soil samples from each pit (the five) was bulked together to obtain composite soil samples for each replicates (the three blocks) in three depth interval and land use types. Soil clods in each composite sample was thoroughly broken to make a uniform mix, and then divided into four equal parts from which two diagonal parts was retained and the other two parts removed. This process was continued several times until the successive quartering reduced the weight of a composite sample to about 0.5 kg.

Samples were air dried ground and passed through 2 mm sieve for analysis. Analysis of soil samples were carried out at Chemistry laboratory of Ambo University based on their standard laboratory procedure. Particle size distribution and bulk density were determined by the hydrometer and the core sample methods respectively. Soil pH was determined in soil to water ratio of 1:2.5 (w/v). OC was determined by wet digestion method. Total N was determined by micro-Kjeldahl wet digestion and distillation method, while available P was extracted by the Bray II method and finally quantified by spectrophotometer. Cation exchange capacity (CEC) and exchangeable bases were extracted by 1M ammonium acetate (pH 7) method. Exchangeable bases (Ca, Mg, Na, and K) were extracted with 1M ammonium acetate at pH 7. Ca and Mg were analyzed by titrations using EDTA method. Exchangeable K and Na were measured by flame photometer.

2.3. Statistical Analysis

Soil physicochemical properties were subjected to analysis of variance using statistical analysis system version 9.0 [13]. Treatment means of the different land use types were compared according to Tukey test.

3. Results and Discussion

3.1. Impact of Land Use on Selected Physical Properties of the Soils along Depths

3.1.1. Soil Particle Size Distribution, Soil Bulk Density and Soil Moisture Content

Silt content across land use was not significant except for middle surface, which is significant (P = 0.05). High silt content was observed under forest land while low silt content was observed under cultivated land in all soil depths (Table 1). The distribution of sand across land use was not significant (P = 0.05) higher in all soil depths of grazing land than both forest land and cultivated land. Silt to clay ratio in the 0-10 cm depth was low and varied from 0.67, 0.32 and 0.49 for soils of forest, cultivated and grazing lands, respectively. While, Silt to clay ratio in the 20-30 cm depth was high and varied from 0.63, 0.99 and 1.09 for soils of forest, cultivated and grazing lands, respectively. Higher and lower silt to clay ratio recorded in the forest land and cultivated land respectively. High silt-clay ratio observed in the top surface horizons of forest land (0.63) may be attributed to the deposition of plant materials (litter) which are still undergoing decomposition. While, lower silt to clay ratio recorded in the cultivated land (0.32) attributed to the impacts of deforestation and farming practices [9].

Table 1. *Comparisons of soil physical properties in different land use types and soil depth. Results expressed as mean ± standard deviation.*

Variables	Depth (cm)	Land use Types		
		Forest land	Cultivated land	Grazing land
Sand (%)	0-10	24.26±4.00^{ab}	22.60±2.20^{b}	26.42±3.78^{a}
	10-20	20.66±4.76^{a}	19.72±5.50^{a}	23.88±1.99^{a}
	20-30	21.40±1.56^{a}	18.93±2.70^{a}	22.56±2.34^{a}
Silt (%)	0-10	30.51±2.26^{a}	18.78± 0.74^{b}	24.21±3.04^{ab}
	10-20	29.67±1.87^{a}	19.42±3.03^{c}	25.65±2.90^{b}
	20-30	33.52±2.94^{a}	19.20±1.64^{b}	20.66±2.67^{b}
Clay (%)	0-10	45.23±1.46^{b}	58.62±6.25^{a}	49.37±0.74^{b}
	10-20	49.67±8.53^{b}	60.86±7.14^{a}	50.47±4.89^{b}
	20-30	45.08±0.93^{b}	61.87±2.95^{a}	56.78±0.33^{a}
BD(g/cm^3)	0-10	1.08±0.03^{b}	1.27±0.03^{a}	1.22±0.01^{a}
	10-20	1.11±0.01^{b}	1.30±0.02^{a}	1.27±0.01^{a}
	20-30	1.13±0.02^{a}	1.35±0.02^{b}	1.29±0.02^{c}
MC (%)	0-10	10.59±0.21^{a}	8.65±0.35^{b}	8.23±0.22^{b}
	10-20	11.80±0.20^{a}	8.66±0.58^{b}	8.46±0.20^{b}
	20-30	12.99±0.16^{a}	9.14±0.08^{b}	8.79±0.25^{b}

Means within rows followed by different letters are significantly different (P =0.05) with land use.

Soil bulk density of both 0-10 cm and 10-20 cm soil depths was not significant (P = 0.05). While soil bulk density was significant at lower surface 20-30 cm of the soil depth. The distribution of bulk density in all soil depths of cultivated land were higher compared to both forest and grazing land. Soil bulk density was higher in the lower compared to 0-10cm and 10-20 cm soil depths in all land use types indicating the tendency of bulk density to increase with depth due to the effects of weight of the overlying soil and the corresponding decrease in soil organic matter content.

The relatively lower bulk density in the top surface than in the lower layer may reflect organic matter concentration [9].

Soil moisture content of cultivated and grazing land was not significantly different (p=0.05) from each other at all soil depths while moisture content of forest land was statistically different from both cultivated and grazing land at all soil depths [14].

3.2. Impact of Land Use on Selected Soil Chemical Properties along Depths

Table 2. Comparison of pH, OM, AV.P and TN in different land use types and soil depth, Results expressed as mean ± standard deviation.

Variables	Depth (cm)	Land use types		
		Forest land	Cultivated land	Grazing land
pH	0-10	5.28 ± 0.11^a	4.24 ± 0.29^c	4.66 ± 0.15^b
	10-20	5.53 ± 0.41^a	4.41 ± 0.17^c	4.47 ± 0.04^b
	20-30	5.82 ± 0.10^a	4.63 ± 0.20^c	4.92 ± 0.08^b
OM (%)	0-10	2.31 ± 0.50^b	1.50 ± 0.62^b	1.93 ± 0.15^b
	10-20	1.71 ± 0.31^a	1.27 ± 1.89^b	1.76 ± 0.04^a
	20-30	1.33 ± 1.18^b	1.22 ± 0.70^b	1.55 ± 0.08^a
Av. P (ppm)	0-10	4.79 ± 0.31^a	2.34 ± 0.01^a	2.51 ± 0.04^a
	10-20	3.82 ± 0.22^a	2.12 ± 0.01^b	2.53 ± 0.03^b
	20-30	2.85 ± 0.06^a	1.71 ± 0.02^b	2.47 ± 0.01^a
TN (%)	0-10	0.12 ± 0.03^a	0.09 ± 0.07^a	0.10 ± 0.01^a
	10-20	0.11 ± 0.01^a	0.08 ± 0.03^a	0.09 ± 0.03^a
	20-30	0.10 ± 0.01^a	0.08 ± 0.02^a	0.08 ± 0.04^a

Means within rows followed by different letters are significantly different (*P* =0.05) with land use.

3.2.1. Soil pH, Organic Mater, Available Phosphorus and Total Nitrogen

The distribution of soil pH was significantly varies across all land use. The highest soil pH in all soil depth was observed under forest land compared to both grazing and cultivated land. On the other hand soil pH increase with increasing soil depths (Table 2). The reason can be the reduction of Ca and Mg ions along soil depth which lowers soil pH from top to down the soil layers [15].

The highest soil OM contents were observed in the forest land use type for both 0-10 cm and 20-30 cm soil depths while its content was high in grazing land for 10-20 cm soil depth (Table 2). At 0-10 cm soil depth the SOM content of forest land was significantly higher than cultivated land and grazing land at similar soil depth. This is attributed partly to the continuous accumulation of undecayed and partially decomposed plant and animal residues in the surface soils. In general, forest clearing followed by conversion into agricultural fields brought about a remarkable depletion of the soil OM stock. Hence findings of this study are in agreement with findings of similar studies by [16, 17].

Total N content was not significant in all land use types (P=0.05). The higher total N (0.31%) was found on the 0-10 cm depth range of forest land, whereas total N content of forest land was 0.28 % and 0.26% at 10-20 cm and 20 30 cm soil depths respectively. It is also known that in forest and grass dominated land, competition for nutrient is higher in the bottom surface layer due to the root biomass on this layer is much denser than the other layers [18].

The distribution of available soil phosphorus content of

forest land use types in all soil depths were found to be higher compared to cultivated and grazing land. However, available phosphorus of grazing land at 10-20 cm soil depth was greater than the top surface 0-10 cm this could be attributable to the tillage practice which leads a relocation of clay separates and soil mix up which could increase phosphorus fixation by soil colloids [19].

3.2.2. Soil Cation Exchangeable Capacity and Exchangeable Cations

In this study, cation exchange capacity at 0-10 and 10-20 cm soil depth was not significant for both cultivated and grazing lands (Table 3). The CEC of the soil of all land use types at 20-30 cm was significantly higher than both 0-10 cm and 10-20 cm soil layers in all land use types, generally the present study found that, CEC of the soil of different land use types was low in the top layer except for forest land which was higher in 10-20 cm depth. The decrease in CEC from the bottom to the top soil layer might be attributed to the increase in clay contents with depth. As per the ratings recommended by [20] overall mean CEC value of forest land was moderate where as grazing land and cultivated land were classified as low status of CEC value.

Distributions of Ex. Ca and Ex. Mg across land use types were significantly influenced by soil depths (P=0.05). Generally, Ex. Mg and Ca distribution was decreased with increasing soil depths. The top 0-10 cm depth was higher in both Ex. Mg and Ca than the other two depths in all land use types. The higher content of Ex. Mg and Ca in the surface is probably due to forest litter and dead plant accumulation. [15] Stated that higher contents of Mg and Ca in the surface soil were due to the association of biological accumulation with biological activity and accumulation from plant. On the other hand, [21] stated that subsoil may be playing an important role as a nutrient storage.

Exchangeable potassium (Ex. K) contents of different land use systems were significantly affected by soil depths across all land use types in the study area while, distributions of exchangeable Sodium across land use type are not significant for both 10-20 cm and 20-30 cm depths [22].

In general, Deforestation, leaching, limited recycling of dung and crop residue in the soil, declining fallow periods or continuous cropping and soil erosion have contributed to depletion of basic cations on the cultivated land as compared to the adjacent forestland and grazing land. Although the farming system in Hursa water shade area of Gindeberet district is predominantly mixed crop-livestock, nutrient flows between the two are predominantly one sided, with feeding of crop residue to livestock but little or no dung returned to the soil.

Table 3. Comparison of CEC (cmol($^-$) kg^{-1}) and Exchangeable cations (cmol($^-$) kg^{-1}) in different land use types and soil depth. Results expressed as mean ± standard deviation.

Variables	Depth (cm)	Land use Types		
		Forestland	Cultivated land	grazing land
Ex. Ca	0-10	14.63±0.15[a]	8.70±0.10[c]	12.53±0.30[b]
	10-20	13.43±0.21[a]	8.03±0.15[c]	11.70±0.20[b]
	20-30	12.60±0.61[a]	7.76±0.15[c]	10.40±0.56[b]
Ex. Mg	0-10	5.60±0.10[a]	3.86±0.06[c]	4.80±0.20[b]
	10-20	5.40±0.30[a]	3.50±0.10[c]	4.50±0.10[b]
	20-30	5.00±0.10[a]	3.26±0.05[c]	3.80±0.20[b]
Ex. K	0-10	0.67±0.03[a]	0.23±0.01[c]	0.51±0.05[b]
	10-20	0.71±0.02[a]	0.29±0.05[c]	0.54±0.05[b]
	20-30	0.75±0.01[a]	0.33±0.03[c]	0.63±0.02[b]
Ex. Na	0-10	0.71±0.01[c]	0.21±0.03[a]	0.40±0.04[b]
	10-20	0.66±0.02[b]	0.41±0.04[a]	0.53±0.01[b]
	20-30	0.89±0.04[a]	0.46±0.01[a]	0.78±0.03[b]
CEC	0-10	16.56±3.06[a]	7.24±0.30[b]	15.16±1.60[b]
	10-20	18.76±2.10[a]	7.48±3.00[b]	14.54±1.80[b]
	20-30	14.28±1.01[a]	8.16±1.60[b]	13.54±1.00[c]

Means within rows followed by different letters are significantly different at (P =0.05).

4. Conclusions

The soil chemical properties of the study area were significantly affected by land use types except Ex. Na, which was insignificantly influenced by land use types. SOM content was observed high in forest land but it was low cultivated land. TN content of cultivated land soil decreased by 15, 27.27 and 20 % at 0-10, 10-20 and 20-30 cm soil depths respectively when compared with soil of forest land. The analysis of soil pH of the cultivated land showed that it was more acidic than the grazing land. The exchangeable cations in the study area did not show similar trends along the land use types. Ex. Ca and Ex. Mg were decreases along with depth from top to bottom surface soil. But, Ex. Na and Ex. K were increase with increasing soil depths. On the other hand Ex. Ca and Ex. K were significantly influenced by different land use types while Ex. Mg and Ex. Na were insignificantly influenced by different land use types. Based on the study on the selected soil physicochemical properties the following recommendations are made. This study indicates that, there is an urgent need to improve soil fertility by developing sustainable land use/cover practices to reduce the rate of soil erosion and to ensure long-term sustainability of the farming system, as a result national efforts are urgently needed to protect the remaining forests and to implement extension programmes to ensure sustainable use of lands and conservation of forested areas.

References

[1] A. Yifru and B. Taye "Effects of land use on soil organic carbon and nitrogen in soils of Bale, Southeastern Ethiopia". *Tropical and Subtropical Agro ecosystems* 14: 229 – 235, 2011.

[2] D. Halvorson and A. Reule, L. Anderson. "Evaluation of management practices for converting grass land back to cropland". *Soil Water Conser*vation. 55: 57–62, 2000.

[3] C. Kennedy and R. Papendick. "Microbial characteristics of soil quality". *Journal of Soil and Water Conservation.* 50: 243-248, 1995.

[4] A. Aluko and J. Fagbenro. "The role of tree species and land use systems in organic matter and nutrient availability in degraded Ultisol of Onne, Southeastern Nigeria". *Annual Conference of Soil Science Society, Ibadan*, Oyo State, 2000

[5] C. Achalu, G. Heluf, K. Kibebew and T. Abi. "Status of selected physicochemical properties of soils under different land use systems of Western Oromia", Ethiopia. *Journal of Biodiversity and Environmental Sciences* 2(3): 57-71, 2012.

[6] G. Majaliwa, R. Twongyirwe, R. Nyenje, M. Oluka, B. Ongom, J. Sirike, D. Mfitumukiza, E. Azanga, R. Natumanya, R. Mwerera and B. Barasa. "The Effect of Land Cover Change on Soil Properties around Kibale National Park in South Western Uganda". *Applied and Environmental Soil Science.* 1(10): 1-7, 2010.

[7]   G. Heluf and N. Wakene. "Impact of Land Use and Management Practices on Chemical and Properties of Some Soils of Bako Area Western Ethiopia". *Ethiopian Journal of Natural Resources.* 8(2): 177-197, 2006.

[8]   S. Mustafa and D. Orhan. "Influence of selected land use types and soil texture interactions on some soil physical characteristics in an alluvial land". *International journal of Agronomy and Plant Production.* 3 (11), 508-513, 2012.

[9]   G. Fikadu, A. Abdu, L. Mulugeta and F. Aramde. "Effects of Different Landuses on Soil Physical and Chemical Properties in Wondo Genet Area Ethiopia". *New York Science Journal* 5(11):110-118, 2013.

[10] B. Jones. "Laboratory Guide for Conducting Soil Tests and Plant Analysis". CRC press LLC. Pp 363-365, 2001.

[11] F. Khormali, M. Ajami, S. Ayoubi, C. Srinivasarao and P. Wani. "Role of deforestation and hillslope position on soil quality attributes of loess-derived soils in Golestan province Iran". *Agriculture, Ecosystems and Environment.* 134: 178-189, 2009.

[12] PEDOWS. "Zonal Atlas of West Shewa Planning and Economic Development Office for West Shewa Administrative zone. 1st edition, Ambo, Ethiopia". Printing section of the Ministry of Economic Development and Cooperation. Pp 34-35, 1997.

[13] SAS. "SAS System Version 9 for Microsoft Windows". SAS Institute Inc, Cary, NC, USA. 2002.

[14] M. Awdenegest, D. Melku and Y. Fantaw. "Land use effects on soil quality indicators: A case study of Abo-Wonsho Southern Ethiopia". *Applied and Environmental Soil Science.* 1(2013):1-9, 2013.

[15] B. Soto and F. Diazfierroz. "Interaction between plant ash leachates and soil". *International Journal of Wild life,* 3:207-216, 1993.

[16] B. Woldeamlak and L. Stroosnijder. "Effects of agro-ecological land use succession on soil properties in the Chemoga watershed, Blue Nile basin, Ethiopia". *Geoderma.* 111: 85-98, 2003.

[17] W. Genxu, M. Haiyan and C. Juan. "Impact of land use changes on soil carbon, nitrogen and phosphorus and water pollution in an arid region of northwest China". *Soil use and Management* 20: 32-39, 2004.

[18] Y. Teshome, G. Heluf, K. Kibebew and B. Sheleme. "Impacts of Land Use on Selected Physicochemical Properties of Soils of Abobo Area, Western Ethiopia". *Agriculture, Forestry and Fisheries.* 2(5): 177-183, 2013.

[19] M. Watkins, H. Castlehouse, M. Hannah and D. Nash. "Nitrogen and phosphorus changes in soil and soil water after cultivation". *Applied and Environmental Soil Science,* Hindawi Publishing. 2012

[20] P. Hazelton and B. Murphy. "Interpreting Soil Test Results: What Do All the Numbers Mean? 2nd Edition". Csiro publishing, Australia. PP 152, 2007.

[21] S. Ohta, S. Effendi, N. Tanaka and S. Miura. "Ultisols of lowland Dipterocarp forest in East Kalimantan, Indonesia, Clay minerals, free oxides and exchangeable cations". *Journal of Soil Science and Plant Nutrition* 39: 1-12, 1993.

[22] S. Sharma. and H. Manchanda. "Influence of leaching with different amounts of water on desalinization and permeability behaviour of chloride and sulphate-dominated saline soils'. *Agricultural and Water Management* 31(3): 225–235, 1996.

3

Effect of plant spacing on the yield and yield component of field pea *(pisum sativum L.)* at Adet, North Western Ethiopia

Yayeh Bitew, Fekremariam Asargew, Oumer Beshir*

Adet Agricultural Research Centre, Amhara Agricultural Research Institute, Bahir Dare, Ethiopia

Email address:

Yayeh_bitew@yahoo.com (Y. Bitew)

Abstract: Field pea is an important low-input break crops throughout the highlands of Ethiopia. The experiment was conducted on effect of spacing on the yield and yield component of field pea cultivars *(pisum sativum L.)* in 2012-213 cropping season at Adet Agricultural research station. Three intra row spacing's (5 cm, 10 cm and 15 cm) and two inter row spacing (20 cm and 25 cm) were evaluated using two released varieties, Sefinesh and Megeri on a plot size of 5 m x 5m (25 m2). The experimental design was a completely randomized block with 12 treatments in three replications. JMP-5 (SAS) software was used to compute the analysis of variance, correlation and regression analyses. Main effects of variety and intra row spacing had significant effect ($P<0.05$) on plant height, number of seeds per pod, seed yield while inter row spacing did not affect all examined attributes. The overall highest seed yield was recorded when Sefinesh was planted in 15 cm intra row spacing followed by Megeri in 5cm intra row spacing. The experiment revealed that average yield of Megeri increased when intra row spacing decreased. The reverse is true for Sefinesh. Similarly, increasing the intra row spacing revealed a peak seed yield at approximately 15 cm intra row spacing in Sefinsh. More importantly, increase in inter and intra row spacing together leads to increase and decrease the seed yield of Sefinesh and Megeri, respectively. Hence, 25 cm inter row with 15 cm intra row and 20 cm inter row with 5 cm intra row spacing, respectively gave the highest mean seed yield, and thereby increase the productivity of filed pea cultivars in West Gojam, but the experiment should be tested under small scale farmers' conditions.

Keywords: Field Pea, Intra Row Spacing, Inter Row Spacing, Megeri, Sefinesh

1. Introduction

Among the high land legume crops, field pea is the second important stable food grain in Ethiopia and mainly grown under rain fed conditions. It is important low-input break crops throughout the highlands of Ethiopia (1800-3000m. a.sl) [1]. About 150 thousand hectare of land is allocated to field pea production every year putting Ethiopia in the list of major filed pea producing countries in the world [2]. Field pea is primarily used for human consumption. It has high levels of amino acids, lysine and tryptophan, which are relatively low in cereal grains. Field pea contains approximately 21-25 % protein and high levels of carbohydrates, are low in fibber [3].

The seed yield obtained by local farmers is quite low and variable. Thus, the national average yield of field pea is 12 qt/ha [4]. Among the many yield limiting factors in field pea production under farmers practice plant population and planting method are important. High yields are realized with optimum plant population and planting method. In the past some observations in plant population and planting methods were made at Holleta. The indications were 500,000plants/ha and row planting at 40 cm spacing were optimum [2].

In Ethiopia, farmers traditionally either broadcast their seeds in isolation or mixed with Faba beans or as intercropping with other cereals and cover it with a local plow [2]. Due to lack of recommendations on inter and intra row spacing of field pea cultivars, plant populations on farmers' fields appear lower or higher than the optimum.

As a result very low yield is obtained. In addition to this seeds sown broadcast seeding is distributed unevenly (which may result in overcrowding). Also this method may not ensure that all seeds are sown at the correct depth. Furthermore, in field pea field broadcast seeding makes difficulty to weeding and other intercultural practices. Hence, the study on plant spacing (Inter and Intra-row planting methods) on field pea was initiated to find out the optimum plant spacing (Inter-row and Intra-row planting) for two selected varieties, Sefinesh (large seeded) and Megeri (Small seeded) of sowing at Adet Agricultural Research station.

2. Materials and Methods

The experiment on the effect of plant spacing (intra and inter-row spacing) on the two varieties of filed pea was conducted for two year (2012-2013) at Adet Agricultural research station located between 11017, N latitude and 370 43, E longitude with an altitude of 2240 m.a.s.l. The mean annual total rain fall is 1257 mm, ranging between 860 mm and 1771 mm and the average annual temperature is ranging from 90c to 25.50c.The highest rain fall in 2012 and 2013 were observed at August and September, respectively. However, the distribution of rain fall during the experimental months is explained in Figure 1. Monthly rain fall were more or less good uniform distribution during 2012 cropping season as compared to 2013 cropping season (Figure 1). Detailed chemical and physical soil characteristics of the experimental area are explained in Table 1 [5]. The experimental area is mainly clay in texture. Three intra-row spacing's (5 cm, 10 cm and 15 cm) and two inter-row spacing (20 cm and 25 cm) were evaluated using two released varieties, Sefinesh (large seeded) and Megeri (Small seeded) on a plot size of 5 m x 5m (25 m2) in RCBD design with 3 replications. The distance between each plot and replication were 0.5 m and 1m, respectively. At planting 100 kg/ha DAP were applied, planting date, weeding and other crop management practices were applied as recommended for the site. All the relevant data's including plant height, number of pods per plant, number of seeds per pod, seed yield (kg/ha) and thousand seed weight (gram), were collected from the net plot size and subjected to analysis of variance using [6] computer software. Comparisons between treatment means were made using Turkey HSD Test at 0.05 probability level. Correlation coefficients were used to determined relationships among the examined traits. Regression analyses were also performed between seed yield and intra row spacing and the interaction between both row spacing's.

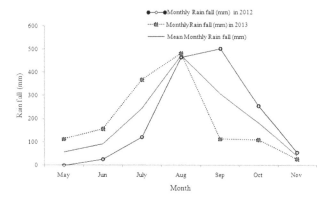

Figure 1. *Monthly Rain fall (mm) during the experimental years (2012-2013) at Adet Agricultural Research station*

Table 1. *Chemical properties of the soil in the research area (Adet Agricultural research area)*

Soil characteristics	Values
Sand	11%
Silt	21%
Clay	69%
Available P (Olsen)	1.688
Total N %	0.0949
Organic Matter %	1.898
cmol/kg K+	0.45
cmol/kg CEC	25.32
PH	5.17

3. Results and Discussion

3.1. Plant Height

The influence of row spacing (Intra and Inter-row spacing) and variety on the yield and yield component of field pea *(pisum sativum L.)* are presented in this paper. Inter row spacing had no significant effect on all growth, yield and yield component of field pea varieties (Table 2). Combined data of the two years demonstrated that variety (Table 2), intra row spacing (Table 2), interaction between variety and intra row spacing

(Table 3) and interaction between variety and inter row spacing (Table 3) had significant effect (P<0.05) on plant height. Results of analysis of variance also showed that combined data presented in Table 4 indicated that interaction between all possible treatments were differed highly significantly (P<0.01) in plant heights.

According to Table 1, results of combined statistical analysis indicated that, Sefinesh cultivar (157.6 cm) recorded the highest plant height as compared to Megeri (134.4 cm). Planting field pea using Intra row spacing of 5 cm (157.7 cm) gave the highest plant height as compared to boarder intra row spacing; 15 cm (143.3 cm) intra row spacing showed the lowest plant height (Table 1). Plant height increased linearly by increasing plant population (decreasing intra row spacing) due to competition of plants in higher densities on light, resulting in taller plants. Similar findings were achieved by [7], who indicated that denser plant population of pea increased plant height due to competition among plants.

The present study also more clarify that planting Sefinesh and Megeri in 10 cm intra row spacing gave the highest (160.2-166.5 cm) and the lowest (122.7-135.6 cm) mean plant height, respectively (Table 3). However, planting Sefinesh and Megeri in 25 cm inter row spacing gave the highest (156.8-161.5 cm) and the lowest (127.3-137.5 cm) mean plant height, respectively (Table 3). Planting Sefinesh in 10 cm intra row spacing and 20 inter row spacing (165-169.3 cm) gave the highest overall plant height as compared to other treatments (Table 4). However, it was on a par with planting Sefinesh in 5 cm intra row spacing and 25 cm inter row spacing (163-168.7 cm).This indicating that taller plant height was recorded when Sefinesh was planted under wider intra row spacing with narrower inter row spacing; and vise versa. The lowest overall plant height was recorded when Megeri was planted in 10/15 cm intra row spacing and 25 cm inter row spacing (112.7-134.8 cm) (Table 4). According to [8] plant height of different cultivars of field pea significantly affected by row spacing's. Higher plant height for Sefinesh planted under higher inter and intra row spacing was supported by [9], who expressed that cultivar x row spacing interactions, affected plant height character which was greater when wide rows was used. Higher plant height for Megeri planted under lower inter and intra row spacing was supported by [7], who indicated that denser plant population of pea increased plant height due to competition among plants. This is might be due to close row spacing, the space for plant spreading was less and hence plant height increased significantly. Findings of [10] found that the tallest plant from closer row spacing.

3.2. Number of Pods per Plant

In the current study, combined analysis of variance showed that both the main and the interaction effect did not significantly affect (P>0.05) the number of pods per plant

(Table2-4). Though, interaction between, variety and each plant spacing were statistically non significant effect on number of pods per plant on field pea cultivars in combined analysis, highest mean values was recorded when Sefinesh and Megeri were planted in 10 cm intra row spacing and 25 cm inter row spacing (Table 4). The increase in the number of pods per plant in wider row spacing may be due to vigorous plants as in wider spacing; plant grew vigorously and produced more branches which resulted in high number of pods per plant. On the other hand, in closer row spacing, the plant growth was decreased which resulted in less number of pods per plant. [10] found the highest number of pods/plant in wider row spacing as compared to closer spacing.

A reduction in the plant population (wider intra and inter row spacing) in the case of Sefinesh significantly increases the number of pods per plant and it was supported by [11] who found that a reduction in the plant population significantly increases the number of pods per plant.

3.3. Number of Seeds per Pod

Table 2. Combined main effects (variety, Intra-row spacing and Inter-row spacing) on the growth, yield and yield component of field pea at Adet (2012 -2013)

Treatment and statistics	Mean of agronomic attributes				
	PH	NPPP	NSPP	SY	TSW
Variety					
Sefinesh (large seeded)	157.6[a]	9.2	4.3[b]	946.8	139.4
Megeri (Small seeded)	134.4[b]	8.3	4.9[a]	1005.9	135.2
LSD (5 %)	**	NS	*	NS	NS
Intra row spacing					
5 cm	152.7[a]	8.3	4.4	913.3	135.1
10 cm	146.2[b]	9.6	4.8	993.1	137.5
15 cm	139.1[b]	8.3	4.7	1022.6	139.3
LSD (5 %)	*	NS	NS	NS	NS
Inter row spacing					
20 cm	146.2	8.6	4.6	975.9	137.1
25 cm	145.8	8.8	4.6	976.8	137.5
LSD (5 %)	NS	NS	NS	NS	NS

Table 3. Combined interactions between variety; intra and inter row spacing on the agronomic attributes of filed pea at Adet research station (2012-2013)

Treatment and statistics	Mean of agronomic attributes				
	PH	NPPP	NSPP	SY	TSW
Variety * Intra row spacing					
Sefinesh* 5 cm	160.9[a]	8.9	4.2[c]	767.4[b]	137.6
Sefinesh*10 cm	163.3[a]	9.8	4.4[abc]	954.0[ab]	140.0
Sefinesh* 15 cm	148.5[ab]	8.8	4.3[bc]	1119.0[a]	140.6
Megeri* 5 cm	144.4[abc]	7.7	4.7[abc]	1059.1[ab]	132.6
Megeri* 10 cm	129.1[c]	9.4	5.1[a]	1032.2[ab]	135.0
Megeri*15 cm	129.8[bc]	7.7	5.0[ab]	926.27[ab]	138.0
LSD (5 %)	*	NS	*	*	NS
Variety * Inter row spacing					
Sefinesh*20 cm	156.0[a]	8.9	4.2[c]	913.5	138.8
Sefinesh* 25 cm	159.1[a]	9.5	4.3[bc]	980.1	140.0
Megeri* 20 cm	136.5[b]	8.4	5.0[a]	1038.2	135.3
Megeri* 25 cm	132.4[b]	8.1	4.9[ab]	973.5	135.1
LSD (5 %)	*	NS	*	NS	NS
Intra-row spacing * Inter row spacing					
5 cm *20 cm	152.8	8.0	4.5	943.0	134.6
5 cm * 25 cm	152.5	8.6	4.4	883.5	135.6
10 cm* 20 cm	150.3	9.3	4.7	976.1	137.6
10 cm*25 cm	142.1	9.9	4.9	1010.2	137.3
15 cm* 20 cm	135.6	8.6	4.7	1008.6	139.0
15 cm* 25 cm	142.6	8.0	4.6	1036.6	139.6
LSD (5 %)	NS	NS	NS	NS	NS

3.4. Seed Yield

The present study also demonstrated that two year combined analysis of variance showed that variety (Table 2), interaction between variety and intra row spacing (Table 3), interaction between varieties and inter row spacing (Table 3) had significant effect (P<0.05) on the number of seeds per pod. Further, as seen Table 5 number of seeds per pod was positively and significantly correlated with seed yield (r=0.35*) and positively correlated with number of pods per plant (r=0.2). In contrary, 1000-seed weight was negatively correlated (r=-0.16) with number of seeds per pod. This indicated that increasing these traits in pea cultivar decrease 1000-seed weight and increasing seed yield of the pea cultivar. These results are in agreement with [8] who stated that negatively correlation between 1000-seed weight and seed per pod; and significant and positive correlation between seed yield and seed per pod indicated that increasing these traits in pea cultivar decrease 1000-seed weight and increase seed yield of the pea cultivar.

Among the field pea varieties, Megeri (4.7-5.2) gave the mean highest number of seeds per pod as compared to Sefinesh (3.8-4.8) (Table 2). The highest mean number of seeds per pod was recorded when Mgeri was planted in 10 cm (5.0) intra row spacing while planting Sefinesh in 5cm (4.2) gave the lowest number of seeds per pod (Table 3).This may be due to the fact that in wider row spacing the pods length was maximum resulting in maximum no seeds per pod in these plots and vice versa. Reports of [10] found that maximum number of seeds per pods in wider row spacing than closer row spacing

It is evident from the combined analysis of variance across two years indicated in Table 2 that the differences in variety and intra-row spaces had caused to significant changes in seed yield of field pea. Based on the statistical results, we detected that interaction effects between variety and intra row spacing was significantly influence (P<0.05) the seed yield of field pea cultivars (Table 3). Inter row spacing (Table 2), interaction between intra and inter row spacing (Table 3), interaction between intra row spacing, inter row spacing and variety (Table 4) had no significant effect (P>0.05) on seed yield of field pea cultivars. In terms of seed yield, data given in Table 5 also showed that positive and significant correlation between seed yield and pod number per plant (r=0.37*), seed number per pod (r=0.35*) functioned as major contributors to seed yield of field pea cultivars. These results represented that selection based on number of pods per plant and seed number per pod increase seed Yield of field pea. Similar results were obtained by [12] and [13].

Although, the mean effect of variety was statistically non significant, Megeri (1005.9 kg/ha) gave the highest grain yield as compared to Sefinesh (946.8 kg/ha) (Table 2). These variations in yield among field pea might be due to difference genetic potential of the variety and environment [14]. The above result was in contrary with the research out puts of [15], who reported that smaller and larger seeds of a same variety of

soybean will have the same yield potential. However, [16] reported that seed size was positive and significant relationship between TSW and seed yield of chickpea and lentil. As well, although, there was no significantly differences among the combined results of intra row spacing, the mean highest seed yield was recorded when field pea cultivars were planted in 15 cm intra row spacing (1022.6 kg/ha) (Table 2). The response of seed yield explained using quadratic equation (Figure 2) showed that increasing the intra row spacing revealed a peak seed yield at approximately 15 cm intra row spacing in the larger seed of field pea cultivar (Sefinsh). When planting Sefinesh from 5 cm to 15 cm intra row spacing the seed yield increased by 31.8%. However, planting Megeri from 5 cm to15 cm intra row spacing seed yield decreased by 12.3%.

Although, the interaction between of variety, intra and inter row spacing were statistically non significant effect on the seed yield of filed pea cultivars in the combination of two years, the combined overall highest seed yield was obtained when Sefinesh was planted in 15 cm intra and 25 cm inter row spacing (1175.8 kg/ha) followed by Megeri was planted in 5 cm intra and 20 cm inter row spacing (1121.7 kg/ha) (Table 4). The response of seed yield explained using quadratic equation illustrated under Figure 3 also showed that increasing the intra row spacing and inter row spacing together revealed a peak seed yield at approximately 15 cm intra 25 inter row spacing in the larger seed of field pea cultivar (Sefinsh) (Figure-3). Further increases intra and inter row spacing together for small seeded field pea cultivars (Megeri) result in yield penalty. This is similar to the previous findings of Ozveren, (2013) who showed that further increases in row spacing causes reduction in seed yield of field pea cultivars. However, further decreases intra and inter row spacing together for large seeded field pea cultivars (Sefinesh) result in seed yield reduction. The former and the later results are disagree and in agree, respectively with results of [17], who reported that the seed yield of dry pea increased with increasing plant population densities from 30 to 80 plant per m2.

3.5. Thousand Seed Weight

Combined analysis of variance showed that both the main and the interaction effects were not statistically significant change for 1000 seed weight (Table 2-4). Similarly, [8], [11] [18] and [19] identified that there were no significant differences in row spacing for 1000 seed weight. It was also reported by [20] who reported that there were no significantly differences among cultivars in terms of 1000-seed weight of field pea. Negatively correlation between 1000-seed weight and number of pods per plant, number of seeds per pod (Table 5) indicated that increasing these traits in pea cultivar decrease 1000-seed weight of the pea cultivar.

4. Conclusions

The present study demonstrated that the overall highest seed yield was recorded when Sefinesh was planted in 15 cm intra row spacing followed by Megeri in 5cm intra row spacing. The experiment revealed that average yield of Megeri increased when intra row spacing decreased. The reverse is true for the large seeded field pea (Sefinesh). Similarly, increasing the intra row spacing revealed a peak seed yield at approximately 15 cm intra row spacing in the larger seed of field pea cultivar (Sefinsh). When planting Sefinesh from 5 cm to 15 cm intra row spacing the seed yield increased by 31.8%. However, planting Megeri from 5 cm to15 cm intra row spacing

seed yield decreased by 12.3%. More importantly, increase in inter and intra row spacing together leads to increase and decrease the seed yield of Sefinesh and Megeri, respectively. Hence, 25 cm inter row with 15 cm intra row and 20 cm inter row with 5 cm intra row spacing, respectively gave the highest mean seed yield, and thereby increase the productivity of filed pea cultivars in West Gojam, but the experiment should be tested under small scale farmers' conditions.

Table 4. Combined interactions between varieties, intra and inter row spacing on the agronomic attributes of filed pea at Adet research station (2012-2013)

Treatment and statistics	Mean of agronomic attributes				
	PH	NPPP	NSPP	SY	TSW
Variety * Intra-row spacing *Inter row spacing					
Sefinesh* 5 cm*20 cm	156.1[abc]	8.0	4.2	764.3	139.8
Sefinesh*5 cm*25 cm	165.8[a]	9.9	4.2	770.5	134.8
Sefinesh* 10 cm*20 cm	167.1[a]	9.2	4.4	914.1	134.8
Sefinesh* 10 cm*25 cm	159.5[ab]	10.4	4.5	994.0	139.8
Sefinesh*15 cm*20 cm	144.8[abc]	9.5	4.2	1062.3	137.3
Sefinesh* 15 cm*25 cm	152.1[abc]	8.2	4.4	1175.8	137.3
Megeri*5 cm*20 cm	149.6[abc]	8.1	4.8	1121.7	134.8
Megeri*5 cm*25 cm	139.2[abc]	7.3	4.6	996.5	139.8
Megeri*10 cm*20 cm	133.5[bc]	9.3	4.9	1038.0	139.8
Megeri*10 cm*25 cm	124.8[c]	9.4	5.3	1026.4	134.8
Megeri*15 cm*20 cm	126.5[c]	7.7	5.2	954.9	137.3
Megeri*15 cm*25 cm	133.2[bc]	7.7	4.9	897.5	137.3
LSD (5 %)	*	NS	NS	NS	NS
CV (%)	6.8	11.1	2.0	16.5	5.8

Note: Means followed by different letters in columns are significantly different at 5 % of probability level according to Tukey HSD Test. On the above table PH , NPPP , NSPP, N0F· SY and TSW refers to plant height in cm, number of pods per plant, number of seeds per pod, seed yield in kilogram/ha and Thousand seed weigh in gram, respectively

Table 5. Estimation of correlation between important agronomic traits of field pea as affected by variety, intra and inter-row spacing at Adete research station average in two years (2012-2013) (n=36).

Variable	by Variable	Correlation	Signif Prob	Plot Corr
NPPP	PH	0.3902	0.0186	
NSPP	PH	-0.2971	0.0785	
NSPP	NPPP	0.2019	0.2377	
SY	PH	0.0136	0.9371	
SY	NPPP	0.3700	0.0263	
SY	NSPP	0.3498	0.0365	
TSW	PH	0.1598	0.3518	
TSW	NPPP	-0.0354	0.8375	
TSW	NSPP	-0.1597	0.3523	
TSW	SY	0.1640	0.3391	

Notes: PH , NPPP , NSPP, N0F· SY and TSW refers to plant height in cm, number of pods per plant, number of seeds per pod, seed   yield in kilogram/ha and Thousand seed weigh in gram, respectively

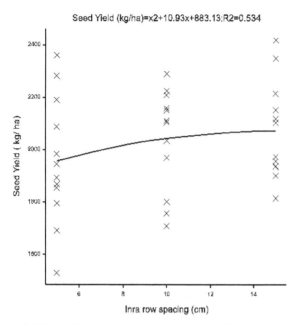

Figure 2. *Relationship between intra row and seed yield in combined of the two years*

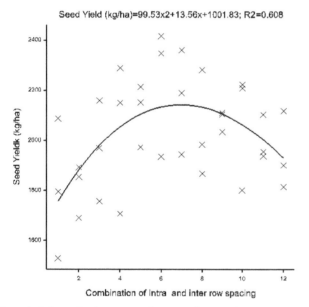

Figure 3. *Relationship between combination of intra and inter row spacing and seed yield in combined of the two years*

References

[1]   Amare G., and Adamu M., (1993). Faba bean and field pea Agronomy research Holleta and Sheno Agricultural research center pages 199-225h. In: Proceeding of the first National Cool-season food legumes review conference ,16-20 Decmber,1993 (Asfaw T., Geletu B., Mohan C.,and Mahmoud B.,eds), Adiss Abeba, Ethiopia

[2]   Beyene D., Alem B., Hailu G. and Beniwal S. (1989). Yield testing of promising field pea lines, Holeta research center. In: Nile Valley Regional program cool season food legumes, Research results, 1989 crop season, Institute ofAgricultural research, Addis Abeba, Ethiopia. akota State University and U.S. Department of Agriculture cooperating, USA

[3]   Central Statistical Agency (2012) Agricultural Sample Survey. 2011/2012 Volume I Report on Area and Production of Major Crops. Statistical Bulletin. May 2012, Addis Ababa.

[4]   Tilahun Tadesse, Alemayehu Assefa, Minale Liben and Zelalem Tadesse, 2013. Effects of nitrogen split-application on productivity, nitrogen use efficiency and economic benefits of maize production in Ethiopia. International Journal of Agricultural Policy and Research Vol.1 (4), pp. 109-115

[5]   SAS Institute Inc., 2002. JMP-5 Statistical Software, Version 5.Cary, NC, USA.

[6]   Inanç S and B. Yıldırım., 2007. The Effect of Different Row Space Applications on the Yield and Yield Components in Pea (Pisum sativum L.). Turkish VII. Field Crops Congress, 25-27 July, Erzurum

[7]   Derya Ozveren Yucel, 2013. Impact of plant density on yield and yield components of pea (pisum sativum ssp. sativum l.) cultivars. arpn Journal of Agricultural and Biological Science 2(8):169-174

[8]   Shirtliffe S.J. and A.M. Johnston. 2002. Yield-Density relationships and Optimum Plant Populations in two Cultivars of Solid-Seeded Dry Bean (Phaseolus vulgaris L.) Grown in Saskatchewan. Can. J. Plant Sci. 82: 521-529.

[9]   Dahmardeh M, M Ramroodi and J Valizadeh. 2010. Effect of Plant Density and Culticars on Growth, Yield and YieldComponents of Faba Bean (Vicia faba L.). Afr J. of Biot. 9(50): 8643-8647.

[10]   Hoorman J.J., R. Islam and A. Sundermeier. 2009. Sustainable Crop Rotations with Cover Crops. OSU, Agriculture and Natural Resources. Fact Sheet.

[11]   Sharma, S.K., 2002. Effect of sowing time and spacing levels on seed production of pea cultivar Arkel. Seed Res. 30(1): 88-91.

[12]   Momoh, E. J. J. and W. Zhou, 2001. Growth and yield responses to plant density and stage of transplanting in winter oilseed rape (Brassica napus L). Journal of Agronomy and Crop Science, 186: 253-259.

[13]   Togay N., Y. Togay, B. Yıldırım and Y. Doğan., 2008. Relationships between Yield and Some Yield Components in Pea (Pisum sativum ssp arvense L.) Genotypes by using Correlation and Path Analysis. African J. of Biotechnology. 7(23): 4285-4287

[14]   Anlarsal A.E., C. Yücel and D. Özveren., 2001. A Research on Determination of Yields and Adaptation of Some Pea (Pisum sativum ssp. sativum L. ve Pisumsativum ssp. avense L.) Lines at Conditions of Çukurova. J. Agric. Fac. Ç.Ü. 16 (3): 11-20.

[15]   Biçer, T., 2009. The effect of seed size on yield and yield components of chickpea and lentil. African Journal of Biotechnology Vol. 8 (8), pp. 1482-1487, Available online at http://www.academicjournals.org/AJB

[16]   Gan Y.T., P.R. Miller, B.G. McConkey, R.P. Zentner,P.H. Liu and C.L. McDonald. 2003. Optimum Plant Population Density for Chickpea and Dry pea in a Semiarid Environment. Can. J. Plant Sci. 83: 1-9.

[17]   Türk M., S. Albayrak and O. Yüksel. 2011. Effect of Seeding Rate on the Forage Yields and Quality in Pea Cultivars of Different Leaf Types. Turkish J. of Field Crops. 16(2): 137-141.

4

Water supply efficiency of brought for phenological stages on a few morpho-physiological parameters of the durum wheat (*Triticum durum Desf.*)

Aïcha Megherbi-Benali[1], Zoheir Mehdadi[2], Fawzia Toumi-Benali[1], Laid Hamel[1], Mohamed Benyahia[1]

[1]Ecodevelopment Spaces Laboratory, Sciences Environment Department Djilali Liabès University, Sidi Bel-Abbès, Algeria
[2]Vegetal Biodiversity, Conservation and Enhancement Laboratory, Sciences Environment Department, Djilali Liabès University, Sidi Bel-Abbès
 Algeria

Email address:

megaicha@yahoo.fr (A. Megherbi-Benali)

Abstract: Our work consists in quantify the benefits effects of a water supply brought to different phenological phases on the durum wheat yield and morpho-physiological traits associated to it. For this purpose, two tests are performed in the field: the first conducted under rainfed conditions and the second with the addition of water at 50 mm tillering and heading, and 60 mm during the filling phase grain. Results obtained for first trial confirm poor performance in the rainfed treatment compared with the irrigated treatment for all measured characters. However, the effect is more or less significant, depending on the caseon account of difference of the period for elaboration of each component of yield.

Keywords: *Triticum durum* Desf., Water Supply, Morpho- Physiological Traits, Yield, Sidi Bel Abbes

1. Introduction

The environmental stresses are very common in Algeria. Drought, cold and hot weather are often presented. In recent years, drought has become common, thus damaging the northern regions, especially those in western Algeria. The production constraint is abiotic aggravates loss of crop yields, especially durum. In Algeria, the practice of rained agriculture represents only 4.8 million hectares, which constitute almost the half of area, so two million hectares annually are not worked due to lack of rain and especially due to its poor distribution in space and time (Smadhi & Mouhouche, 2000).rainfall deficiency cause an important abiotic stress (frost, high temperatures, soil salinity, etc.) that results a significant yield losses (Baldy, 1992).

In fact, for stability or increased production, two alternatives are presented and must be moreover carried out:

- Further investigation on the identification and definition of morpho-physiological traits of adaptation, resistance, tolerance or escape to water stress,
- Develop a reasoned approach that fits the needs of the plant that help better characterize the water variable (precipitation), identify periods of stress and provide additional water for irrigation (Slama, 2002). This can be done through the identification of drought events, their occurrences, their intensities and durations, identifying likely risks of coincidence of these occurrences with sensitive phases of the plant and finally the estimated contributions of additional water and the responses of the plant in terms of efficiency, which are measured at harvest.

The objective of stabilizing or increasing the yield by additional inflows during critical phases of the growth cycle must be sought in the light of optimum utilization of this water. It is in this context that our work is to characterize six durum wheat genotypes vis-a-vis water stress in two different trials (*Triticum durum Desf.*): The first is conducted under rainfed conditions and the second with a supply of water at different phenological stages by measuring certain morphological, physiological and phenological characters.

The objective of stabilizing or increasing the yield by additional inflows during critical phases of the growth cycle must be sought in the light of optimum utilization of this water. It is in this context that the aim of our work is to characterize six durum wheat genotypes vis-à-vis water stress in two different trials (*Triticum durum Desf.*): The first is conducted under rainfed conditions and the second with a supply of water at different phenological stages by measuring certain morphological, physiological and phenological characters.

2. Material and Methods

2.1. Biological Material

Six durum wheat genotypes have been the subject of this study (Table 1).

Table 1. Code, name, pedigree and origin of genotypes tested

Code	Name	Pedigree	Origin
1	Oued Zénati	Sélection dans la population locale Bidi 17 T *Durum leucomelan*	GUELMA (Algérie)
2	Vitron	JO 's' //fg's'	Espagne
3	Oum Rabie 9	Haurani 27/Joc69	ICARDA (Syrie)
4	Chen 's'	Shwa's'/Bit's'	CIMMYT (Mexique)
5	Waha	Waha's'PLC's'/Ruff//Gta's'/3/Rolette	ICARDA (Syrie)
6	Boussalem	Heider/Marte//Huevo De Oro	ICARDA – CIMMYT

2.2. Study Area

The study was performed during the year 2010/2011 in the experimental station of the Technical Institute of the Field crops in the Sidi Bel Abbes region (Western Algeria), located west longitude 0 ° 38 'to latitude 35 ° 11 'and at an altitude of 486 meters, where the inter-annual and seasonal variability of rainfall is considered the major cause of changes in cereal yields which remain very low, ranging from 3 to 13 quitals / ha as indeed most arid areas where water is a limiting factor (Mourret & al., 1990).

The soil of study area is characterizing by moderate alkaline pH (8.92), non-saline (0.100 µmhos / cm), rich in limestone (24.83%), rich in phosphorus (33.00 ppm), low in organic matter (1.93%) and have a moderate concentration in nitrogen (0.06 %). The climate is semiarid lower-cost winter; precipitations are poorly distributed in space and time, situated between 200 and 400 mm / year, often resulting in significant water deficit (Benseddik & Benabdelli, 2000).

Our experiment was practicing in an environment with a total rainfall of 262.7 mm, and an overall deficit of - 62.3 mm compared to the average Seltzer (1946). This deficit is even greater if we consider the period of settlement stand and tillering which corresponds to december (-7.6 mm) and January (- 28 mm) and the period of maximum development of dry matter which is in February (- 26 mm) (ONM, 2011). In addition, the number of days in winter frost is important during the months of January (13 days) and February (18 days) periods coincided with the installation of the stand and tillering stages.

2.3. Methodology

We have adopted supplementary irrigation system, made by sprinkling. A factorial arrangement is adopted when the main factor is the irrigation system and the secondary factor is all genotypes thus established, planted in randomized blocks with three replications. The unit area is 6 m². The plots are irrigated with the aforementioned irrigation system and soil moisture is measured by gravimetry each decade.

We have applied a phosphorus and nitrogen fertilization background and burndowns anti-cotyledons tillering stage. Sowing was made december 17, 2010 with a population of 250 plants / unit area.

For this study, both treatments were adopted:

1 Treatment 1: control of the water supply is carried out using rain gauges. The amounts of water have been made of two irrigations of 50 mm, made during the two stages tillering and heading, and a third 60 mm irrigation during the grain filling stage, or 160 mm in total.

2 Treatment 2: it is the trial in dry (under rained conditions).

2.4. Evaluation of the Measured Parameters

2.4.1. The Morpho-Phenological Parameters Evaluated are

- The number of days from sowing to heading: number of days from emergence to the release date of 50% of ears per variety and per plot (DHE)
- The number of semi-mature days: number of days from emergence to the date when the envelope spike lose their green color (DMA)
- The plant height (PH): Average height in centimeters plants, measured from the ground to the top of the spikes (barbs not included)
- The spike cervical length (SCL)
- The length of the last inter-node (LLIN)
- The spike length (S.L)
- The beards length (B.L)
- Senescence of the flag leaf (LAD).

2.4.2. The Physiological Parameters are

- Leaf area (SF) by the method of Paul et al. (1979).
- Relative Water Content (RWC) by the method of Barrs (1978).
- Rate water loss (cuticular transpiration) (CT), by the method of Clarke (1990).
- Chlorophyll content (Chl.C) by method Hicox Israelstam & (1978).

2.4.3. Yield Components are

- The tillers number per square meter (Tillers / m²)
- The spikes number per square meter (Spikes / m²)
- The rate of tillers regression (RT.R)
- The grains number per spike (Grs. /spike)
- The thousand kernel weight in grams (TKW)
- Biological yield in quintals per hectare (Bio. yield)

- The crop yield in quintals per hectare (yield grains.)
- Harvest index (HI).

2.5. Statistical Analyses

We have treated our data by the Test of Variance to show means and correlations between our results, for this we have using STATISTICA version 6

3. Results

The observations of our study demonstrate the positive effect of supplemental irrigation on all yield components and show that earliness at heading is a key character in the adaptation of wheat durum to water stress and water intake during this period, and enable it to better express its growth and development, to obtain high yields. In fact, it is therefore imperative to understand the climate risk management, often the cause of low agricultural production. As earliness at heading is a key character in the adaptation of durum wheat to water stress and water intake during this period, and enable it to better express its growth and development, to obtain high yields. It is therefore imperative to understand the climate risk management, often the cause of low agricultural production.

3.1. ANOVA for the Various Traits Measured

The response of the different genotypes tested, providing natural or artificial water, shows an improvement in most of the parameters measured. In our study, we note a significant early genotypes Chen's' and Waha in both treatments with 102.33 respectively, 103 days in rainfed and irrigated 160 and 165.33 (Table 2); and the best Straw yields correspond to greater heights irrigated. Statistical analysis revealed a significant genotype effect and irrigation on all yield components in both treatments (Table 3).

Table 2. Variance analysis of the morpho-phenological caracters.

	DHE (days)		DMA (days)		HP (cm)		SCL (cm)	
	Rainfed	Irrigated	Rainfed	Irrigated	Rainfed	Irrigated	Rainfed	Irrigated
- Oued Zénati	119.67	168.67	123.67	174.67	71.25	113.75	20.3	29.20
- Vitron	116.67	165.33	122.33	174.00	69.58	115.92	18.58	22.10
- Chen's'	102.33	160.00	117.00	170.00	55.00	81.67	26.70	32.10
- Boussalem	109.67	167.00	112.33	164.33	67.50	90.00	21.33	28.60
- Waha	103.00	165.33	109.00	164.33	61.25	95.83	21.33	32.93
- Oum Rabie 9	111.67	160.33	105.00	167.00	68.33	88.33	25.24	35.00
- C.v (%)	2.00	1.40	1.00	1.00	8.30	7.20	8.16	14.20
- Standart variation	2.17	1.62	1.65	1.74	4.45	7.02	1.94	4.26
- Génotype effect	S	S	S	S	S	S	S	S
- Average experimenti	110.50	114.89	164.78	169.06	65.49	97.58	22.30	29.99

Table 2. Continued

	L.LIN (cm)		Spike L (cm)		Barbs L. (cm)	
	Rainfed	Irrigated	Rainfed	Irrigated	Rainfed	Irrigated
- Oued Zénati	9.32	13.36	4.73	4.90	12.46	14.23
- Vitron	10.48	15.05	5.10	4.23	12.77	14.2
- Chen's'	10.1	13.01	5.83	6.13	11.92	13.83
- Boussalem	10.59	13.85	5.63	6.47	12.07	15.7
- Waha	10.63	11.67	6.57	5.90	13.07	14.83
- Oum Rabie 9	8.61	13.37	5.20	5.63	12.46	13.27
- C.v (%)	10.00	9.40	6.90	11.10	7.50	13.80
- Standart variation	1.69	2.20	0.38	0.62	0.93	1.98
- Génotype effect	NS	NS	S	S	NS	NS
- Average experimenti	9.95	13.39	5.51	5.54	12.46	14.34

Table 3. Variance analysis of yield and its components.

	Tillers / m²		Spikes / m²		RTR (%)		Grs / épi	
	Rainfed	Irrigated	Rainfed	Irrigated	Rainfed	Irrigated	Rainfed	Irrigated
- Oued Zénati	465.00	551.67	212.50	273.50	61.48	41.18	19.93	24.87
- Vitron	569.17	582.50	206.67	271.67	64.52	52.26	16.67	22.20
- Chen's'	505.33	509.83	287.50	303.33	43.55	40.03	30.60	44.53
- Boussalem	456.17	579.17	247.50	415.00	45.74	28.34	23.73	31.27
- Waha	600.00	726.33	271.67	320.83	54.72	55.85	23.80	30.13
-Oum Rabie 9	529.17	589.17	238.33	335.00	54.96	43.14	27.73	41.07
- C.v (%)	7.00	7.10	8.10	14.10	5.20	7.40	17.40	16.30
- Standart variation	37.66	71.89	19.66	45.07	7.88	6.3	4.12	5.28
- Génotypeeffect	S	S	S	S	S	S	S	S
- Mean experiment	538.14	572.53	244.03	320.72	54.16	43.46	23.74	32.34

Table 3. Continued

	TKW (grs)		Bio. yield (qx/ha)		HI (%)		Grains yield (qx/ha)	
	Rainfed	*Irrigated*	*Rainfed*	*Irriauted*	*Rainfed*	*Irrigated*	*Rainfed*	*Irrigated*
- Oued Zénati	47.65	56.75	32.18	62.33	17.93	25.61	5.77	15.98
- Vitron	52.00	57.36	30.39	39.65	21.25	32.53	6.46	12.90
- Chen's'	41.60	44.09	51.37	56.85	10.53	42.14	5.41	23.96
- Boussalem	50.33	53.16	41.89	59.49	19.33	28.10	6.28	16.72
- Waha	45.44	49.41	49.04	55.67	12.80	37.11	8.10	20.66
-Oum Rabie 9	44.27	49.60	39.81	44.43	14.01	35.98	5.58	15.99
- *C.v (%)*	*7.20*	*6.80*	*11.40*	*9.50*	*8.70*	*7.60*	*7.85*	*8.50*
- *Standart variation*	*3.44*	*3.47*	*5.06*	*6.29*	*8.00*	*7.90*	*0.74*	*3.27*
- *Génotypeeffect*	*S*	*S*	*S*	*S*	*S*	*S*	*S*	*S*
- *Mean experiment*	*47.71*	*50.90*	*40.78*	*53.06*	*15.97*	*33.57*	*6.27*	*17.70*

Table 4. Variance analysis of the physiological parameters.

	SF (cm²),		RWC (%)		RWL (gr.10⁻³/cm2/min)		Chl. C (μg/gMF)	
	Rainfed	*Irrigated*	*Rainfed*	*Irrigated*	*Rainfed*	*Irrigated*	*Rainfed*	*Irrigated*
- Oued Zénati	29.42	39.86	33.22	65.95	1.54	1.98	26.23	42.47
- Vitron	36.83	40.05	37.24	67.24	1.04	1.66	37.00	60.02
- Chen's'	32.05	37.46	40.63	68.75	1.38	1.91	26.60	47.35
- Boussalem	29.53	33.96	53.21	70.17	1.02	1.80	29.35	52.24
- Waha	26.04	42.18	47.96	71.25	2.34	3.04	38.73	59.47
- Oum Rabie 9	27.01	34.24	41.02	70.96	1.98	2.51	36.58	60.29
- *CV (%)*	*9.00*	*12.30*	*4.02*	*8.90*	*9.20*	*12.30*	*9.50*	*10.20*
- *standart variation*	*4.56*	*5.36*	*7.9*	*6.45*	*0.35*	*0.42*	*2.31*	*4.84*
- *Genotype effect*	*S*	*NS*	*S*	*NS*	*S*	*S*	*S*	*S*
- *Mean experiment*	*30.14*	*37.95*	*42.21*	*69.05*	*1.55*	*2.15*	*32.41*	*53.64*

S: significant effect; NS: insignificant effect; CV: coefficient of variation

The relative water contents of six genotypes is higher in irrigated trial conducted (Table 4). Genotypes Boussalem, Waha and Umm Rabie recorded the highest values in both treatments (53.21, 47.96, 41.02, 70.17, 71.25, and 70.96% respectively in dry and irrigated). Also, genotypes lose more water and recorded a rate greater than total chlorophyll irrigated stress conditions (Table 4).

3.2. Correlation between Performance and the Parameters Measured

1 *In the trial conducted under rainfed conditions, yields are correlated:*
- Positively to the spike length (+ 0.621 +) and spike cervical length (+ 0.723); negatively and DHE(- 0.931) and DMA (- 0.747),
- Positively to the spikes number per unit area (+ 0.955) and grains number / spike (+ 0.771); and negatively correlated with the thousand kernels weight (- 0.803)
- Positively to all the physiological parameters measured.
2 *In the trial conducted under irrigation, yields are correlated:*
- Positively to the barbs length (+ 0.894), the harvest index (+ 0.704), the tillers number (+ 0.255) and of spikes per unit area (+ 0.311)
- Negatively total chlorophyll content (- 0.404).

4. Discussion

In the six genotypes, we note a clear difference in all measured parameters, not only between genotypes and between treatments.

Heading date is often used as an important character that influences grain yields, especially in areas where the distribution of rainfall and temperature variability affect the length of the development cycle (Hadjichristodoulou, 1987). It is at this stage that the plant architecture becomes apparent and has a maximum reaches; it often gives an indication of the differential capacity of genotypes (Kirby et al., 1999). In our study, there is a significant negative correlation between not only DH E and final yield (- 0.931) against dry - 0.397 in irrigated), but also with the spike length, grains / spike and TKW (- 0.858, - 0.790, and + 0.762, - 0.853, - 0.705 and + 0.779), respectively in rainfed and irrigated). Given the random distribution of rainfall in semi-arid to arid regions, thus the adoption of genotypes in a relatively short cycle is required. Fisher & Maurer (1978) found that daily gain in early generates an efficiency of 30-85 kg / ha.

In the environment where late frost is a constraint to cereal production (10 days of frost in March), excessive precocity is of no use; rather, it may be a source of instability in grain yields. Too much early is not always advantageous following the negative effects of low spring temperatures on fertility of the pollen grain and that of the ovary (Bouzerzour & Benmahammed 1994). Dodging allows the plant to reduce or cancel the effects of water stress by a good fit of the crop cycle the length of the rainy season (Amigues & *al.,* 2006). Performance of many genotypes was improved by shortening the life cycle in almost all annual crop species (Turner et al., 2001), legume (Subbarao, 1995), as the corn (Fokar & *al.,* 1998).

In fact   by using irrigation, we note a significant correlation between plant height, fertile spikelets number, grains number / spike, and the DHE and DMA with a coefficient respectively - 0.815 - 0.877 + 0.705 and + 0.711. By cons in trial in rainfed conditions, the height recorded a low correlation with all the traits measured; water deficit recorded during the run causes regression of large, lower stems and the surface of the outer leaves (Gate *et al.,* 1990). Similarly, it should be noted that the best yields straw correspond to greater heights in irrigated (Oued zenati with a height of 113.75 cm and recording a biological yield of 62.33 quintals / ha). According to Ben Abdellah and Ben Salem (1993), genotypes with high straw have better adaptation to water deficit.

For beards, their length is strongly correlated with irrigated yield (+ 0.894), whereas it is very weakly correlated dry (+ 0.318). Their presence, in cereals, increases the possibility of use of water and the development of dry matter during the maturation of seeds. The photosynthesis in bearded to genotypes hairless show that genotypes are less sensitive to the inhibitory action of high temperatures during grain filling (Fokar & *al.,* 1998). Comparing three durum wheat genotypes by Slama (2002) found that the genotype with the beard most developed under water stress, has the best performance. Indeed, the barbs can improve performance under drought conditions by increasing the photosynthetic area of the spike (Slama & *al.,* 2005).

The reduction in leaf area, when water stress is very important as mechanism for reducing the need for water (Perrier & Salkini, 1987). In rainy conditions, this surface is highly correlated with Chl.C (+ 0.871) and L.LIN (+ 0.975); while it is insignificant in irrigated (+ 0.005).

Researchers conducted under conditions of water restriction on the response of wheat to an addition of natural or artificial water show an improvement in most components of performance that strongly depends on the grains number / spike , weight grains / spike and spikes number / m² (Assem & *al.,* 2006), except for those who are sensitive to the phenomenon of compensation or interaction, which particularly affects the TKW for a high number of grains per spike and / or HI for high biomass production. After the vegetative period, for the development of the biomass used for ensuring the achievement of the breeding season, their development is carried out in several stages. During the first stage, it is to ensure a high rate of heading which ensures better grains / spike. Thereafter, it will, on the one hand to maintain the high number of maximum grain avoiding the phenomenon of seed abortion implemented. On the other hand, it will properly meet these seeds giving good assimilate transfer from vegetative organs, which allows a high TKW.

Factually, tillering is a factor in determining grain yield in cereals (Hucl & Baker 1989) and early during water stress reduces the number and size of tillers in wheat vegetative phase (Davidson & Chevalier, 1990; Stark & Longley, 1986; Blum & *al.,* 1990). Water stress occurring at the vegetative stage causes a reduction in the number of tillers fertileset early during water stress reduces the number and size of tillers in wheat vegetative phase. Waha genotype had the best coefficient of tillering (600 and 726.83 tillers / m², respectively in dry and irrigated). Under rainfed conditions, this component is positively correlated with spikes / m² (+ 0.716) and L. barbs (+ 0.889); by against irrigated, it is correlated to spikes / m² (+0. 5.72).

The ability to produce high-ground biomass indicates a better adaptation to the production environment (Austin *et al.,* 1980). As against an excess of foliage can also lead to a waste of water, following a greater leaf area at a time when the plant has more need (Araus & *al.,* 1998). The appearance of tillers seems to be related to the general characteristics of the genotype and environmental conditions; while their disappearance is linked to deficiencies caused by competition between tillers to environmental factors (Baldy, 1992) and water (Day & *al.,* 1978).

Spikes number / m² comes from the number of tillers emitted by the plant during tillering and tiller number lost during the next phase, and water deficit in the run results in regression of spikes number / m² (Megherbi & *al.,* 2012). This component is highly correlated with the number of spikelets fertile in both treatments with a correlation, coefficient of + 0.723 and + 0.933 respectively in irrigated and dry. The supply of water is a significant way to the formation of the ears stand. Boussalem genotype produced the best number spikes irrigated (415 spikes / m²) and Chen's' dry (287.50 spikes / m²).

For Grs.N / spike, the determination of this component is more complex because it depends on the number of spikelets / spike and number of grains per spikelet that occur at different times of the cycle; and the final amount of each component depends on the number of differentiated organs. Thus, genotypes Oum Rabie and Chen's' maintain a high number grains / spike in both treatments (27.73 and 30.60 in dry, 41.07 and 44.53 respectively irrigated). Under rainfed conditions, this parameter is positively related to S.C.L. (+ 0.980), spikes N. / m² (+ 0.917), and grains yieldns (+ 0.771); while in irrigated, it is positively related to SCL (0751) and spikes N./ m² (+ 0.624) and negatively with plant height (- 0.877).

To make the best bet of the water, increase the useful plant biomass where each party involved in grain filling and thus creating a high TKW (Riou, 1993).

The results show that the difference is not significant between the two treatments for TKW (gain = 3.19 gr). The weather is a key during the maturation phase vis-a-vis the TKW; However, you should know that stress occurring before heading may limit the weight acting on the envelope size (Couvreur, 1985). In rainy conditions, this component is positively correlated with DHE (+ 0.762) and DMA (+ 0.931) and negatively Grs. N. / spike (- 0.884) and grains yield (- 0.804); which irrigated it is positively correlated with plant height (+ 0.630) and DHE (+ 0.779), negatively to Grs. N./ spike (- 0.804) and biological yield (- 0.799). The lack of water after flowering, combined at elevated temperature results in a decrease of TGW by alteration of the rate of grain filling and / or the filling time (Triboi 1990). During this phase, the lack of water results in a reduction of the grain size (scalding), thus is reducing the yield.

Irrigation is very significant effect on biological yield. Irrigated, it is strongly correlated with spike length (+ 0.823), spikes N. / m² (+ 0.855), Grs. N. / spike (+ 0.719) and S.C.L. (+ 0.666). In this context, the genotype Oued zenati recorded the fastest straw yield (62.39 quintals / ha), genotype (at considerable height) seems better use the water supply; followed Boussalem (59.49 quintals / ha).

Genotypes Chen's' and Waha have low H.I. under rainfed conditions (10.80 and 12.53%) and higher in irrigated (42.14 and 37.11%) compared to others. That irrigated, it is positively correlated with the spike length. (+ 0.500) and SCL (+ 0.710), characters highlighted in the selection (Ferrera *et al.,* 2004), spikes N. / m² (+ 0.628), Grs. N. / spike (+ 0.825) and biological yield (+ 0.841); and negatively for TKW (- 0.882) and DHE (- 0.798). In rainy conditions, it is positively correlated with tillers N. / m² (+ 0.796), the spike L. (+ 0.629), to SCL (+ 0.692) and Grs. N. / spike (+ 0.652).

Leaf area is an important determinism sweating thus one of the first responses of plants to drought stress is to reduce the leaf area in order to keep their water resources (Lebon *et al.,* 2004). The results reflect the existence of a positive relationship between the S.F and the TKW, HI, RWC, RWL and L. LIN under rainfed conditions, with respective correlation coefficients of + 0.958 + 0.956 + 0.764 , + 0.965 + and + 0.975 ; and negatively with the tillers number / m². Irrigated, it is positively correlated with the height plant (+ 0.660) and negatively with tillers N. / m² (- 0.725) and grs. N. / spike (- 0.605).

Water scarcity induced in stressed plants, we look a decrease in the RWC, which gives an indication of the water status of the plant at a time when beginning to determine grain yield (Clark & Mac Caig, 1982). Richards and Townley (1987) mentioned that this characteristic is positively correlated to the efficiency of water use. The water content of the sheets decreases proportionally with durum the reduction of water in the soil (Bajji & *al.,* 2001). The decrease in the RWC is faster in susceptible genotypes than in resistant genotypes (Scofield & *al.,* 1988).

Our results also show the existence of a positive correlation between RWC and height (+ 0.715), PMG (+ 0.690), the grains yield (+ 0.538) and HI (+ 0.865) and negatively with tillers N. / m² (- 0.613) under rainfed conditions. Irrigated, there is a positive relationship between RWC and Grs. N. / Spike (+ 0.465) and negatively with DHE (- 0.550), the DMA (- 0.943) and plant height (- 0.819).

For RWC, several researchers have shown that the leaves that come from stressed plants lose more water than non-stressed plants (Clark and Mac Caig, 1982). Kirkham *et al.,* (1980) directly link the loss of water to the leaf surface, the more it is wider, the attrition rate increases as in the case of our study. Some genotypes have the feature to roll their leaves during water deficit, allowing the reduction of water loss by cuticular transpiration. Clark and Romagosa (1991) also note that the rate of loss of leaf water is indicative of the ability of drought tolerance.

RWC, under rainfed conditions, correlates positively with Grs. N. / spike (+ 0. 519), TKW (+ 0.948), the biological yield (+ 0503), the grains yield (+ 0.518), HI (+ 0.907 +) and the RWC (+ 0.669); and negatively tillers N. / m² (- 0.660). In irrigated, the RWL is positively correlated with spikes N. / m² (+ 0531), the RWC (+ 0.670), the Chl. C. (+ 0607) and SCL (+ 0.706) and was negatively correlated with DMA (- 0.572) at tillers N. / m² (- 0.779) and L.LIN (-0837). In rrigated, this parameter is positively correlated to the RWC (+ 0.607) and negatively biological yields (- 0.780); under rainfed conditions, it is positively correlated with height plant (+ 0.601), the TKW (0813 +) to HI (+ 0.930), RWC (+0.915), the RWL (+ 0.803), SF (+ 0.871) and L.LIN (+ 0.830) and negatively correlated to the number of tillers / m² (-0.630).

Quantitative differences in the total chlorophyll content noted between genotypes are related to drought tolerance (Gummuluru and Hobbs, 1989). Hireche (2006) shows in his work on alfalfa, the genotype Dessica tends to fight against water stress by lowering its chlorophyll content. Falling Chlorophyll is the consequence of the reduction in stomatal aperture to limit water loss through evapotranspiration and increased resistance to the input of atmospheric CO_2 for photosynthesis (Bousba *et al.,* 2009). The amount of chlorophyll of the leaves can be influenced by many factors such as age of the leaves, the leaves and the position of the environmental factors such as light, temperature and water availability.

5. Conclusion

In the Mediterranean region, drought is a major cause of yield losses ranging from 10-80% depending on the year.

As we have pointed out, much of the biomass production is developed before the breeding season, which accounts for the bulk of our two experiments designed to study the effect of supplemental irrigation brought at different phenology stages of the breeding season on yield components of durum wheat. However, the function of development and growth continues to be achieved during the first part of the breeding season, but with less intensity as the results (development of tillers and spikes number per unit area) shows The beneficial effects of supplemental irrigation are apparent for most of the parameters studied. We see that the most important seed weight components are represented by Grs N./ spike and TKW, respectively with a relative production of grains +8.60 / +3.19 gr spike over the trial in terms storm.

The results obtained by the storm treatment especially show that lack of water is much more negatively on the TKW on the number of seeds / spike.

Comparison of results for the first test confirms the poor performance of rainfed treatment versus irrigated treatment for all yield components. Indeed for the second test, a make-up water for each of the three phenological phases had a beneficial effect on all components of performance. However, this effect is more or less important, as appropriate, due to the shift of the period of development of each component of return that actually corresponds to a specific phase of the breeding season. This confirms the interest for focused the most sensitive phases that promote better irrigation water.

Although understanding the mechanisms developed by the cereal in response to additional water supplies is a useful approach, analysis in terms of efficiency only, is less cumbersome to implement and may be recommended to users.

References

[1] Amigues JP, Debaeke P, Itier B, Lemaire G, Seguin B, Tardieu F, Thomas A, 2006. Sécheresse et agriculture. Réduire la vulnérabilité de l'agriculture à un risque accru de manque d'eau. Expertise scientifique collective, *Rapport, INRA (Fr)*.

[2] Araus, JL, Amaro T, Voltas J, Nakhoul H, and Nachit MM. 1998. Chlorophyll florescence as selection criteria for grain yield in durum wheat under Mediterranean conditions. *FCR*, 55: 209-223.

[3] Assem N, El Hafid L, Haloui B, El AtmaniK 2006. Effets du stress hydrique appliqué au stade trois feuilles sur le rendement en grains de dix variétés de blé cultivées au Maroc oriental. Science et changements planétaires. *Sécheresse*. 17 (4): 499-505.

[4] Austin RB, Bingham J, Blackwell RD, Evans LT, Ford MA, Morgan CL, Taylor M, 1990. Genetic improvements in winter wheat yields since 1900 and associated physiological changes. *The journal of agricultural science* 94: 675-689.

[5] Bajji M, Lutts S, Kinet JM, 2001. Water deficit effects on solute contribution to osmotic adjustment as a function of leaf ageing in three durum wheat (*Triticum durum* Desf.) cultivars performing differently in arid conditions. *Plants Sciences*. 160: 669 - 681.

[6] Baldy, C. 1992. "Effet du climat sur la croissance et le stress hydrique des blés en Méditerranée occidentale". In : Tolérance à la sécheresse des céréales en zones méditerranéennes. Colloque Diversité génétique et amélioration variétale, Montpellier (France), 15-17 décembre 1992. *Les colloques* 64: 83-93.

[7] Ben Abdellah N, Ben Salem M, 1993. Paramètres morpho-physiologiques de sélection pour la résistance à la sécheresse des céréales. In : Tolérance à la sécheresse des céréales en zones méditerranéennes. Colloque Diversité génétique et amélioration variétale, Montpellier (France), 15-17 décembre 1992. *Les colloques* 64 : 173-190.

[8] Benseddik B, Benabdelli K, 2000. Impact du risque climatique sur le rendement du blé dur (Triticum durum Desf.) en zone semi-aride : approche éco-physiologique. *Cahiers sécheresse* 11 (1): 45-51.

[9] Blum A, Ramaiah S, Kanemasu ET, and Paulsen GM, 1990. Recovery of wheat from drought stress at the tillering developmental stage. *Field Crop Res* 24 : 67-85.

[10] Bousba R, Yekhlef N, Djekoun A, 2009. Water use efficiency and flag leaf photosynthetic in response to water deficit of durum wheat (*Trticum durum Desf*). *World Journal of Agricultural Sciences* 5: 609 -616.

[11] Bouzerzour H, Benmahammed A, 1994. Environmental factors limiting barley grain yield in the hight plateux of eastern Algeria. *Rachis* 12: 11-14.

[12] Couvreur F, 1985. Formation du rendement d'un blé et risques climatiques. *Perspectives Agricoles* 95 : 12-15.

[13] Clarck JM, MacCraig P, 1982. Excised leaf water relation capability as an indicator of drought resistance of Triticum genotypes. *Canadian Journal Plant Sciences 62*: 571-576.

[14] Clark JM, Romagosa I, 1991. Evaluation of excised leaf water loss rate for selection of durum wheat for dry environments. *Les colloques* 55: 401-414.

[15] Davidson DJ, and Chevalier P, 1990. Pre-anthesis tiller mortality in spring *wheat. Crop Sciences 30*: 832-6.

[16] Day W, Legg BJ, French BK, Johnston AE, Lawlor DW, and Jeffers C, 1978. A drought experiment using mobile shelters: the effect of drought on barley yield, water use and nutrient uptake. *Journal Agricultural Sciences. Camb.* 91: 599-623.

[17] Ferrera R, Sellés G, Ruiz RS, Sellés IM, 2004. Effect of water stress induced at different growth stages on grapevine cv. Chardonnay on production and wine quality. *Acta Horticulturae 664*: IV International Symposium on Irrigation of Horticultural Crops: 233- 236.

[18] Focar M, Nguyen HT, Blum A, 1998. Heat tolerance in spring wheat. Grain filling. *Euphytica* 104: 9-15.

[19] Fischer RA, Maurer R, 1978. Drought resistance in spring wheat cultivar. Grain yield responses. *Australian Journal of Agricultural Resarch* 29: 897-912.

[20] Gate Ph, Boutier A, Woznica K, Manzo MO, 1990. Drought resistance of winter wheat. The first results. *Perspective Agricoles* 145: 17-23.

[21] Gummuluru S, Hobbs LA, 1989. Genotype variability in physiological cbaracters an dits relationship to drought tolerance in durum wheat. *Canadian Journal Plants Sciences* 69 : 703 - 711.

[22] Hadjichristodoulou A, 1987. The effect of optimum heading date and its stability on yield and consistency of performance barley and durum wheat in dry areas. *Journal Agricultural Sciences. Camb.* 108: 599-608.

[23] Hireche M, 2006. *Réponse de la luzerne Médicago sativa (L.) au stress hydrique et à la profondeur du semis*. Thèse de magister, université de Batna (Algérie).

[24] Hiscox JO, and Israelstam JF, 1978. A method for the extraction of chlorophyll from leaf tissue without maceration. *Canadian Journal. Bot.*, 57, 1332-1334.

[25] Hucl P, and Baker RJ, 1989. Tillering patterns of spring wheat genotypes grown in a semiarid environment. *Canadian Journal Plants Sciences* 69: 71-79.

[26] Kirby EGM, Spink JH, Frost DL, Evans EJ, 1999. A study of wheat development in the field: analysis by phases. *European journal of Agronomy* 11: 63-82.

[27] Kirkham MB, Smith EL, Danasobhon C, Draket TI, 1980. Resistance to water loss of winter wheat flag leaves. *Cereal Research Communications* 8: 393-399.

[28] Lebon E, Pellegrino A, Tardieu F, Lecoeur J, 2004. Shoot development in grapevine is affected by the modular branching pattern of the stem and intra and inter-shoot trophic competition. *Annals of Botany* 93: 263 -274.

[29] Megherbi A, Mehdadi Z, Toumi F, Moueddene K, Bachir Bouadjra SE, 2012.Tolérance à la sécheresse du blé dur (*Triticum durum* Desf.) et identification des paramètres morpho-physiologiques d'adaptation dans la région de Sidi Bel-Abbès (Algérie occidentale). *Acta Botanica Gallica* 159: 137-143.

[30] Mourret J C, Conesa AP, Bouchier A, Ould Saîd H, Gaîd M, 1990. Identification des facteurs de variabilité du blé dur en condition hydriques limitantes dans la région de Sidi Bel-Abbès. *Céréaliculture* 23 : 1-10.

[31] Office National de la Météorologie. - Données 1975-2011

[32] Perrier ER, Salkini AB, 1987. Supplemental irrigation in the Near-east and North Africa. *Proceedings of a Workshop on Regional Consultation on Supplemental Irrigation.* ICARDA and FAO, Rabat, Morocco, 7–9 December.

[33] Paul MH, Planchton C, Ecochard R, 1979. Etude des relations entre le développement foliaire, le cycle de développement et la productivité chez le soja. *Amelioration des plantes 29 : 479 -492.*

[34] Riou C, 1993. L'eau et la production végétale. *Sécheresse* 2: 75-83.

[35] Scofield T, Evans J, Cook MG, Wardlow IF, 1988. Factors influencing the rate and duration of grain filling in wheat. *Australian Journal of Plant Physiology* 4: 785 - 797.

[36] Seltzer P, 1946. *Le climat de l'Algérie.* Alger: Institut de Météorologie Physique du globe de l'Algérie.

[37] Slama A, Ben Salem M, Bennaceur M, Zid E, 2005. Les céréales en Tunisie : production, effet de la sécheresse et mécanismes de résistance. *Sécheresse* 16 : 225-229.

[38] Slama A, 2002. *Étude comparative de la contribution des différentes parties du plant du blé dur dans la contribution du rendement en grains en irrigué et en conditions de déficit hydrique.* Thèse de doctorat, faculté des sciences de Tunis.

[39] Stark JC, and Longley TS, 1986. Changes in spring wheat tillering patterns in response to delayed irrigation. *Agron J* 78: 892-6.

[40] Smadhi D, Mouhouche B, 2000. Etude comparée de l'évapotranspiration et des besoins en eau des cultures céréalières de trois étages bioclimatiques. *Pub., Prem., Symp., Intern.*, filière Blé, O.A.I.C, Alger, Algérie. 239-246.

[41] Subbarao GV, Johansen C, Slinkard A, Nageswara E, Rao RC, Saxena NP, and Chauhan YS, 1995. Strategies for improving drought resistance in grain legume. *Crit Rev Plant Sci* 14 : 469- 523.

[42] Triboï E, 1990. Modèle d'élaboration du poids du grain chez le blé tendre. *Agronomie* 10: 191- 200.

[43] Turner NC, Wright GC, Siddique KHM, 2001. Adaptation of grain legume to water-limited environments. *Advagron* 71: 193-231.

5

In vitro germination and direct shoot induction of Yeheb (*Cordeauxia edulis* Hemsl.)

Yohannes Seyoum[1], Firew Mekbib[2, *]

[1]Dry land Crop Research Department, Somali Region Pastoral and Agro-pastoral Research, Jijiga, Ethiopia
[2]School of Plant sciences, Haramaya University (HU), Dire Dewa, Ethiopia

Email address:

yohanes1195@gmail.com (Y. Seyoum), firew.mekbib@gmail.com (F. Mekbib)

Abstract: 'Yeheb' (*Cordeauxia edulis* Hemsl) is a multipurpose and evergreen shrub and endemic to southeastern corner of Ethiopia and Somalia. It is adapted to low and irregular rainfall and survives a very long dry season. It has enormous economic and food security role to the pastoralist of Somali in Ethiopia. However, the plant is threatened with extinction due to overexploitation and its' poor natural regeneration capacity. In addition, 'yeheb' is usually reported having limited reproductive capacities and often have very specific and limited conditions for seed germination, flowering and seed shelf life. Therefore, to overcome these propagation challenges, an experiment was conducted with the aim of developing a protocol for the *in vitro* regeneration of 'yeheb' from cotyledonary node. The result of these studies revealed that seed was washed by 5% sodium hypochlorite for ten min in aseptic condition found to be more effective in surface sterilization. The sterilized seed cultured on half strength of Gamborg (B5) medium was found to be the most suitable medium for germination (26.67%).The highest shoot initiation percentage (89 % of explants produces shoots), number of shoots per explant and number of leaf per shoot were obtained from cotyledonary node explants cultured on Murashige and Skoog (MS) media supplemented with 2.00 mg. l^{-1} N^6-benzylaminopurine (BAP) within nine weeks. While, the highest shoot length and shoot fresh weight were recorded from control (free BAP) and 6.00 mg. l^{-1} BAP, respectively. The highest shoot multiplication (4.56 number of shoot induced) and elongation (2.97cm) were obtained from the induced shoot were cut and placed on MS media supplemented with 2.00 mg. l^{-1}BAP+6.00 mg. l^{-1}of gibberellic acid (GA$_3$) and free BAP+6.00 mg. l^{-1} of GA$_3$, respectively. The elongated shoots were transferred to different media supplemented with various types and levels of hormones but none of them induced root. As a conclusion, this is the first attempt for direct *in vitro* regeneration of *C. edulis* and permissible result for cryopreservation.

Keywords: BAP, Cotyledonary Node, Germination, In Vitro, GA$_3$ Shoot Induction, Shoot Elongation

1. Introduction

Cordeauxia edulis Hemsl, belongs to the family *Leguminosae* and the subfamily *Caesalpinioideae* and is locally known as 'Yeheb' [1]. It is among the most important edible wild food plants in Ethiopia [2]. It is a multi-stemmed ever green shrub or small tree up to 2.5 m height, and is endemic to restricted localities in eastern Ethiopia and parts of central Somalia [1 and 3].

The shrub thrives well in frost-free climatic conditions with 28 °C of mean annual temperature with two rainy seasons; one more reliable in March-May and another one in October–November, gives an annual rainfall of 85-400 mm [4]. The species has resistance to normal drought periods of 4-5 months, and up to 10-15 months in irregular drought periods [5 and 6].

'Yeheb' is a multi-purpose plant where most parts of the plant are used. The seeds are edible and eaten fresh, roasted, boiled or dried. The seed of the species is potentially a valuable protein source with high sugar and fat contents. It has high energy value (0.39-1.87 MJ/Kg). The leaves are also rich in energy (5.59-5.86 MJ/Kg dry matter) [3]. In semi-arid and arid areas, the species represents a viable economical interest. As it is adapted to lower and irregular rainfall and survives a very long dry season, it could represent an enormous advantage in the fight against hunger. The development of cultivation of such plants for the semi-desert region like Sahelian zone could also constitute an interesting food supplement in an area poor in protein supply [7]. It constitutes the staple food of the pastoralist of Somali region in Ethiopia. Moreover, the nut is sold on the market and even exported to the coastal cities of Somalia.

Another major use of the species is its contribution of up to half of the biomass in the area that makes it important dry season browse to camel and goat. The estimated average forage production is 325-450 kg ha^{-1} (1.4-2 kg/plant) [8]. Fodder value of the leaves is comparable to other tropical tree legumes but some mineral levels (P, Mg, Mn and partly Zn) would not satisfy the demands of animals if 'yeheb' were the only source of fodder [4]. Leaves have been used to dye cloths, calico and wool, since the cordeauxiaquinone forms vividly colored and insoluble combinations with many metals [9].

Even if the species has such and many other uses and has a potential to play a role in ensuring food security in the region, the plant is threatened with extinction due to overexploitation of the shrub by long term heavy grazing pressure, harvesting of seeds, cutting and fire. In addition, erosion, drought and war in the region has led to poor or none natural regeneration [9, 10, and 11].

Some Reference [5 and 12] reported the decline and progressive destruction of the stands of C. edulis due to over grazing, and recommended protection from use of the plant. Likewise [4] reported that C. edulis plant is in great danger of extinction and speedily narrowing distribution area because of the increase in population and their herds. Unlike many other plants, yeheb shrubs flowers just before the onset of rains and the seeds mature when the plant moisture content is at its peak [8]. Yeheb seeds have been reported not to retain viability for more than a few months, even if they are stored under ideal conditions and the recommendation has therefore been to sow them immediately [13 and 14].

Studies on C. edulis seed storage behavior and germination indicated characteristics of intermediate storage behavior and 70-84% of germination when seed moisture is above 24%, and achieved a germination percentage of 58% and 41% when seed moisture content was 12.3% and 9.6%, respectively [13]. Acid treatment (Gibberellic acid and Potassium nitrate) tests did not indicate a positive result for seed germination. The same author studied the desiccation tolerance of 'Yeheb'. The results indicated that germination percentage was dependent on seed moisture content, i.e., there was a reduction in germination percentage from 70 to 57.5% when seed moisture content dropped from 24.4 to 12.3%. Further drying to 9.6% moisture content reduced germination percentage even less than 41.3%.Some pilot studies had been made regarding vegetative propagation but so far without greater success [14].

Due to poor germination and death of young seedlings under natural conditions, propagation through seeds, as with most leguminous trees, is unreliable. Hence, rapid in vitro propagation method is required for mass production of healthy and excellent C. edulis planting materials to rehabilitate the ecology of 'Yeheb' grown area and save the species from extinction. On the contrary, there is no cost effective in vitro regeneration protocol developed for C. edulis mass production anywhere. Therefore, it is imperative to develop and/or optimize a tissue culture protocol for effectively and efficiently carryout in vitro regeneration and mass multiplication of C. edulis for regeneration of the species to maintain the ecology and enhance its economic importance. Thus, the objective of this study was to develop a protocol for the in vitro germination and shoot induction of C. edulis.

2. Materials and Methods

2.1. Description of the Experimental Area

The experiment was conducted at the Plant Biotechnology Laboratory of Holetta Agricultural Research Center (HARC). The center is located 29 km west of Addis Ababa at an altitude of 2400 meter above sea level, 90 00'N latitude, 380 30'E longitude.

2.2. Experimental Material

As the shrubs did not produced seed during the experimental period due to recurrent moisture stress in the region, the seeds that were full size and dried were collected from local market of 'Boh', Warder Zone of Somali regional state in June, 2011. Healthy seeds were selected carefully and used as explants for this study.

2.3. Explant Sterilization Experiment

The seeds used for this experiment were washed with 30g l^{-1} kocide under running tap water for different duration (thirty and sixty min) and used as treatment. This was followed by immersing 70% (v/v) ethanol for three min, and later rinsed three times (three min each) with sterile distilled water. After sterilization, seeds were soaked in sterile distilled water for twelve hours. The seed coats were then removed and subjected to surface disinfection with 5.00 % sodium hypochlorite for different duration (5, 8, 10 and 15 min) also used as treatment, and then rinsed three times (three min each) with sterile distilled water.

2.4. Germination Induction Experiment

After surface sterilization, seeds were directly inoculated on full and half strength Murashige and Skoog (MS; M499, PhytoTechnology) [15] and Gamborg (B5; G398, PhytoTechnology) [16] media. The half strength media were supplemented with 1.00 mg. l^{-1} of N^6-benzylaminopurine (BAP; B9395, Sigma), while the other treatments were not supplemented with BAP. The media contained 3% sucrose (S5390, Sigma) and 0.7% agar. The media was adjusted at pH 5.7 after addition of the plant growth hormone, prior to adding agar. Then after, it was dispensed into magentas, and later autoclaved at 120^{0}C for fifteen min. Finally, the seeds (45 seeds used for each treatment) cultured on the media were incubated for two months at 25±2 ^{0}C with a 16 h photoperiod. The number of germinated seeds per treatment was recorded after three and six weeks of culture. The combined data were used for statistical analysis.

2.5. Direct Shoot Regeneration Experiment

2.5.1. Shoot Induction

Cotyledonary nodal explants of *Cordeauxia edulis* from twenty-one days old in vitro raised seedlings were planted on MS media, supplemented with BAP and kinetin (kin; K750, PhytoTechnology) at 0, 0.50 1.00, 2.00, 3.00, 4.00 and 6.00 mg. l^{-1} concentrations separately. The media contained 500 mg. l^{-1} of casein hydrolysate (C7290, Sigma), 3% sucrose and 100 mg. l^{-1} of activate charcoal for prevention of browning of cultures and 0.7% agar. The media were adjusted to pH 5.7 after addition of the plant growth hormone, but prior to adding agar. Later the media were distributed to magenta before autoclaving at $120^{0}C$ for 15 minutes. The culture was maintained at $25\pm2^{0}C$ with a 16 hour photoperiod at a light intensity of 2700 lux from cool white florescent 40 watt bulbs. Data on number of days to shoot initiation, number of shoots per explant, shoot length, number of leaves per explant and shoot fresh weight were recorded after 3 and 6 weeks of culture. The combined data were used for statistical analysis.

2.5.2. Shoot Multiplication and Elongation

After six weeks, those individual shoots (1 cm long) harvested from each explants were cultured in MS medium supplement 0.00, 1.00 and 2.00 mg. l^{-1} BAP + 2.00, 4.00 and 6.00 mg. l^{-1} gibberellic acid (GA$_3$; G7645, Sigma) alone or in combination for shoot multiplication and elongation. Media constituent and preparation were similar to shoot induction media, and also culture condition. After harvesting the shoots, the original explants were transferred to fresh treatment medium for further shoot proliferation and elongation. Data on number of shoot and shoot length were recorded after three and six weeks of culture on shoot multiplication and elongation. The combined data were used for statistical analysis.

2.6. Experimental Design and Data Analysis

Treatments in all the experiments were arranged in a completely randomized design (CRD) with three replications. The data was subject for analysis of variance (ANOVA) using SAS (version 9.0) [17] and significant differences among mean values were compared using Duncan's Multiple Range Test (DMRT) at p<0.05. Logarithmic transformation was used for percentages data to attain normality, before doing analysis of variance.

3. Results and Discussion

3.1. Optimizing Sterilization Technique

Analysis of variance (ANOVA) revealed that the sterilization treatment had highly significant effect on level of contamination, survival and germination percentage of yeheb seed *in vitro* culture. The highest contamination (62.96%) and lowest survival percentage (37.04 %) were recorded on treatment five: 5.00% sodium hypochlorite (NaOCl) solution for five min, while the lowest contamination (0.00%) and highest survival percentage (100.00 %) was recorded on treatment four: 30 g l^{-1} kocide for sixty min with 5.00% NaOCl solution for ten min (Table 1).

Table 1. Effect of disinfectants and time of exposure on contamination, survival and germination percentage.

Sterilization treatment	Time of exposure (min)		Contamination %	Survival %	Germination %
	Kocide*	NaOCl**			
1	30	8	37.04 ± 6.42^c	62.96 ± 6.42^d	11.11 ± 0.00^b
2	60	8	25.93 ± 6.42^d	74.07 ± 6.42^c	7.41 ± 6.42^{bc}
3	30	10	11.11 ± 0.00^e	88.89 ± 0.00^b	0.00 ± 0.00^c
4	60	10	0.00 ± 0.00^f	100.00 ± 0.00^a	0.00 ± 0.00^c
5		5	62.96 ± 6.42^a	37.04 ± 6.42^f	11.11 ± 0.00^b
6		8	48.15 ± 6.42^b	51.85 ± 6.42^e	7.41 ± 6.42^{bc}
7		10	14.81 ± 6.42^e	85.19 ± 6.42^b	18.52 ± 6.42^a
8		15	7.41 ± 6.42^{ef}	92.59 ± 6.42^{ab}	0.00 ± 0.00^c
Mean			25.93	74.07	11.67
CV (%)			5.43	1.79	8.91

Means with same letter (s) in the same column are not significantly different at 1% according to Duncan's Multiple Range Tests (DMRT). CV= coefficient of variation (%), *=30 g l^{-1} of kocide used before hood, **= 5% of NaOCl used after the seed coat removed, Three-min with 70% ethanol was used after Kocide and before NaOCl in aseptic condition, and in each steps the seed was rinsed three times for three min by double distilled sterilized water.

The highest seed germination percentage (18.52%) and the second lowest contamination percentage were recorded on treatment seven (5.00% NaOCl solution for ten min), while poor or no germination percentage was recorded on treatment three, four and eight. Treatments have lowest contamination and highest survival percentage. This indicated that the chemical and time used to sterilize or disinfect the seed from microbial affected the germination percentage. Specially, germination and contamination percentage were dramatically reduced when both disinfectant agents with long time exposure used together as one treatment, and the survival percentage was increased (Table 1).

Generally, considering all parameters and the aim of sterilization, treatment-seven (5.00% NaOCl solution for ten min) was the most effective sterilization treatment among tested, which had highest germination percentage (18.52%), moderate contamination (14.81%) and high survival percentage (85.19%). As time of exposure increased, so also did the level of disinfection, whereas the germination percentage significantly reduced. Several protocols for seed disinfection were carried out using a sodium hypochlorite solution [18 and 19], which is preferred for its simplicity and lower cost [20]. Similarly, our work have determined that 5% NaOCl for ten min is more effective to control contamination from *C. edulis* seed explants with minimum mortality effect. This sterilization technique is easy, locally available, less costly and less toxic compared to other sterilization agents (eg. HgCl$_2$), i.e. does not require special handling and waste disposal precaution [21]. Similar result was reported by [22] on Kinnow bud culture disinfection using 5% NaOCl for 10–15 minutes.

3.2. In Vitro Seed Germination of Yeheb (Cordeauxia Edulis)

The seeds which were sterilized through optimized procedure were cultured on germination media. The half and full strength of MS and B5 media were tested for germination percentage. The analysis of variance revealed that different types of germination media had significant effect on germination percentage. Highest germination percentage (26.67%) was recorded on half strength of B5 media, while poor germination was recorded on full strength of MS media (Table 2; Fig. 1a and 1b, respectively). This result is similar with [23] who reported that B5 media gave 20% germination on *Commiphora wightii*.

Low *C. edulis* germination percentage (17%) was obtained from *ex vitro* experiment [14]. Several species of dry land plants have also been reported to exhibit similar low germination rates. For example, *Temarindus indica*, *Acacia auriculiformis* and *Chamaecytisus palmensis* have been reported to have a physical or chemical inhibitor for germination so that the seed will only germinate when conditions are favorable [24].

| (a) | (b) |

Figure 1. *In vitro germination C. edulis seed on various germination media (a) half strength of B5 media; and (b) full strength MS media.*

Table 2. *Effect of different media on germination of 'yeheb' seed*

Treatment	Germination percentage (%)
Full strength of MS media	15.56 ± 3.85^d
Half strength of MS media	17.78 ± 3.85^{cd}
Half strength of MS media +1mg. l^{-1} BAP	20.00 ± 0.00^{bcd}
Full strength of B5 media	24.44 ± 3.85^{ab}
Half strength of B5 media	26.67 ± 0.00^a
half strength of B5 media +1mg. l^{-1} BAP	22.22 ± 3.85^{abc}
Mean	21.11
CV (%)	11.5

Means with same letter (s) in the same column are not significantly different at 1% according to Duncan's Multiple Range Tests (DMRT). CV= coefficient of variation (%).

The effect of BAP had no significant impact compared to other treatments on germination percentage, which contradicted with that reported by [23]. This might be due to the probable reduction of seed viability during the experiment period, since 'yeheb' seeds can be stored for only 3-4 month only [13].

Even if the mean germination percentage obtained from various media had not statistical difference for *in vitro* germination half strength of B5 media was found to be the permissible medium for *in vitro* germination (26.67%) for six month stored seed of *C. edulis*.

3.3. Direct Method of Regeneration

Induction media supplemented by different types and levels of concentrations of cytokinin were tested for shoot induction experiment on cotyledon node explant which was excised from 3 weeks old seedlings of *C. edulis*. Among tested cytokinin; BAP was able to induce considerable numbers of shoots compared to Kinetin. The superiority of BAP had also been reported on different Acacia species [25-27] and castor bean [28].

3.3.1. Effect of BAP Hormone on Shoot Induction

Results of the analysis of variance revealed that different level of BAP had a highly significant (p≤0.01) effect on shoot initiation percentage, number of shoot, shoot length, number of leaf and shoot fresh weight. The higher mean shoot initiation percentage (89% of explant) was obtained on media supplemented by 2.00 mg. l^{-1} of BAP within six weeks cultured compared to control (Table 3, Fig. 2a). On the other hand, poor shoot initiation percentage (22%) was recorded from BAP free and 6.00 mg. l^{-1} BAP (Table 3, Fig. 2b and 2c). This result indicated that the addition of BAP promoted the initiation of more shoots, i.e. the exogenous BAP for the shoot initiation is indispensable for cotyledonary nodal explant of *C. edulis*. In consistent with this result, [29] had reported that BAP induce more shoot on cotyledonary nodal explants of *Acacia sinuate*.

The maximum mean number of shoots per explant (3.00±0.33) was obtained on media supplemented by 2.00 mg. l^{-1} BAP, followed by 1.78±0.19 and 1.55±0.19 shoots from 3.00 and 1.00 mg. l^{-1} BAP, respectively. While lower number of shoots of 0.78 ± 0.19 was obtained at 6.00 mg. l^{-1} BAP. In both low concentration of BAP (0.50 mg. l^{-1}) and high concentration of BAP (6.00 mg. l^{-1}) the shoot number was reduced. In addition, regenerated shoots exhibited slightly different morphology at high concentration of BAP (6.00 mg. l^{-1}) and the explant gave dense clump of non-elongated new shoots (Fig. 2c). Reference [30] similarly reported that higher concentration of BAP reduce the number of shoots on *Ricinus communis* L.

The longest mean shoot length (1.72 ± 0.15) was measured on control (free BAP media), followed by 1.52±0.22 cm length from 0.50 mg. l^{-1} BAP, and both means were not statistically different; with the shortest shoot length (0.32 ± 0.16 cm) recorded from 6.00 mg. l^{-1} BAP. As BAP concentration increased, the mean shoot length decreased significantly. Reference [31] reported similar result on *Ceratonia siliqua*. The reduction in shoot length at high concentration of BAP might be due to the toxic effects of ethylene, produced at high cytokinin concentration. This result is in accordance with [32] Thomas and Blakesley reported that the production of ethylene by the excessive cytokinin application caused the inhibition of internodes elongation and number of regeneration of tobacco disc.

The maximum number of leaf and shoot fresh weight was recorded on media supplemented with 2.00 and 6.00 mg. l^{-1} BAP, respectively. Similarly, [33] reported that maximum number of leaf and shoot fresh weight was obtained from cotyledonary nodal explant of 'korarima' on medium supplemented with high concentration of BAP, respectively.

Generally, increasing BAP concentration up to certain level (2.00 mg. l^{-1}) increased shoot initiation percentage and shoot number. After maximum production, it starts declining with further increase in the BAP concentration. Therefore, selection of proper concentration of plant growth regulator is critical to shoot induction. To this end, we found that BAP at a concentration of 2.00 mg. l^{-1} the most suitable growing condition with regard to shoot initiation percentage (89 % of explant produces shoots within six weeks) and number of shoots per explant (three shoots per explant) with mean length of 0.97 cm. (Table 3).

3.3.2. Effect of Kinetin Hormone on Shoot Induction

None of the treatments induced shoot, rather large and green calli were observed after six weeks (Figure 2d). Similar results were observed on castor bean by [28].

Table 3. *Effect of various level of concentration of BAP on different morphogenetic responses of C. edulis cotyledon node on MS medium after six week of cultured.*

BAP (mg. l^{-1})	Shoot initiation percentage (%)	Number of shoots per explant (n) (Mean ± SE)	Shoot length per explant (cm) (Mean ± SE)	Number of leaf per shoot (n) (Mean ± SE)	Shoot fresh weight per explant (g) (Mean ± SE)
0.00	22^d	1.22 ± 0.19^c	1.72 ± 0.15^a	2.22 ± 0.19^{cd}	0.266 ± 0.02^f
0.50	33^{cd}	1.33 ± 0.00^c	1.52 ± 0.22^a	2.44 ± 0.19^{bc}	0.339 ± 0.03^e
1.00	55^{bc}	1.55 ± 0.19^{bc}	1.07 ± 0.17^b	2.56 ± 0.19^b	0.402 ± 0.06^d
2.00	89^a	3.00 ± 0.33^a	0.97 ± 0.23^{bc}	3.22 ± 0.19^a	0.446 ± 0.02^d
3.00	67^{ab}	1.78 ± 0.19^b	0.77 ± 0.10^{cd}	2.22 ± 0.19^{cd}	0.517 ± 0.04^c
4.00	33^d	1.33 ± 0.00^c	0.67 ± 0.10^d	2.00 ± 0.00^d	0.576 ± 0.00^b
6.00	22^d	0.78 ± 0.19^d	0.32 ± 0.16^e	1.67 ± 0.00^e	0.642 ± 0.03^a
Mean	50	1.62	1.10	2.33	0.46
CV (%)	1.25	8.95	2.48	6.97	7.14

Means with same letter (s) in the same column are not significantly different at 1% according to Duncan's Multiple Range Tests (DMRT). CV= coefficient of variation (%)

(a)   (b)

(c)   (d)

Figure 2. *Shoot induction of C. edulis from cotyledon node explants on shoot induction media supplement with different types and various level of cytokinin: a) 2.00 mg. l^{-1} BAP; b) free hormone; c) 6.00 mg. l^{-1} BAP.; and d) 2.00 mg. l^{-1} kin*

3.3.3. Shoot Multiplication and Elongation

The analysis of variance revealed that BAP and GA$_3$ hormones had highly significant (p<0.01) effects on shoot multiplication and elongation; and their interaction had also significant (p<0.05) effect on shoot multiplication but not on shoot elongation. This interaction effect indicated that the two factors are dependent on each other for shoot multiplication but not for shoot elongation of *C. edulis* shoots.

The maximum mean number of shoots (4.56 ± 0.20) was obtained on MS medium supplemented with 2.00 mg. l^{-1} BAP + 6.00 mg. l^{-1} GA$_3$ (Table 4; Fig. 3a), while lower shoots (1.11 ± 0.19 and 1.22 ± 0.19, respectively) were recorded from PGR free (control) and BAP free + 2.00 mg. l^{-1} GA$_3$. Shoots number increased when the BAP concentration increased along with increased of GA$_3$ concentration in media.

(a)   (b)

Figure 3. *Shoot multiplication and elongation of C. edulis on media supplemented with: a) 2.00 mg. l^{-1} BAP + 6.00 mg. l^{-1} GA$_3$; and b) free BAP+6 mg. l^{-1} GA$_3$.*

The mean longest shoot (2.97 ± 0.04) was obtained on

media supplemented with 6.00 mg. l^{-1} GA$_3$ (Table 4; Fig. 3b), while shortest shoot was recorded from PGR free (control). The shoot length increased with the increasing GA$_3$ concentration but it decreased when BAP concentration increase in culture media. In addition dwarf shoots were observed on higher concentration of BAP. Similar result was obtained by [34] on Walnut trees.

Generally, these results showed that shoot proliferation was influenced by the combination of BAP with GA$_3$ than individually (Fig. 3a) This is due to the effect of GA$_3$ on shoot elongation which resulted in longer shoot having a potential to induce more number of bud than shorter shoot [35]. When the concentration GA$_3$ increased within similar level of BAP in the media, it increased shoot proliferation. However, increasing BAP concentration within similar level of GA$_3$ in the media decreased the shoot length (Table 4).This illustrated that the combination of BAP with GA$_3$ enhanced shoot proliferation but BAP alone had negative effect on shoot elongation. Similar result was reported by [36 and37].

Table 4. Effect of BAP and GA3 on number of shoots induces and shoots length during shoot multiplication

Plant growth hormone (mg. l^{-1})		Number of shoots/ explant (Mean ± SE)	Shoots length/ explant (cm) (Mean ± SE)
BAP	GA$_3$		
0.00	0.00	1.11 ± 0.19^g	2.39 ± 0.10
0.00	2.00	1.22 ± 0.19^g	2.59 ± 0.05
0.00	4.00	1.56 ± 0.20^f	2.72 ± 0.09
0.00	6.00	1.67 ± 0.00^{ef}	2.97 ± 0.04
1.00	0.00	1.67 ± 0.00^{ef}	2.25 ± 0.07
1.00	2.00	1.78 ± 0.19^{ef}	2.52 ± 0.07
1.00	4.00	1.89 ± 0.19^e	2.60 ± 0.10
1.00	6.00	3.78 ± 0.19^c	2.74 ± 0.08
2.00	0.00	2.22 ± 0.19^d	1.39 ± 0.15
2.00	2.00	3.89 ± 0.19^{bc}	1.61 ± 0.13
2.00	4.00	4.11 ± 0.19^b	1.98 ± 0.17
2.00	6.00	4.56 ± 0.20^a	2.03 ± 0.12
Mean		2.45	2.32
CV (%)		7.51	4.15

Means with same letter (s) in the same column are not significantly different at 1% according to Duncan's Multiple Range Tests (DMRT). CV= coefficient of variation (%).

4. Conclusion

Yeheb (*Cordeauxia edulis* Hemsl.) is a multi-purpose plant where most parts of the plant are usable. Even if the species has multitude uses and has a potential to play a role in ensuring food security in the region, the plant is threatened with extinction due to overexploitation. This, in turn, has led to poor or none natural regeneration. Generally, this study found a permissible result to rescue rare, endemic, and endangered species through mass and continuous plantlet production within short period of time. In addition it may be used as a baseline point for ex situ conservation through cryopreservation. This is the first attempt *in vitro* regeneration of C. edulis.   Hence, the aforementioned potential benefits of the outputs of this study can be reaped in the areas of future 'Yeheb' conservation, research, and development.

References

[1] Ali, H.M., 1988. Cordeauxia edulis: Production and Forage Quality in Central Somalia. Thesis for the degree of Master of Science in Rangeland Resources, National University of Somalia, Somalia.

[2] Teketay, D. and Eshete, A., 2004. Status of indigenous fruits in Ethiopia. In: Chikamai B, Eyog-Matig O, Mbogga M (eds.) Review and Appraisal on the Status of Indigenous Fruits in Eastern Africa: A Report Prepared for IPGRI-SAFORGEN in the Framework of AFRENA/FORENESSA, Kenya Forestry Research Institute, Nairobi, Kenya, pp 3-35.

[3] Miège, J. and Miège, M.N., 1978. Cordeauxia edulis a Caesalpinaceae of Arid Zones of East Africa, Caryologic, blastogenic and biochemical features. Potential aspects for nutrition. Economic Botany, 32: 336-345.

[4] Drechsel, P. and Zech, W., 1988. Site conditions and nutrient status of Cordeauxia edulis (Caesalpiniaceae) in its natural habitat in central Somalia. Economic Botany, 42: 242–249.

[5] Hemming, C.F., 1972. The vegetation of the northern region of Somalia Republic. Proceeding of Linnaeus Social London, 177:173-250.

[6] Watson, R.M., Tippett, C.J., Becket, J.J. and Scholes, V., 1982. Somali Democratic Republic, Central Rangelands Survey, London. Resource Management and Research, 1:3–10

[7] N.A.S (National Academy of Science), 1979. Tropical legumes: Resource for the Future, Nat Acad. Sci. Washington DC, pp261.

[8] Brink, M., 2006.Cordeauxia edulis Hemsl Record from Protobase. PROTA (Plant resources of tropical Africa / Ressourcesvégétales de l'Afriquetropicale), Wageningen, Netherlands http://database.prota.org/search.htm (Accessed on September 14, 2011)

[9] Booth, F.E.M. and Wickens, G.E., 1988. Non-timber uses of selected arid zone trees and shrubs in Africa. FAO Conservation Guide 19, 52-58.

[10] FAO, 1988. Traditional food plants, Food and nutrition paper 42:224-27.

[11] Assefa, F., Bollini, R. and Kleiner, D., 1997. Agricultural potential of little used tropical legumes with special emphasis on Cordeauxia edulis (Ye-eb nut) and Sphenostylisstenocarpa (African yam bean). Giessener Beiträge zur Entwicklungsforschung, 24:237–242.

[12] Bally, P.R.O., 1966. Miscellaneous notes on the flora of Tropical East Africa, 29. Enquiry into the occurrence of the Yeheb nut (Cordeauxia edulis Hemsl.) in the Horn of Africa.Candollea 21 (1), 3-11.

[13] Liew, J., 2003. Desiccation tolerance of yeheb (Cordeauxia edulis Hemsl.) seeds.Thesis for the degree of Master of Science in Agriculture, SLU, Ultuna, Sweden.

[14] Mussa, M., 2010.Cordeauxia edulis (yeheb): resource status, utilization and management in Ethiopia. Thesis for the degree of Philosophiae Doctor, University of Wales.

[15] Murashige, T. and Skoog, F., 1962.A revised medium for rapid growth and bioassays with tobacco tissue cultures. Physiologia Plantarum, 15: 473-497.

[16] Gamborg, O.L., Murashige, T., Thorpe, T.A. and Vasil, I.K., 1976.Plant tissue culture media. In Vitro, 12:473-478.

[17] SAS Institute Inc., 2002. Statistical Analysis Software, Version 9.0. Cary, North Carolina, USA.

[18] Pierik, R.L.M., 1997. In vitro culture of higher plants. Kluwer Acadamic publishers. Wageningen Agricultural University, The Netherland, pp 1-72.

[19] Alvarez, V.M., Ferreira, A.G., and Nunes, V.F., 2006. Seed disinfestation methods for in vitro cultivation of epiphyte orchids from Southern Brazil. HorticulturaBrasileira 24: 217-220.

[20] Vujanovic, V., Arnaud, S.T.M. and Barabé, D., 2000.Viability testing of orchid seed and the promotion of coloration and germination.Annals of Botany, 86:79-86.

[21] George, E. F., Hall, M.A. and Klerk, G.D., 2008. Plant Propagation by Tissue Culture 3rd Edition, Publication by Springer, Netherlands.

[22] Altaf, N., 2006. In vitro bud culture of Kinnow tree, Pak. J. Bot. 38(3): 597-601.

[23] Kant, T., Prajapati S. and Parmar, A.K., 2010. Efficient micropropagation from cotyledonary node cultures of Commiphora wightii (Arn.)

[24] Roshetko, J.M., 1995. Albiziasaman: pasture improvement, shade, timber and more (NFTA 95-02). Winrock International, Morrilton, Arkansas. http://www.winrock.org/ forestry/factpub/factsh/a_saman.htm. (Accessed on September 14, 2011).

[25] Mittal, A., Agarwal, R., Gupta, S.C., 1989.In vitro development of plantlets from axillary buds of Acacia auriculiformis. Plant Cell, Tissue and Organ Culture, 19: 65–70.

[26] Dewan, A., Nanda, K., Gupta, S.C., 1992. In vitro micropropagation of Acacia nilotica sub sp. Indica Brenan via cotyledonary nodes. Plant Cell Reports, 12: 18–21.

[27] Singh, H. P., Singh, S., Saxena, R.P., Singh, R.K., 1993. In vitro bud break in axillary nodal segments of mature trees of Acacia nilotica. Indian Journal Plant Physiol. 36:21–24.

[28] Aha, Y.J., and Chen, G.Q., 2008. High frequency regeneration through adventitious shoot formation in caster bean (Riceinus communis L.). In Vitro Cellular and Developmental Biology -Planta, 43:9-15.

[29] Vengadesan, G. and Pijut, M.P., 2009. In vitro propagation of northern red oak (Quercusrubra L.)In Vitro Cellular and Developmental Biology – Planta, 45:474–482.

[30] Alam, I., Sharmin, S.A., Mondal, S.C., Alam, M., Khalekuzzaman, M. and Anisuzzaman, M., 2010. In vitro micropropagation through cotyledonary node culture of caster bean (Riceinus communis L.). Australian Journal of crop science, 4(2):81-84.

[31] Naghmouchi, S., Khouja, M.L., Rejib, M.N. and Boussaid, M., 2008. Effect of growth regulators and explant origin on in vitro propagation of Ceratonia siliqua L. via cuttings. Biotechnol. Agron. Soc. Environ.:12 (3):251-258.

[32] Thomas, T. and Blakesley, D. 1987 .Practical and potential uses of Cytokinin in agriculture and horticulture. In: Bonga, J. M., and Durzan, D.J.,(eds) Cell and tissue culture in trees, vol. 1 Amsterdam: Martinus Nijhoff Publishers.37p.

[33] Eyob, S., 2009.Promotion of seed germination, subsequent seedling growth and in vitro propagation of korarima Aframomum corrorima Braun. Journal of Medicinal Plants Research, 3(9):652-659.

[34] Payghamzadeh, K. and Kazemitabar, S.K, 2011. In vitro propagation of walnut. African Journal of Biotechnology, 10(3):290-311.

[35] Chitra D.S.V. and G., Padmaja, 1999. Clonal propagation of mulberry through invitroculture of nodal explants. Scientia Hort. 80: 289-298.

[36] Deore, A.J. and Johnson, A.T., 2008. High frequency plant regeneration from leaf-disc culture of Jatropha curcas L.: an important biodiesel crop. Plant Biotechnology Rep., 2:7-11.

[37] Najafabadi, A.J. and Hamidoghli, Y., 2009. Micropropagation of thornless trailing blackberry (Robus sp.) by axillary bud explant. Australian Journal of crop science, 3:191-194.

6

The efficacy of *Trichoderma* spp. and *Bacillus* isolates in the control of chickpea wilt pathogens

Hanan Ibrahim Mudawi[1, *]**, Mohamed Osman Idris**[2]

[1]The Environmental, Natural Resources and Desertification Research Institute, National Centre for Research, Ministry of Science and Technology, Khartoum, Sudan
[2]Department of Plant Protection, College of Agriculture, Khartoum University, Khartoum, Sudan

Email address:
hananmodawi@gmail.com (H. I. Mudawi), shid_agic@hotmail.com (I. O. Mohamed)

Abstract: Dual experiments were carried out in 2007 at the laboratories of the National Center of Research, to test the antagonistic efficacy of three *Trichoderma spp* and *23 Bacillus* isolates, for the control of chickpea wilt and root- rot pathogens: *Fusarium oxysporum f. sp. ciceris* and *F. solani* adopting CRD. *Trichoderma harzianum* was found highly antagonistic compared to *Trichoderma viride* isolates as it inhibited the mycelial growth of *F. oxysporum f. sp. ciceris* and *F. solani* by 85.29% and 86.21% after 12 days of *in-vitro* incubation, whereas *T. viride* (isolate Tv1) gave an inhibition percentage of 81.88% and 76.64%. Antagonistic hyphae of *T. harzianum* showed parasitic behavior against Fusarium spp. The parasite reached and recognized *F. oxysporum f. sp. ciceris* by coiling around the hyphae of the pathogen and disintegrating the hyphae and spores. Only 17 out of 23 Bacillus isolates from 130 colonies of bacteria screened showed significantly antagonistic properties against wilt pathogens. Only B3, B16, B2, B15and B20 proved to be the most effective among the rest of isolates and were considered strongly antagonistic against *F. oxysporum f. sp. ciceris* and *F. solani in-vitro*, with an inhibition percentage range of 57.57% - 64.65%. The management of Chickpea root/rot wilt complex disease incited by *F. oxysporum f. sp. ciceris* and *F. solani* could be achieved successively by the use of bioagents derived from various fungal and bacterial isolates.

Keywords: Antagonisms, *Bacillus* spp, Biocontrol, *Cicer arietinum* L, *Trichoderma harzianum*, *T. viride*

1. Introduction

Chickpea, *Cicer arietinum* L., is one of the best legumes for human consumption and widely grown in Northern Sudan. *Fusarium oxysporum f. sp. ciceris* and *F. solani* are the wilt and root- rot pathogens causing severe damage wherever this crop is grown [1, 2].The use of bioagents in the control of pests is a result of the change in the public attitude towards the use of chemical pesticides and fumigates [3]. In this respect, *Trichoderma* spp. has been studied as biological control agents against soil-borne plant pathogenic fungi and nematodes. They have been investigated for over 80 years and recently used as biological control agents and their isolates have become commercially available. The mechanisms suggested to be involved in their bio-control are antibiosis, lysis, competition, mycoparasitism and promotion of plant growth [4].

Plant growth promotion by Plant Growth Promoting Rhizobacteria (PGPR) may also be an indirect mechanism of biological control, leading to disease escape when the growth promotion results in shortening the time that a plant is in a susceptible state [5]. It was originally reasoned that endophytic bacteria which could colonize vascular tissues of plants would be potential antagonists of vascular-invading pathogens, such as *F. oxysporum* and *Verticillium* spp. [6]. One of the advantages of using entophytes is that, once inside the host, they are better protected against environmental stress and microbial competition [7]. Strains of Bacillus are among the most common bacteria found to colonies plants endophytically and are plant growth promoters [8]. *Bacillus* spp. has the characteristics of being widely distributed in soils, having high thermal tolerance, showing rapid growth in liquid culture, and readily form resistant spores. Abeysinghe [9] screened four isolated rhizobacteria in dual Petri dishes assay as antagonistic against *F. solani f.sp. phaseoli* and reported that *Bacillus subtilis* CA32 effectively antagonized the pathogen growth by 55.05% mycelium inhibition mean. Harlapur et al.[10] evaluated eight bio-agents under *in-vitro* conditions against *Exserohilum turcicum*, that affect maize leaf, and among the bio-agents tested *T. harzianum* caused significantly maximum inhibition (65.17%) followed by *T. viride* (56.95%) and *B. subtilis* (49.57%), while *Pseudomonas fluorescens* was found to be the least effective (19.30%).

The objective of this study is to test the antagonistic ability of three *Trichoderma* spp and two *Bacillus* isolates obtained from chickpea growing areas at Northern Sudan against *F. oxysporum f. sp. ciceris* and *F. solani* the causal agents of wilt root rot disease complex. This disease is highly sever at chickpea grown areas, that leads the farmers to alter the crop with other profitable alternative crops. The morphological antagonistic behaviour of *Trichoderma* spp *in- vitro* were also observed.

2. Materials and Methods

2.1. Source of Pathogens

Soil samples were randomly collected at March 2007 from the rhizosphere of chickpea grown on a sick plot at Shambat Research Station Farm (ARC), heavily infected with wilt/ root-rot disease complex incited by *Fusarium* spp and the lesion nematode *Pratylenchus* spp. The sick plot is repeatedly used for chickpea Fusarium wilt resistance screening experiments. The soil was maintained in paper bags and left to dry at room temperature (25-30^{0}C).

2.2. Isolation and Identification of Fusarium spp

Fusarium spp were isolated from air-dried soil samples following the dilution plate method [11]. The samples were thoroughly mixed, and a suspension of 1g (dry weight equivalent) in 9 ml of sterilized distilled water was prepared from each sub-sample. A serial dilution of the soil suspensions was prepared (ten-folds), and inoculated on a Fusarium selective Spezieller Nährstoffarmer Agar media (SNA) that composed of KH2PO4 1.00 g, KNO3 1.00 g, MgSO4.7H2O 0.50 g, KCl 0.50 g, Glucose 0.20 g, Sucrose 0.20 g, Agar 20.00 g, Distilled water 1.00 L, supplemented with 0.05g chloramphenicol as anti-bacterial. And 1-2 pieces of sterile filter paper (Whatman № 1) were placed, approximately 1 cm² on the agar surface to enhance sporulation. The media were allowed to dry for 3 days [15]. Individual colonies resembling *Fusarium* spp were transferred to Potato Dextrose Agar (PDA) media (Unpeeled potato 200g, Dextrose 20g, Agar 20g, Distilled water 1 L) and incubated for 10– 15 days at 25°C in the dark. Production of pigmentation was observed on SNA media. The isolates were maintained on sterilized soil-agar at 4°C and / or PDA medium at 25°C and sub-cultured every three months. Identification of isolates was based on cultural, microscopic characteristics with reference to Leslie and Summerell's Fusarium laboratory manual [12].

2.3. Isolation of Antagonistic Bacteria (SciencePG-Level3-Multiple-Line)

Twenty- three bacteria isolates were used in this study.

Thirteen strains were isolated from the rhizosphere of the main chickpea growing areas of Adu Hamad (northern Sudan). Eight grams of soil samples were transferred to a 100 ml beaker and filled with 40ml of sterilized water. The beaker was heated in a water path for 10 minutes at approximately 80^{0}C and agitated during the process. The soil suspension was serially diluted, spread on Nutrient Agar (NA), on Petri plates using a spreader and inoculated at 28^{0}C for 48 h. Five replicated plates were prepared for each dilution (1×10^5 and 1×10^7). Colonies were isolated on the basis of their different visual characteristics. After isolation, all colonies were purified by single colony isolation after triple re-streaking on NA medium. Colonies morphologically resembling *Bacillus* species were subcultured and maintained on NA tubes according to Claus and Berkeley [13]. The bacterial isolates were stored in NA tubes at 4^{0}C and sub-cultured every three months. Active cultures were prepared in NA tubes 48 h before application.

2.4. Isolation of Trichoderma spp and Inoculum Preparation

The fungal antagonists were isolated from soil samples brought from Shambat area (Central Sudan), using dilution plate techniques [11] on Trichoderma selective medium, cultures were incubated for 2 days on SNA medium in the dark, followed by incubation under ambient laboratory conditions of light and temperature (about 23°C). After an incubation period, colonies determined to be *Trichoderma* spp., were purified and identified on the basis of their morphological characters [14]. *Trichoderma* spp. were maintained on PDA media and stored at 4 °C. Two isolates were isolated and identified in this study *Trichoderma harzianum* (Ts) and *T. viride* (Tv1), the other isolate were obtained from Khartoum Crop Protection Department (isolated from northern Sudan soil) and were identified as *T. viride* (Tv2).

The inocula of *Trichoderma* spp were prepared from 8 – 10 days old culture grown on PDA media. Ten (10) ml of sterilized distilled water were added to each Petri dish, and the surface of the culture was scraped with a glass spatula to dislodge the spores. The spore suspension derived from six Petri dishes was transferred to 100 ml sterilized flask. One ml of each isolate was poured into PDA Petri dishes, allowed to dry for 3 days, and incubated at room temperature for one day.

2.5. In-Vitro Antagonistic Tests

The *in-vitro* antagonistic properties of bacteria isolates and *Trichoderma* spp were investigated against *Fusarium* spp. Assay were performed in Petri plates (90 mm) containing 20 ml of PDA, allowed to dry for 3 days. An entire pure culture, 4 mm in diameter obtained from one day old *Trichoderma* spp and 7-10 days old cultures of *F. oxysporum f. sp. ciceris* and *F.solani* were cut using a cork-borer. Pathogens disks were transferred 10 mm from the edge of each Petri dish. Each *Trichoderma* spp was placed 10 mm from the other edge, opposite to the pathogens. The inoculated plates were then incubated upside-down at 25^{0}C, and were observed for inhibition or otherwise of their growth for 12 days. Each bacterial isolate were spotted with a sterile tooth-stick (four spots per plate), 10 mm from the opposite edge of the Petri dish and opposite to the pathogens [15] and were observed for inhibition for 7 days. The control Petri dishes were inoculated each alone with *F. oxysporum f. sp. ciceris* and *F.solani*. The radial growth of the two pathogens in the control and the treated Petri dishes were measured every 24 h, and the inhibition percentage of the antagonism was calculated according to the following formula:

The percentage inhibition = R1 - R2 / R1 ×100

Where R1 is the value of radial growth of pathogen in control plates and R2 is the radial growth of the pathogen in the treated plates [16].

2.6. Analysis of Data

A completely randomized design (CRD) with five replicates for the effect of Trichoderma spp and three replicates for bacterial effect was adopted in this part of the study. The percentage of inhibition was arc sin transformed and the data means were analyzed according to Duncan multiple range test at $P \leq 0.05$ to test significant differences between treatments [17].

3. Results

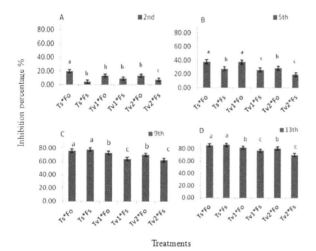

Figure 1. *Mean inhibition percentages of the radial growth of F. oxysporum f.sp. ciceris and F.solani treated with three Trichoderma isolates on PDA media from the 2nd day after inoculation. .Vertical bars represent standard errors. Bars with the same letter are not significantly different at P ≤ 0.05.*

Using the serial dilution isolation technique described above, thirteen isolates were preliminary characterized as member of the genus Bacillus based on its colony morphology, gram-positive reaction, spore forming and the presence of the bacillus-shape under the electron microscope [13]. The fungi isolated from the sick plot (Shambat Research Station Farm) were identified as *F. oxysporum f. sp. ciceris* and *F. solani* according to Leslie and Summerell's Fusarium laboratory manual [12]. Whereas three species of Trichoderma were isolated from Shambat area (Central Sudan), one isolate was identified as *T. harzianum* (Ts), and

two were identified as *T.viride* (Tv1 and Tv2) according to [14]

Antagonistic properties of *T.viride* (Tv1 and Tv2) and *T. harzianum* (Ts) were tested against the *F. oxysporum f. sp. ciceris* and *F. solani*, using dual Petri plate method. It appeared from the data presented in Figure 1 that all the antagonistic fungi significantly (*P≤0.05*) inhibited the growth of *F. oxysporum f. sp. ciceris* and *F.solani* against the control plates during all days of treatments. *T. harzianum* were found significantly superior in antagonizing the two pathogens than *T. viride* isolates, inhibiting the mycelial growth of *F.oxysporum f. sp. ciceris* and *F. solani* with a range of 19.78% to 85.29% and 4.52% to 86.21% respectively during the incubation days (Fig.1) with no significant difference at the13th day after inoculation. The effect of *T.harzianum* was highly superior from the second day of inoculation recording an inhibition percentage of 19.78% in comparison with the other species. After 5 days of inoculation, *T. harzianum* resulted in inhibiting the two pathogens by 37.93%, 75.90%, 85.29% and 27.83%, 77.70 %, 86.21% during the 5th, 9th and 13th days after inoculation; respectively (Fig.1.A, B, C).

T. viride isolates (Tv1 and Tv2) gave an inhibition percentage of 37.62%, 28.67% and 26.17%, 19.52% after 5 days; respectively (Fig.1A) and 72.79%, 69.84% and 63.42%, 61.80% after the 9th day; respectively (Fig.1B) and 81.88%, 80.29% and 76.64%, 69.67% after the 13th day after inoculation; respectively (Fig.1C).

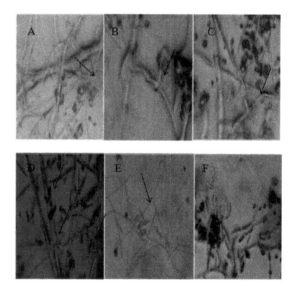

Figure 2. *(A) The parasitic hyphae of T. harzianum reached and recognized the host hyphae F.oxysporum f.sp. ciceris (B) coiling, (C) encircling, (D) penetrating the mycelia and disintegrating the hyphal cell wall, (E) Normal septation of F.oxysporum f.sp. ciceris spores, (F) invaded and disintegrated spores by T. harzianum with no or faint septation as shown by the arrows.*

From day five onward, the inhibition percentage of the mycelial growth of pathogens by *Trichoderma* isolates increased (Fig.1). From day 9 to the end of the experiment *F.oxysporum f. sp. ciceris* and *F. solani* were more sensitive. *T. viride* isolates were never observed to overgrow the tested fungus. Among the two *T. viride* isolates isolate Tv1 was significantly (*P≤ 0.05*) more effective in inhibiting the mycelial growth of the two pathogens.

Observations of hyphal interaction indicated that antagonistic hyphae of *T. harzianum* showing parasitic behavior against Fusarium spp. The parasite reached and recognized the pathogen (Fig. 2A), by coiling around the hyphae of pathogen and disintegrating them (Fig. 2B and C). Occasionally *T. harzianum* hyphae formed a hook or bunch like structures around the hyphae of the pathogen from where penetration took place (Fig. 2C and D). Hyphae of antagonist either coiled around the hyphae of *F. oxysporum f. sp. ciceris* before penetration (Fig. 2B) or entered directly without the formation of appressorium-like structures suggesting mechanical activity (Fig. 2D). The host hyphae eventually disintegrated (Fig.2E).The spores of *F. oxysporum f. sp. ciceris* was disintegrated shown no or faint septation (Fig. 2F). The antagonistic mycelium of *T. harzianum* overgrew on the mycelium of F. oxysporum f. sp. ciceris after day 10, whereas *T. viride* isolates only arrested the mycelium growth of *F.solani*.

Twenty three bacterial isolates were tested *in- vitro*, out of 130 colonies isolated from the infected soil, for inhibition of mycelium growth of *F.oxysporum f.sp. ciceris* and *F. solani* (Fig. 3). Sixteen isolates suppressed mycelium growth of *F.oxysporum f.sp ciceris* but nine isolates, B3, B16, B2, B15, B22, B20, B11 and B18 were significantly (*P≤0.05*) superior to the rest of isolates (Fig. 3A), with an inhibition percentage of 64.65%, 63.95%, 62.60%, 62.60%, 58.22%,57.57% 51.64%, and 50.96%, respectively. Also the mycelial growth of *F. solani* was significantly (*P≤0.05*) suppressed by nine isolates, B3, B4, B6, B10, B11, B14, B16, B19, and B22, but the most effective isolates were B16, B3 and B22 (Fig. 3B) with an inhibition percentage of 55.76%, 51.06% and 49.17% respectively.

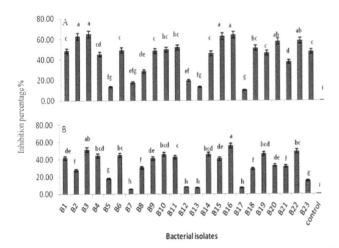

Figure 3. *Mean inhibition percentage of (A). F.oxysporum f.sp. ciceris and (B). F.solani treated with 23 bacterial isolates on PDA media after 7 days of inoculation. Vertical bars represent standard error from the means. Bars with the same letter are not significantly different at P ≤ 0.05.*

These three isolates were considered highly inhibitory against the two pathogens. *F.oxysporum f.sp ciceris* was more sensitive to bacterial isolate than *F. solani* evidenced by the inhibition percentages recorded "Fig. 3". However, for the rest of the bacterial isolates, *F.oxysporum f.sp ciceris* and *F. solani* over grew the bacterial isolates. In the control treatments the phytopathogens completely covered the Petri dishes.

4. Discussion

The results obtained suggested that *Trichoderma* spp. is capable of producing a range of metabolites, which have antifungal activity. *T. harzianum* and *T. viride* significantly (*P≤0.05*) inhibited the growth of fungal pathogens and reduce the mycelial growth indicating antagonistic properties against the pathogens. Somasekhara et al. [18] reported that bioagents such as *T. viride*, *T. harzianum* and *T. hamatum* are effective in controlling pigeonpea wilt caused by *F. oxysporum f. sp. udum*.

The minimum growth inhibition percentages of *Fusarium* spps (8.45%-34.43%), incited by *Trichoderma* isolates were at 4 and 5 days after incubation. This may be due to high pathogenic virulence of the pathogens, which resisted the inhibitory action of *Trichoderma* species. From day 6 a gradual increase in the inhibition action was recorded, attaining its maximum of 86.21% at day 13 after inoculation. Overgrowing behaviour of the pathogen was observed when *F. oxysporum f.sp. ciceris* was treated with *T. harzianum*. The *Trichoderma* spps used in this study were found more aggressive antagonisms than that used by Nikam et al. [19] who reported that *T. harzianum* was more effective in inhibiting the mycelial growth of *F. oxysporum f.sp. ciceris* by 83.33% than *T. viride* (76.66%). Production of chitinolytic and glucanolytic enzymes from *T. harzianum* may have direct significance in the parasitism on *F. oxysporum f.sp. ciceris* as these enzymes function by breaking down the polysaccharides, chitin and β-glucan that are responsible for the rigidity of fungal cell walls and interact synergistically to achieve a high level of antifungal activity, thereby destroying cell wall integrity, which could explain the strong antagonistic results of *T. harzianum* [20,21].

It was observed that the antagonisms were expressed as arresting the mycelium growth of *F.solani*. One of possible mechanisms is fungistasis, in which inhibition of germination has been considered a survival mechanism which arrest fungal growth and biologically controls the pathogenic fungi [22]. There could be other reasons for this result; for instance, the increase was favored through antibiotic production by the mature mycelium. Such antibiotics had indeed been reported by workers like Jayaswal et al. [23], Mcloughlin et al. [24] and Montealegre et al. [25] who reported that *Trichoderma* spp. secreted chitinase and B-1, 3 glucanase in supernatants.

The *in vitro* culture of *Fusarium* spp and *Trichoderma* spp in culture media led to a variety of interactions. *Fusarium* spp growth was generally inhibited; the host cell contents disorganized and the hyphae were intensively parasitized by *Trichoderma* spp. similar reactions have been reported on other fungal pathogens [26].

Mycoparasitism behaviour displayed by *Trichoderma* spp. was evidenced by the microscopic observations of the interaction regions between *F.oxysporum f.sp. ciceris* and *T.*

harzianum. The mycelia of *T. harzianum* grew on the surface of the pathogen always coil round their mycelia and later penetrate their cell walls directly without formation of appressorium structures. The mycelia of the pathogen then disintegrate suggesting an enzyme action [27, 28]. Lorito et al. [20], Metcalf and Wilson [29] and Sharon et al. [30], demonstrated the possible role of chitinolytic and/or glucanases enzymes in the biocontrol exhibited by Trichoderma. These enzymes function by breaking down the polysaccharides, chitin, and glucans that are responsible for the rigidity of fungal cell walls, thereby destroying cell wall integrity limiting the growth of the pathogen. A mixture of several enzymes might be necessary for efficient cell wall lysis. *T. viride* and *T. harzianum* were reported by several workers as the best antagonists for growth inhibition of several soil and seed borne plant pathogens [31, 32].

From the 23 bacterial isolates used, only nine isolates were able to antagonize *F. oxysporum f.sp. ciceris* in dual Petri plate assay. Isolate B3, B16, B2, B15, B 22, B20, B11, B18 and B10 significantly inhibit *F. oxysporum f.sp. ciceris* radial growth recording a percentage inhibition range of 64.65% - 50.96%, on the other hand, only three isolates were highly significantly able to inhibit *F.solani* percentage radial growth namely B16, B3 and B22, with a percentage inhibition of 55.77%, 51.05% and 49.17%; respectively. These results were in agreement with that reported by Abeysinge [11]. Some bacteria isolates change the media colour especially near the inhibition zone resulting in mycelial growth retardation. This behavior was observed when *F. oxysporum f.sp. ciceri* was treated with B3 and B16, which assumed to be a result of excretion of inhibitory substances or enzymes. Reduction of fungal growth by certain PGPR and formation of inhibition zones were presumably due to the materials; antifungal substances and/or cell wall degrading enzymes; released by the bacteria into the culture medium [33]. Also, Sarhan et al. [34] and Montealegre et al. [25] pointed that the cell free culture filtrate of *B. subtilis* inhibited the mycelial growth, radial growth, and spore germination and germ-tubes length of *F. oxysporum f.sp. ciceri*. Many strains of *Bacillus* strains have been found to be potential biocontrol agents against fungal pathogens. This antifungal action involves the production of antibiotics, especially within soil microsites [35]. However, it is likely that several mechanisms act in concert to achieve control, including the production of volatiles, which have a significant effect on soil microbiology, including the soil-borne plant pathogens such as *Rhizoctonia solani* and *Pythium ultimum* [36]. The isolates B3 and B16 were the strongest antagonistic isolates against the two pathogens.

The protection exerted by the Trichoderma isolates against the fungal pathogens was pronounced than *Bacillus* isolates conformed by the high mycelial radius growth inhibition percentages, which confirms the results of Harlapur et al. [10].This difference may be due to more than one mode of mechanisms exerted by the *Trichoderma* spp which may have an additive effect in plant protection. Moreover, *Trichoderma* spp is a well-known producer of cell wall-degrading enzymes and antibiotics thus could act synergistically with other mechanisms [37]. *F.oxysporum f.sp ciceris* was more sensitive against fungal and bacterial antagonisms than *F. solani* evidenced by the inhibition percentages recorded.

In conclusion, the present study clearly demonstrated that antagonistic *Bacillus* isolates and *Trichoderma* spp can be used as biological control agents in order to protect chickpea plants from wilt/root-rot pathogens. The combined use of these biocontrol agents and the evaluation of the biological control efficacy under pot trail conditions are underway.

Acknowledgements

The authors like to express their thanks to Dr Suad Algam at the department of crop protection, Khartoum University for kindly providing ten isolates of *Bacillus* spp.

References

[1]   R. S. Singh, "Plant diseases, 6th ed., Oxford and IBH Publishing Co. Pvt. Ltd. 1996, pp. 98-102.

[2]   P. Castillo, M. P. Mora-Rodríguez, J. A. Navas-Cortés and R. M. Jiménez- Díaz "Interactions of *Pratylenchus thornei* and *Fusarium oxysporum f. sp. ciceris* on chickpea". Phytopathology, 1998, 88:828-836.

[3]   Y. Elad, I. Chet, and J. Katan, "*Trichoderma harzianum*: a biocontrol agent effective against *Sclerotium rolfsii* and *Rhizoctonia solani,*" Phytopathology, 70, 1980, pp. 119–121.

[4]   L. G. Hjeljord, A. Stensvand, and A. Tronsmo," Antagonism of nutrient-activated conidia of *Trichoderma harzianum* (*atroviride*) P1 against *Botrytis cinerea,*" Phytopathology, 91, 2001, pp. 1172-1180.

[5]   J. W. Kloepper, R. Rodriguez-Ubana, G. W. Zehnder, J. F. Murphy, E. Sikora, and C. Fernandez, "Plant root-bacterial interactions in biological control of soilborne diseases and potential extension to systemic and foliar diseases" Australasian Plant Pathology , 28, 1999, pp.21-26.

[6]   J. Hallmann, A. Quadt-Hallmann, W.F Mahaffee, and J.W. Kloepper," Bacterial endophytes in agricultural crops," Canadian Journal of Microbiology, 43, 1997, 895-914.

[7]   R. Law and D.H. Lewis," Biotic environments and maintenance of sex: Some evidence from mutualistic symbioses," Biological Journal Linnean Society, 20: 1993, pp. 249-276.

[8]   W.F. Mahaffee, and J.W. Kloepper, "Temporal changes in the bacterial communities of soil rhizosphere and endorhiza associated with field- grown cucumber (Cumcumis sativus L.)," Microbial Ecology, 34, 1997, pp. 210- 223.

[9]   S. Abeysinghe. "Biological control of *Fusarium solani f. sp. phaseoli* the causal agent of root rot of bean using *Bacillus subtilis* CA32 and *Trichoderma harzianum* RU01". Ruhuna Journal of Science. Vol. 2, 2007, pp. 82-88,. http://www.ruh.ac.lk/rjs/rjs.html.

[10]   S.I.Harlapur, M.S. kulkarni, M.C. Wali, and Srikantkulkarni, "Evaluation of Plant Extracts, Bio-agents and Fungicides against *Exserohilum turcicum* (Pass.) Leonard and Suggs. Causing Turcicum Leaf Blight of Maize," Karnataka J. Agric. Sci., 20(3) 2007, pp. 541-544.

[11] H.W. Seeley, and P.J. Van Demark, Microbes in action. A laboratory manual of Microbiology, 3rd ed., W.H Freeman and Company U.S.A. 1981, pp. 350.

[12] J.F. Leslie and B. A. Summerell, Fusarium laboratory manual, 1st ed., Blackwell Publishing, Asia, 2006, pp. 387.

[13] D. Claus and, R. C. W. Berkeley. "Genus Bacillus," in, Bergey's Manual of Systamic Bacteriology, vol. 2, P. H. A. Sneath, N. S. Mair and, N. S. Sharp, Eds. Williams and Wilkins, Baltimore, MD, 1986. pp. 1105-1139.

[14] W Gams, and J. Bissett, Morphology and identification of Trichoderma. In, Trichoderma and Gliocladium. Basic biology, taxonomy and genetics, vol 1. G. E. Harman, and C. P Kubicek, Eds. Taylor and Francis Ltd, London, 1998, pp. 3-31.

[15] F. Besson, F. Peypoux, G. Michel, and, L. Delcambe. "Identification of antibiotics of iturn group in various strains of Bacillus subtillus". Journal of Antibiotics 1978, 31: 284-288.

[16] A.C. Odebode, "Control of postharvest pathogens of fruits by culture filtrates from antagonistic fungi. Journal of plant protection research, vol. 46, No. 1. 2006

[17] K. A Gomez and A. A. Gomez, Statistical Procedures for Agriculture Research. 2nd ed. Wiley, New York, U.S.A. 1984, pp. 680.

[18] Y.M. Somasekhara, T.B. Anilkumar, and A.H. Siddarad, "Biocontrol of pigeonpea wilt Fusarium udum," Mysore J. Agric., 30, 1996, pp. 159-163.

[19] P. S. Nikam, G. P. Jagtap, and P. L. Sontakke, "Management of chickpea wilt caused by Fusarium oxysporum f. sp. ciceris," African Journal of Agricultural Research, vol. 2 (12), 2007, pp. 692-697.

[20] M. Lorito, SL. Woo, I. Garcia Fernandez, G. Colucci, G.E. Harman, J.A. Pintor-Toro, E. Filippone, S. Mucciflora, C.B. Lawrence, A. Zoina, S. Tuzun, and F. Scala, "Genes from mycoparasitic fungi as a source for improving plant resistance to fungal pathogens," USA Proc. Natl. Acad. Sci., 95, 1998, pp. 7860-7865.

[21] C.R. Howell, "Mechanisms employed by Trichoderma species in the biological control of plant diseases: The history and evolution of current concepts," Plant Disease, 87, 2003, pp. 4-10.

[22] D. Klein, and D.E. Evenigh, "Ecology of Trichoderma," in Trichoderma and Gliocladium. Basic biology, taxonomy and genetics, vol1, G. E. Harman; and C. P Kubicek, Eds. Taylor and Francis Ltd, London, 1998, pp. 57- 74.

[23] R. K. Jayaswal, M. A. Fernandez, and, R. O. Schroeder, "Isolation and characterization of a Pseudomonas strain that restricts growth of various phytopathogenic fungi," Appl. Environ. Microbiol., 56, 1990, pp. 1053–1058.

[24] T.J. Mcloughlin, J.P. Quinn, A. Betterman, and R. Bookland, "Pseudomonas cepacia suppression of sunflower wilt fungus and role of antifungal compounds in controlling the disease," Appl. Environ. Microbiol., 58, 1992, pp. 1760–1763.

[25] J.R. Montealegre, R. Herrera, J.C. Velasquez, P. Silva, X. Besoain, and L.M. Perez, "Biocontrol of root and crown rot in tomatoes under greenhouse conditions using Trichoderma harzianum and Paenibacillus lentimorbus. Additional effect of solarization," Electronic Biotech., 8, 2005, pp. 249-257.

[26] L. E Hanson, and, C. R. Howell, "Elicitors of plant defence responses from biocontrol strains of Trichoderma virens". Phytopathology, 94, 2004, pp.171-176.

[27] B. Padmodaya, and H.R. Reddy, "Screening of Trichoderma spp. against Fusarium oxysporum f. sp. lycopersici causing wilt in tomato," Indian J. Mycol. Plant Pathol., 26, 1996, pp. 266–270.

[28] D. Kumar, and S.C. Dubey," Management of collar rot of pea by the integration of biological and chemical methods," Indian Phytopath., 57, 2001, pp. 62–66.

[29] D.D. Metcalf, and C.C. Wilson," The process of antagonism of Sclerotium cepivorum in white rot affected onion roots by Trichoderma koningii," Plant Pathol., 50, 2001, pp. 249-257.

[30] E. Sharon, M. Bar-Eyai, I. Chet, A. Hewrra-Estrella, O. Kleifeld, and Y. Spiegal, "Biological control of the rootknot nematode Meloidogyne javanica by Trichoderma harzianum," Phytopathol. 91, 2001, pp. 687-693.

[31] S.C. Dubey, "Integrated management of web blight of mung bean by bio-seed treatment," .Indian Phytopath., 56, 2003, pp. 34–38.

[32] R.K. Poddar, D.V. Singh, and S.C. Dubey, "Integrated application of Trichoderma harzianum mutants and carbendazim to manage chickpea wilt (Fusarium oxysporum f. sp. ciceris)," Indian J. Agric. Sci., 74, 2004, pp. 346–348.

[33] F. Zarrin. M. Saleemi, M. Zia, T Sultan, M. Aslam, R, U. Rehman, and M, C. Fayyaz, "Antifungal activity of plant growth-promoting rhizobacteria isolates against Rhizoctonia solani in wheat," African Journal of Biotechnology, vol. 8 (2), 2009, pp. 219-225.

[34] M. M. Sarhan, S.M. Ezzat, A.A. Tohamy, A.A. El-Essawy, and F.A. Mohamed, "Biocontrol of Fusarium tomato wilt diseases by Bacillus subtilis" Egypt. J. Microbiol., 36, 2001, pp. 376-386.

[35] D. R. Fravel, "Commercialization and Implementation of Biocontrol," Annual Review of Phytopathology, 43, 2005, 337-359.

[36] H. Kim, J. Park, S. Choi, K. Choi, G.P. Lee, S. J. Ban, H. C. Lee, and S. C. Kim, Isolation and Characterization of Bacillus Strains for Biological Control," The Journal of Microbiology, vol. 41, no. 3, 2003, pp.196-201.

[37] F. Vinale, K. Sivasithamparam, E. L. Ghisalberti, R. Marra, S. L. Woo, and M. Lorito, "Trichoderma–plant–pathogen interactions. Soil Biology and Biochemistry, 40, 2008, pp. 1–10. http://www.elsevier.com/locate/soilbio.

7

Adaptability study of black cumin (*Nigella sativa* L.) varieties in the mid and high land areas of Kaffa zone, South West Ethiopia

Ermias Assefa[*], Addis Alemayehu, Teshom Mamo

Southern Agricultural Research Institute, Bonga Agricultural Research Center, Department of Crop Science Research Process, Bonga, Ethiopia

Email address:

ethioerm99@gmail.com (E. Assefa)

Abstract: The bases of the idea to conduct this field experiment are the uses of the spices and the suitable agro-ecology of Kafa zone. There was no any research activity conducted in the Kafa zone in relation to highland seed spices. It is important to evaluate the adaptability of improved black cumin varieties in the Kafa zone in order to diversify their production and to maximize the income of the farmers in the area. Based on this fact, a field experiment was conducted using three improved Black cumin (*Nigella sativa* L.) varieties; Dirishaye, Eden, and Deribera with the local check. The activity was conducted in the 2012 /13 cropping season at Alarigeta and Kaya Kela experimental sites of Bonga Agricultural Research Center. The objective of this study was to test the adaptability of improved Black cumin varieties to the representative areas Kafa zone. These varieties were evaluated for yield, plant height, pods per plant, emergence and flowering dates. The test varieties were used as experimental treatments and arranged in a randomized complete block design with five replications. The grain yield recorded in gram per plot was converted to kilogram per hectare. There were no significant differences ($p < 0.05$) in most parameters evaluated at Alarigata, unlike Kaya Kela site. The local check showed significantly higher grain yield (612.98 Kg ha[-1]) than Eden, Dirishaye, and Deribera (473.06, 451.9, and 449.62 Kg ha[-1], respectively) at Kaya Kela. Based on the results obtained under this study variety Eden could be used for demonstration, popularization and pre-scaling up of the technology at Alarigata and the surrounding areas. On the other hand, this experiment showed the huge potential of the local variety at both experimental sites. Thus, such a potential suggests that the local check or the land race could be used for variety development program which would later be supported by agronomic and pathological studies (fertilizer rate, sowing date, and reaction to insect pests and diseases). This would give rise to the production of adaptive improved black cumin seed spices with specific quality traits at different agro-ecologies of the zone that fulfill the specific international market demands.

Keywords: Black Cumin Variety, Grain Yield, Plant Height, Pods per Plant, Emergence Date, Flowering Date

1. Introduction

Black Cumin (*Nigella sativa* L.) is a member of Apiaceae (Umbelliferae). This species is originated in Egypt and East Mediterranean, but is widely cultivated in Iran, Japan, China and Turkey (Shewaye, 2011). Black cumin grows on a wide range of soils. Sandy loam soil rich in microbial activity is the most suitable for its cultivation. The sloppy soils of heavy rainfall areas and leveled and well drained soils of moderate rainfall areas are quite suitable for its cultivation. Soil pH of 7.0 to 7.5 is favorable for its production (Orgut, 2007).

Black Cumin has a long history of uses for food flavors, perfumes and medicinal values. Oil has been used for bringing smell to some medicines, sterilizing of surgical operation fiber, production of some veterinary and agricultural medicines and plastic components (Aminpour and Karimi, 2004).

Black Cumin seeds have an aromatic odor and bitter taste. They are used as an essential ingredient in soup component, sausages, cheese, cakes and candies. The Ethiopian variety of cumin seed accumulate up to 50% thymol, a monocyclic phenolic compound. The presence of this compound makes cumin valuable source for health care Industry (Black et.al, 2005) and medicinal purposes (Ashraf and Orooj, 2006). In Ethiopia, it is commonly used in Amharic "*Berbere*" in which it tends to reduce its hotness (Hedberg et al., 2003), for preparation of curries, bread, katikala (Jansen, 1981),"*Shamita*" (Mogessie and Tetemke, 1995), traditional Ethiopian stews, "*Wot*" and preservation of butter.

Black cumin is used principally to flavor food, either as whole grain, in powdered form or as an oleoresin extract. It is also used in gripe water and other herbal medicines. Within Ethiopia its main use is as a spice, which is typically ground and mixed with other spices. There is also some use in traditional medicine. The vast majority of Ethiopia's black cumin exports go to Arabic countries, which, together with other predominantly Muslim countries, accounted in 2008 for some 98% of national exports. Sudan overtook Saudi Arabia as the main export destination in 2007 and by 2008 it accounted for almost one half of all official exports. It is uncertain how reliable this market is and whether exports can be maintained at current levels. Value-adding to cumin in Ethiopia is low, with all exports being made in the form of whole grain (Orgut, 2007).

Taking into consideration of its use and the suitable agro-ecology of Kafa zone, there was no any research activity conducted in relation to highland spices. In order to diversify its production and increase the income of the farmers, it is important to evaluate the adaptability of improved black cumin varieties to the area. Therefore, this study was initiated with the objective of selecting the best adaptive black cumin varieties to the area.

2. Materials and Methods

2.1. Description of the Experimental Area

The experiment was conducted at Kaya Kela and Alarigata experimental sites of Bonga Agricultural research Center (BARC), which was located at,Kaffa zone, Southern Nations Nationalities and People's Region (SNNPR). It is found within the southwestern plateau of Ethiopia and 450km and 725km far from Addis Ababa and Hawassa respectively. The Kaya Kela site is located at 07°00'- 7°25'N Latitude and 35°55'-36°37'E Longitude at an elevation of 1753 meters above sea level. The area experiences one long rainy season, lasting from March /April to October. The mean annual rainfall ranges from 1710 mm to 1892mm.The mean minimum and maximum daily temperature ranges from 18.10C to 19.40C.

The topography is characterized by slopping and rugged areas with very little plain land. Whereas, Alarigata experimental site is located at07°17' 316"N latitude and longitude of 36°22'243"E at an elevation of 2400 meters above sea level.

2.2. Land Preparation, Treatments and Design

Land was ploughed three times with oxen, which is similar to farmers' practice. The experimental field was divided into five blocks each having four plots. The width and length of individual plot was 1.8m and 4m, respectively, with each plot sub divided into six rows. The spacing between plots and blocks was maintained to be 1 and 2 m, respectively. The study was conducted using randomized complete block design with five replications. The varieties were assigned as treatments. The improved seeds were collected from national research center and the local seeds of black cumin were taken from farmers of the respective testing sites. The seeds were sown by drilling in rows as soon as the rains started. The spacing between each row was 30 cm.

2.3. Data Collection

Yield and yield component data such us Grain yield, plant height, pods per plant, flowering and emergence dates data were collected.

The Grain yield were recorded in gram per plot and converted in to kilogram per hectare. The plant height data measure the height of the plant from the base "close to ground level" to the shoot tip of the main axis. Pods per plant also collected randomly selected five plants per plot and recorded by counting the number of pod per plant. Flowering date recorded first open flower 50% of the plants per plot by counting the number of days and Days to maturity also recorded when 95% of flowers per plot was matured.

2.4. Statistical Analysis

The analysis of variance was done using statistical analysis system, SAS 9.1 software (SAS, 2007). Wherever F values were found to be significant at a 5% level of probability, least significant difference (LSD) values were computed for making comparisons among the treatment means.

3. Results and Discussion

Table 1. Mean separation.

Mean Values										
Alarigata						**KayaKela**				
Varieties	**DF**	**DM**	**PH**	**PP**	**GY**	**DF**	**DM**	**PH**	**PP**	**GY**
Dirshaye	87[c]	163.2[cb]	51.28	6	997.5	78[b]	133.8[b]	40.08[b]	6	451.9[b]
Eden	96[b]	166.4[ab]	52.2	5.04	1097.9	79.6[b]	136.6[b]	40.96[b]	5.96	473.06[b]
Deribera	85[c]	160[c]	45.88	4.76	979.3	70.6[c]	132.4[b]	39.12[b]	6.88	449.62[b]
Local	114.4[a]	168.4[a]	48.36	4.24	957.8	91[a]	146[a]	51.36[a]	6.78	612.98[a]
Mean	95.33	164.5	49.43	5.01	1008.1	79.8	137.2	42.88	6.41	496.89
CV	4.33	1.74	14.91	36.18	26.04	5.57	2.28	16.27	33.8	18.2
LSD	5.56	3.83	Ns	Ns	Ns	5.96	4.2	9.35	Ns	121.27

*= Significant at 5% probability level, CV= Coefficient of variation, LSD= Least significant difference, DF= Days to flowering, DM= Days to maturity, PH= Plant height, PP= Pods per plant, GY= Grain yield; Values with the same letter(s) are not significantly different

The results of analysis of variance indicated that there were no significant differences (p < 0.05) in most parameters compared at Alarigata site, unlike Kaya Kela site.

3.1. Days to Flowering

Significant variations were observed in days to 50% flowering in the tested varieties at both locations. The local check took the highest mean days to flower at both Alarigata and Kaya kela testing sites (114.4 and 91 days, respectively). Whereas, Dirishaye and Deribera took the shortest average days to flower (87 and 85 days, respectively) at Alarigata; and likewise, Deribera took the shortest mean days (70.6 days) to flower at Kaya kela.

3.2. Days to Maturity

The experimental varieties at both locations showed significant variations (p < 0.05) in days to maturity (Table 1). The longest average duration (168.4 days) was recorded by the local check to mature, while variety Deribera took the shortest duration (160 days) at Alarigata. Similarly, the local check at Kaya kela took the mean highest duration (146 days) to mature over the rest of the improved ones.

3.3. Plant Height

The mean values (Table 1) revealed that there were no statistically significant differences (p < 0.05) in plant height among the tested varieties at Alarigata. On the contrary, the local check showed the highest plant height (51.36 cm) over the improved varieties at Kaya kela.

3.4. Pods per Plant

The LSD result 2.43 at alpha 0.05 indicated that there was no significant difference among the tested varieties in pods per plant at both locations (Alarigata and Kaya kela).

3.5. Grain Yield

The LSD result 352.03 at alpha 0.05 indicated that there was no significant difference among the varieties in grain yield at Alarigata site. Whereas, the highest grain yield (612.98Kg ha^{-1}) was scored by the local check at Kaya kela. The released varieties (Dirishaye, Eden, and Deribera) did not show their grain yield potential at Kaya Kela site as compared with the Alarigata site. This might be associated with the occurrence of wilt disease on the experimental varieties at Kaya kela during the growth periods.

3.6. Quality Evaluation

The laboratory evaluation was made for the experimental varieties under this study. Moisture content, essential oil content, and oleoresin yield of the seeds of the tested black cumin varieties were taken as parameters. Based on the laboratory results (Table 2), the average moisture content, essential oil content both in dry and wet bases, and oleoresin yield of the tested varieties were equal or greater than the national average content (8, 0.6, and 26.3%, respectively). These laboratory results revealed that the tested varieties fulfill the international market demands.

Table 2. Laboratory results of Black cumin varieties tested at Alarigata and Kaya kela testing sites during 2012/13 crop season.

Variety	Moisture Content (%)				Oleoresin Yield (%) Hexane Extract			
	Rep 1	Rep 2	Rep 3	Mean	Rep 1	Rep 2	Rep 3	Mean
Dirishaye	9.09	8.14	8.57	8.6	25.2	24	25..20	24.79
Eden	9.17	8.85	9.65	9.22	28	28.8	28.4	28.4
Deribera	9.55	9.68	9.65	9.62	29	27.5	28.4	28.31
Local	9.34	9.87	10.5	9.89	26.6	28.4	27.88	27.64

Table 2. Laboratory results of Black cumin varieties tested at Alarigata and Kaya kela testing sites during 2012/13 crop season. (continue)

Variety	Essential oil Content (%) WB				DB			
	Rep 1	Rep 2	Rep 3	Mean	Rep1	Rep 2	Rep 3	Mean
Dirishaye	0.52	0.61	0.6	0.58	0.57	0.66	0.66	0.63
Eden	0.43	0.42	0.42	0.42	0.48	0.46	0.46	0.47
Deribera	0.57	0.47	0.47	0.5	0.63	0.52	0.52	0.56
Local	0.53	0.57	0.57	0.56	0.59	0.64	0.64	0.62

Key: WB = wetness base; DB = dry base

4. Summary and Conclusions

The adaptability of improved black cumin varieties (Dirishaye, Eden, and Deribera) and the local checks of the respective locations was evaluated at Alarigata and Kaya kela testing sites of Bonga Agricultural Research Center during 2012/13 cropping season. The experiment was conducted based on the protocol required for black cumin varieties.

Days to 50% flowering, days to maturity, plant height, pods per plant and grain yield were taken as parameters of the evaluation. The results of the experiment indicated that the local check scored higher grain yield than the improved varieties at Kaya kela testing site. However, no statistically significant variations were recorded in grain yield among the tested varieties at Alarigata. Because of the occurrence of wilt disease on the experimental black cumin varieties at

Kaya Kela, the improved varieties yielded less than the national average (9-16 Q ha-1). The materials were also studied under laboratory for moisture content, essential oil content in dry and wet bases, and oleoresin yield. The results of the laboratory studies indicated that the tested materials scored much greater values than the national average. In general, the outcomes of the experiment in all parameters studied showed that the testing areas have huge potential for black cumin production. The performance of the local varieties at both locations also indicated that there is a need for further investigation.

Based on the results of the experiment, Eden could be used for demonstration, popularization and pre-scaling up of the technology at Alarigata and the surrounding areas and future research activities should focus on the further evaluation of the released black cumin varieties with respect to the local checks under various agro-ecologies. The due attention should be given towards the collection, characterization and evaluation of the landrace black cumin in order to help the variety development program. Agronomic and pathological aspects of the crop should also be area of concern under different agro-ecologies in the Kafa zone.

References

[1]   Aminpour and Karimi (2004). Underutilized medicinal spices. Spice India. 17 (12): 5-7.

[2]   Ashraf M, Orooj A (2006). Salt stress effects on growth,ion accumulation and seed oil concentration in an arid zone traditional medicinal plant ajowan (*Trachypermum ammi* [L.] Sprague). J. Arid Environ., 64(2):209-220.

[3]   Black M, Bewley D, Halmer (2005). The Encyclopedia of seed science, technology and uses. wallinoford. CAB P 7.

[4]   Hedberge I, Edwards S, SileshiNemomissa (eds) (2003). Flora of Ethiopia and Eriteria. Vol 4 (2), Apiaceae to Dipsaceae. The Natural Herbarium. Addis Ababa University, Addis Ababa and Uppsala. P. 21.

[5]   Jansen PCM (1981). Spices, condiments and medicinal plants in Ethiopia. their taxonomy and agricultural significance. Addis Ababa: Center for Agricultural Publishing and Documentation pp. 111-120.

[6]   Mogessie A, Tetemke M (1995). Some microbiological and nutritional properties of Borde and Shamita. Traditional Ethiopian fermented beverages. Ethiop. J. Health Dev. 9(1): 105-110.

[7]   Orgut, Market Assessment Study, Ethiopian Nile Irrigation and Drainage Project, Main Report and Annexes, Ministry Of Water Resources, Addis Ababa, June 2007

[8]   SAS (2007) Statistical Analysis Systems SAS/STAT user'sguide Version 9.1 Cary NC: SAS Institute Inc. USA

[9]   Shewaye L (2011).Antifungal Substances from Essential Oils. M.Sc. Thesis. Addis Ababa University. p. 8

Beeswax production and marketing in Ethiopia: Challenges in value chain

Gemechis Legesse Yadeta

Oromia Agricultural Research Institute, Holeta Bee Research Center (HBRC), Holeta, P. O. Box 22, Ethiopia

Email address

gemechislegesse@gmail.com

Abstract: Beeswax is one of the most valuable and oldest bee products to be used by mankind and still being used in the development of new products in various fields such as cosmetics, foods, pharmaceuticals, engineering and industry. Ethiopia has huge apicultural resources that made it the leading beeswax producer in Africa, and one of the important beeswax exporter to the world market. In Ethiopia apicultural research is being conducted in a coordinated manner under the national agricultural research system. Hence, a lot of information have been gathered on different aspects of the beekeeping. This work is a review of various research results from published and unpublished data over a long period of time in the area of beeswax production, chemical analysis, marketing and value chain studies in Ethiopia. Despite the country's huge potential for production of high quality beeswax, only less than 10% of the beeswax produced is exported. The beeswax production and processing practices use traditional and inefficient techniques that leave significant amount of beeswax resource unutilized. The marketing channel for beeswax in the country is also entangled with challenges related to uneasy traceability and adulteration that are affecting both the local and international trade.

Keywords: Beekeeping, Beeswax, Ethiopia, Marketing, Production

1. Introduction

The term beeswax is often limited to wax produced by honeybees (*Apis* species) and many would specify *Apis mellifera* L. as a source [1]. It is one of the most valuable bee products and it is also one of the oldest items used by mankind [2]. Beeswax, with its unique characteristics, is now being used in the development of new products in various fields such as cosmetics, foods, pharmaceuticals, engineering and industry [3]. Specifically, most of the wax produced nowadays are used in the manufacture of cosmetics, such as hand and face creams, lipsticks and depilatory wax and many other uses. Moreover, the pharmaceutical industry uses the wax in various ointments, for coating pills and suppositories and other miscellaneous industrial products [4].

Beeswax is secreted in small wax platelets form by worker honeybees from four pairs of wax glands on the underside of the abdomens which are functional when the bees are about 9–17 days old after being engorged with honey and resting suspended for 24 hours together [5]. Construction of combs saps the colony's energy supplies, through the costly production of wax from the sugars in collected honey [6], and through thermoregulation of the building site by the surrounding festoon of bees [7]. Honeybees fed with sugar syrup during dearth periods couldn't produce more beeswax emphasizing the need of nectar/honey for beeswax production (Gemechis, Holeta Bee Research Center, Ethiopia, unpublished data). The platelets are scraped off by the bees, masticated several times into pliable pieces with the addition of saliva and a variety of enzymes to form part of the comb of hexagonal cells [8]. Wax is used to cap the ripened honey, and when mixed with some propolis protects the brood from infections and desiccation and also employed for sealing cracks and covering foreign objects in the hive [2]. Worldwide in general and in Ethiopia in particular, a lot of research activities had been conducted and data regarding beeswax production, physical and chemical characterization , processing and value addition and marketing and problems related to marketing are documented. However, in particular in Ethiopia, the progresses in different aspects of research in beeswax have not been reviewed and all the available information are found scattered and in inaccessible situation. Therefore, this review work was executed to review the progress of research in production, value addition and marketing of beeswax mainly in Ethiopia.

2. Production

Assessments indicate that Ethiopia has got potential for production of beeswax because of huge number of honeybee colonies being kept in traditional hives [9]. The migratory behavior of the tropical honeybees also contributes to high beeswax production leaving combs behind every time colonies search for new nests [10]. The beeswax production in traditional beehives is 8–10% of the honey yield [2]. In 2005, Ethiopia produced about 4300 tones of beeswax [11]. This made Ethiopia stand first in Africa and third in the world. In the same year there were about 4.55 million hived colonies [12] which, based on FAO data for national production, is equivalent to 0.95 kg wax per hive per year. However, with the current increase in production of honey that is estimated to be around 54,000 tones [13], the annual beeswax production is expected to be more than 5000 tones. [10] indicated that this can be optimized to 9000 tones.

2.1. Characterization of Beeswax

2.1.1. Beeswax Quality

The composition of beeswax is very complex to identify, but it is relatively constant for beeswax from a single species of honeybee [2]. Pure beeswax from *A. mellifera* consists of at least 284 different compounds [1]. Quantitatively, the major compounds are saturated and unsaturated monoesters, diesters, saturated and unsaturated hydrocarbons, free acids and hydroxy polyesters, each consisting of a series of long carbon-chain compounds [1,2]. There are 21 major compounds, each making up more than 1% and together accounting for 56% of the pure unfractionated wax. The other 44% of diverse minor compounds probably account for beeswax's characteristic plasticity and low melting point [1]. Among the physical and chemical features of beeswax, melting point, relative/ specific density, electrical resistance, thermal conductivity, saponification cloud point, ester and acid values and the ratio of ester to acid values serve in determining the quality of beeswax [1]. Quality standards for beeswax are set in most countries according to their pharmacopoeias [8].

Ethiopia has set its standards for beeswax after investigating the physical and chemical properties of samples of beeswax collected from different parts of the country at farm gates and at different beeswax processors and exporters' stores [10]. The physical and chemical properties that are relevant to beeswax quality like melting point, saponification cloud point, acid value, ester value and ester to acid ratio were tested based on the protocols of American Beeswax Importers and Refiners Association INC, 1968 as cited in [14]. Generally, the purified beeswax collected from different parts of Ethiopia met the world standards [10]. The saponification cloud point ranged between 57.9 °C and 65.0°C, while the melting point lied between 61.0 °C and 63.9 °C. Acid value of 18.0 to 32.7 and ester value ranging between 66.4 and 98.0 were recorded, while the ratio of ester to acid values was found 4.2 to 4.0 [10].

2.1.2. Adulteration

The quality of beeswax could deteriorate and its natural composition could alter because of adulteration and prolonged overheating [1]. Under local conditions deterioration of beeswax quality due to overheating from processing is highly likely to happen; some of the processing facilities are not suitable to regulate the optimum temperature during processing [10]. Similarly, adulteration of beeswax with cheaper materials like animal fats, plant oils and paraffin has become a problem for beeswax quality and its marketing, especially adulteration of beeswax with paraffin is a major one [1,2, 15]. But in Ethiopia, animal tallow is highly suspected to be the major adulterant to be mixed with beeswax because of availability and cheapness, actually many times cheaper than beeswax [10].

To detect adulteration, a number of tests may have to be conducted. The simplest is to determine the melting point by measuring the temperature at which the first liquid wax appears during very slow heating. It should be between 61 and 66°C or preferably between 62 and 65°C [8]. In addition, determining the saponification cloud point is an officially accepted sensitive method for determining adulteration. The method is limited to detecting quantities greater than 1 % of high melting (80-85°C) paraffin waxes, or more than 6% of low melting (50-55°C) paraffins. The test measures the amount of hydrocarbons which saponify (turn into soap) in a specific amount of ethanol and give a clear solution. If the solution becomes clear at or below 65°C, the wax is probably unadulterated with paraffin. If it is adulterated, the solution will turn clear only at a higher temperature [8].

An investigation conducted following the previously mentioned protocols of American Beeswax Importers and Refiners Association INC, 1968 cited in [14] to look into the melting point and saponification cloud point of mixture of 10 gm of pure beeswax and animal tallow prepared in the proportion of 1%, 2.5%, 5%, 7.5%, 10%, 12.5%, 15%, 17.5%, 20%, 30%, 40%, 50%, 60%, 70%, 80%, and 90% by weight of animal tallow showed that melting point and saponification cloud point could equally serve in detection of as small as 2% animal tallow used in beeswax adulteration [14].

Beeswax samples adulterated with 1% animal tallow melted at slightly lower temperature at an average of 61°C, which was lower by 1°C than the lower limit of most pure beeswax melting point standards. Beeswax samples mixed with 2.5-7.5% animal tallow melted at lower temperature between 60°C-59°C. Above this adulteration level, the melting point was further below 59°C and as the proportion of animal tallow increased the melting point approached to 46°C, which was the melting point for pure animal tallow [14]. For saponification cloud point, the result followed the same trend in which beeswax samples mixed with 1% animal tallow saponified at 60°C and beeswax samples adulterated with 2.5-7.5% animal tallow saponified at lower temperatures between 59°C-58°C. At adulteration of 7.5% the saponification cloud point fell below 58°C and as the proportion of animal tallow increased the saponification cloud point approached to 44°C, which was the saponification cloud point of pure animal tallow [14].

2.2. Beeswax Processing

In many parts of the world much of the beeswax produced by bees that could be harvested by beekeepers is wasted. The beeswax is left or thrown away because beekeepers do not bother to collect and render it into marketable blocks. As a result only a limited proportion, may be at most one-half, of the world's production of beeswax comes on to the market, the rest being thrown away or lost [2]. From the total amount of beeswax produced annually in Ethiopia, only less than 10% of it is used for export [11,16,17]. The remaining large proportion of it is believed to be wasted at different levels. In Ethiopia, there are so many factors attributing to this wastage of beeswax including consumption of honey in crude form and discarding the crude beeswax at every crude honey consuming points and "tej" (local wine made from honey) makers [18]. The majority of beekeepers in Ethiopia practice beekeeping using dominantly traditional beehives and hence huge amount of beeswax is produced [10]. Observations indicate that these beekeepers do not know the use of beeswax, the rendering techniques or the existence of market for this product [19].

The sources of crude beeswax for rendering could be from cappings removed during honey extraction, which produces a very high quality, light colored wax. Light colored broken combs provide the next quality of wax, whereas old black brood combs yield the smallest proportion and lowest quality of wax. Dark combs contain propolis resin, coccon and pollen packs that lower the quality of the beeswax [2]. In areas with traditional and top bar hive beekeeping, different qualities of wax can be produced by separating new white honey combs from darker ones or from those with portions of brood [8]. In Ethiopia, the majority of the crude beeswax is collected by the local tej makers in the form of the beverage byproduct called "sefef". Sefef is produced in a straw form and sometimes difficult to assume it as a source of beeswax because of its impurities and discoloration. Beeswax blocks produced from this material are of low quality (in its sensory properties) and the color is not as light as beeswax sourced from crude honey due to many ingredients used in tej, like the leaves of plant called "Gesho" (Rhamnus prinoides), and the fermentation process.

To begin any processing of beeswax, the first step is to separate the honey from beeswax. In the case of box frame hives extracting honey leaves the empty combs attached to the frames. Therefore, the combs can be directly taken to prepare purified blocks of beeswax by scraping all the combs from the frame and boiling in water. However, crude honey from traditional hives has more non honey and non beeswax foreign materials due to poor harvesting practices [10]. After the crude honey is strained, the beeswax is left with many types of impurities of pollen packs, cocoon sheath, propolis, dead bodies and parts of honeybees and hive bodies [2]. As expected, the proportion of these impurities in the beeswax greatly affects the amount of pure beeswax recovered during rendering [2, 18].

The amount of pure beeswax produced from crude beeswax is dependent on both the quality of source material and the techniques used in processing. An amount of crude beeswax ranging from 5% to 65.6%, with mean of 27.5%, can be recovered from crude honey produced in traditional hives and collected from beekeepers; while the average percentage of pure beeswax obtained from crude beeswax deriving from the aforementioned source was 73.6% [18]. Recently, (2012)separate study at Holeta Bee Research Center (HBRC), Ethiopia, unpublished data) revealed that the average pure beeswax yield using three extracting methods (manual sack extraction method, submerged and solar extraction) was the highest (67.7%) for crude beeswax sourced from crude honey. The average recovery rate of pure beeswax using these three techniques was 26.8% for old and dark combs and 25.9% for sefef. However, the techniques by themselves significantly affected the amount of pure beeswax recovered from crude beeswax of the three sources (crude honey, old combs and sefef) as manual and submerged methods yielded 44.2% and 49.6%, respectively compared to solar method that gave only 26.4% pure beeswax. This reveals the high and wide variations inefficiency in the existing wax rendering techniques specially for small scale producers and processors.

Several methods of rendering wax are possible and may be adapted to various circumstances. Wax can be separated in solar wax melters, by boiling in water then filtering, or by using steam or boiling water and special presses [2]. However, for small and medium scale producers and processors the existing inefficient beeswax extraction technologies are believed to be one of the factors for the wastage of beeswax in Ethiopia [18]. The manual sack pressing method could recover about 34.2% of the crude beeswax content sourced from crude honey which is far less than 64.7%, the percentage of the pure beeswax recovered from the same material by simple machines applying mechanical and hydraulic pressure developed at HBRC [18]. Similarly, the average pure beeswax recovered from sefef by the manual sack method was about 25%, while mechanical and hydraulic pressure applying simple machines could increase this efficiency by 50% more than that of manual pressing [18].

3. Marketing

In Ethiopia, beeswax is one of the important exportable agricultural commodities [20]. Currently, the annual production of beeswax is expected to be more than 5000 tones. [10] indicated that this is around one tenth of the world annual beeswax production that is estimated to be around 50,000 tones. Because of its pliability, yellow coloration and other physical properties, the Ethiopian beeswax has been highly demanded and mostly used to blend beeswaxes from other sources. It is dominantly yellow in color though white beeswax is produced in southeast and southwest parts of the country. Yellow coloration is mainly due to the pollen stored in combs and propolis polishing [10].

The smallholding beekeepers are the primary sources of beeswax in Ethiopia who sell the majority of crude honey to the *tej* brewers, hence most of marketable crude beeswax comes from them as a byproduct of the beverage [19]. After the beverage production, the *tej* makers collect the crude beeswax and store it as it is in the crude form of "*sefef*" or partially strained form of "*keskes*" [10]. The *sefef* or the partially processed *keskes* is collected from the *tej* makers [19]. Traditional beeswax extractors are also the other intermediate sources who process the *sefef* partially to rough beeswax blocks. Recently, many private firms collect *sefef* and *keskes*, process and export beeswax [20]. The channel of crude beeswax collection, processing and marketing in Ethiopia is very complex and the issue of traceability is a big concern. This is one of the major challenges that is attributing to the increasing adulteration of beeswax with cheap materials like animal fat in addition to the ever increasing price that draws attention of the people involved in the mischief.

Beeswax has good domestic market in Ethiopia. The traditional religious practice of the Ethiopian Orthodox Church followers to burn candle sticks called "*tuaf*" made from pure beeswax is believed to consume a significant amount of the beeswax produced locally even though not quantified so far. Moreover, the intensity of the improved beekeeping extension in the main beekeeping potential areas of the country launched by government and NGOs has created a huge demand of beeswax for foundation making for frame box hives. Currently, a kilogram of purified blocks of beeswax cost about 250-300 ETB (25-30 USD) in the local markets. Generally, the beeswax price at the domestic market is mostly higher than the international beeswax price which makes beeswax export less profitable in Ethiopia.

Most of the world beeswax is supplied by the developing countries. China is the leading producer and exporter [10]. The EU is the major market for beeswax in the world, accounting for more than half of global imports [21]. The price of beeswax exported to EU is also on continuous rise since 2003. For instance, the price of beeswax exported from Ethiopia had risen from 2430 €/ton in 2003 to 3200 €/ton in 2009, with the highest 3630 €/ton in 2006, in main European ports [22]. In EU, most of the beeswax refining and re-exporting is done as the importers in EU are in a better position to enjoy the competitive advantage through advanced technological and market reputation they have already established. They don't actually encourage their suppliers in the developing countries to refine their beeswax.

Exports of beeswax from Ethiopia have increased spectacularly and reached 402 tones of beeswax (1.2% share in world market), destined to different countries (USA, Japan, Greece, Great Britain and Netherlands etc.), generating USD 936 thousands in 2003 [11]. After 2003, the export volume is not far from 400 tones annually [16,17]. However, similar to the local beeswax market, the export of Ethiopian beeswax has threats as adulteration with cheaper materials has become a challenge for its quality and marketing. Exporters complain that significant proportion of the exported beeswax is refused by recipient companies because of compromised quality mainly due to adulteration with cheap materials (personal communication, Ethiopian honey and beeswax processors and exporters association, 2013).

4. Conclusions

The Ethiopian beekeeping is characterized by traditional production system that created an opportunity for high beeswax production potential. However, even the total production expected each year is not properly collected, processed and marketed. In fact, little is known about the beeswax produced but utilized in other applications except used for export market. Challenges at the beekeepers' level that in some potential areas they don't have awareness of the importance of beeswax, skill to collect, process and market beeswax are among many. Moreover, the existing beeswax processing technologies are very limited and traditional of very low efficiency.

The market demand for beeswax both in the domestic and international trade is very high. Beeswax from Ethiopia has higher demand and also earns higher price in EU, that is mainly used for blending low quality beeswax from different sources. The local beeswax market is always short of supply even for the expanding improved beekeeping. Hence, prices have always been rising continuously. This, along with many others is a driving factor for adulteration of beeswax with other cheap materials like animal fats. The *tej* making process from which *sefef* is collected and supplied in the beeswax value chain is one of the important factor in deteriorating the quality of beeswax even though *tej* makers are still the major source of crude beeswax. In addition, the informal and complex nature of beeswax market channel in the country is another serious problem in the production and marketing of the product for both domestic and international trades. This has contributed for the prevailing adulteration practices of beeswax as traceability is hardly possible.

Generally, increasing the production level of beeswax through improved processing technologies of higher efficiency is important step to be taken. Moreover, formalizing the market system and establishing clear path of the products' movements in the market from producers to the consumers will increase the traceability and hence minimizes the problem of adulteration.

Acknowledgments

I would like to acknowledge all the previous investigators who produced valuable research articles and availed for this review paper. My deep appreciation also goes to Nuru Adgaba (PhD) who provided me with some papers for free.

References

[1]   Tulloch, A.P. 1980. Beeswax – composition and analysis. Bee Wor 61: 47-62.

[2]   Crane, E. 1990. Bees and beekeeping: science, practice and world resources. Heinnmann Newness, London. pp. 614.

[3]   Kameda, T. 2004. Molecular structure of crude beeswax studied by solid-state ^{13}C NMR. J. Insect Sci. 4:29

[4]   Brown, R.H. 1988. Honey bees: a Guide to management: The Crowood press Ltd London. pp. 128.

[5]   Brwon, R.H. 1981. Beeswax. Butler and Tanner, LTD, Frome. pp. 74.

[6]   Hepburn, H.R. 1986. Honeybees and wax. Springer-Verlag, Berlin. pp. 205.

[7]   Hepburn, H.R., Hugo, J.J., Mitchell, D., Nijland, M.J.M. and Scrimgeour, A.G. 1984. On the energetic costs of wax production by the African honeybee, *Apis mellifera adansonii*, S. Afr. J. Sci. 80: 363– 368.

[8]   Krell, R. 1996. Value added products from beekeeping. Agricultural Services Bulletin No 124. Food and Agriculture Organization of the United Nations: Rome, Italy. http://www.fao.org/docrep/w0076e/w0076e00.htm., accessed on 13/05/2010.

[9]   Gemechis, L. 2014. Review of progresses in Ethiopian honey production and marketing. Lives. res. for rur. dev. 26(1)

[10] Nuru, A. 2007. Atlas of pollen grains of major honeybee flora of Ethiopia. Holeta Bee Research Centre. Commercial Printing Enterprise. Addis Ababa, Ethiopia. pp 152.

[11] FAO, 2005. Statistical yearbook, FAOSAT.

[12] CSA, 2006. Statistical Abstracts. Central Statistical Agency. Addis Ababa, Ethiopia.

[13] CSA, 2012. Statistical Abstracts. Central Statistical Agency. Addis Ababa, Ethiopia.

[14] Nuru, A. 2000. Physical and chemical properties of Ethiopian beeswax and detection its adulteration. E. J. Ani. Prod. 7: 39-48.

[15] Anam, O.O. and Gathuru, E.M. 1985. Melting point and saponification cloud point of adulterated beeswax. pp. 222-223. Proceedings of 3rd International conference on apiculture in tropical climate, 1984, Nairobi, Kenya.

[16] EEPA, 2010. Ethiopian Export Promotion Agency. Addis Ababa, Ethiopia.

[17] EEPA, 2012. Ethiopian Export Promotion Agency. Addis Ababa, Ethiopia.

[18] Nuru, A. and Eddessa, N. 2006. Profitability of processing crude honey. Pp79-84. Proceedings of 13th Annual Conference of Ethiopian Society of Animal production (ESAP). August 25-27, 2004. Addis Ababa, Ethiopia 244pp.

[19] MoARD, 2003. Honey and beeswax marketing and development. IN DEVELOPMENT, M. O. A. A. R. (Ed.) Plan 2003. Ministry of Agriculture and Rural Development. Addis Ababa, Ethiopia.

[20] Mengistu, A. 2011. Pro-poor value chains to make market more inclusive for the rural poor: Lessons from the Ethiopian honey value chain. pp. 35- 50. Danish Institute for International Studies, Copenhagen, Denmark.

[21] FAO, 2011. Statistical yearbook, FAOSAT.

[22] CBI, 2009. CBI market survey: the honey and other bee products market in the EU.

9

Evaluating the Efficacy of Pituitary Gland Extracts and Ovaprim in Induced Breeding and Fry Quality of *Clarias gariepinus*, Burchell (Pisces: Claridae)

Efe Okere, Ebere Samuel Erondu, Nenibarini Zabbey[*]

Department of Fisheries, Faculty of Agriculture, University of Port Harcourt, PMB 5323 Choba, Port Harcourt, Rivers State, Nigeria

Email address:

nenibarini.zabbey@uniport.edu.ng (N. Zabbey), zabbeyn@yahoo.com (N. Zabbey)

Abstract: This study compared the effectiveness of Ovaprim and pituitary gland extract (PGE) in induced spawning of the African mud catfish, *Clarias gariepinus*, using reproductive output and fry quality indices. At a mean temperature of 26.0 $\pm 0.70^0$C, latency period for Ovaprim and PGE were 613 and 745 minutes, respectively. Workers fecundity was significantly higher ($p<0.05$) for brooders treated with Ovaprim (36086.00 ± 7215.50eggs) than PGE induced spawners (20978.00 ± 6782.15 eggs). Hatching rates also followed the same trend, in which significantly higher hatching success was recorded for Ovaprim ovulated eggs (83.5%) than PGE induced eggs (63.7%). Fry survival rate was 81.90 ± 1.10% for Ovaprim treated fish, while PGE induced fish fry had 77.73± 1.33%; percentage deformed fry was significantly minimal for Ovaprim treated. However, all Ovaprim-treated spent fish died few hours post stripping, contrary to PGE spent brooders that were fully recovered. Production cost analyses revealed that the use of Ovaprim resulted in about 25% cost reduction. It is thus concluded that Ovaprim is superior to PGE in induction of breeding in *Clarias gariepinus*. This notwithstanding, the mortality suffered by all the spent fish treated with Ovaprim raises food safety concerns. This however, needs to be validated.

Keywords: Hatching Success, Induced Breeding, Fecundity, Spawners, Latency

1. Introduction

The African mud catfish, *Clarias gariepinus* (Burchell, 1822) is an important food fish in Nigeria. It is well relished and has high market value (Oladosu *et al.*, 1993; Ayinla *et al.*, 1994). *C. gariepinus* is the most important aquaculture fish species in Nigeria because of its hardiness, ability to survive hypoxic condition, ability to accept pelleted feed, fast growth in captivity and high market value (Ekunwe and Emokaro, 2009; Adewolu and Adeoti, 2010).

Until the late 1980s, most of the fish seeds, including *C. gariepinus*, used for aquaculture in Nigeria were collected from the wild (Osuigwe and Erondu, 1997). But aquaculture that is dependent on wild-bred fish seeds is fraught with several disadvantages such as inadequate supply of fish seed required to meet the production target of the farmer. Meanwhile, adequate supply of quality seeds is an essential prerequisite to successful aquaculture production (Rottmann *et al.*, 1991). Furthermore, stunted individuals are often procured from the wild as it is difficult to differentiate

between siblings and cohorts at early life history stages. To guarantee adequate supply of fingerlings with known age and genetic background, several studies have been carried out, which recent findings have led to improved techniques in induced breeding of *C. gariepinus* (Viveen *et al.*, 1986; Nwokoye *et al.*, 2007; Akinwande *et al.*, 2009; Ataguba *et al.*, 2009).

In order to bridge fingerling demand-supply gap, hatchery techniques have been developed for seed production of some culturable fish species. These are either through natural breeding in captivity, or by induced breeding using exogenous hormone (Ndimele and Owodiende, 2012). Induced breeding involves the use of some exogenous ovulating agents, which trigger the ripening of mature eggs (Ajah, 2007). In Africa, various hormonal substances (Ovaprim, carp pituitary, human chorionic gonadotropin, frog pituitary extracts, etc,) have been used to induce breeding in fish with varying magnitudes of success (Nwadukwe 1993; Okoro *et al.*, 2007). There is, therefore, the need to carry out comparative studies on the effectiveness of these induction agents in order to define the viable options. This will provide valuable information that will promote sustainable hatchery propagation of fish in Nigeria.

This study compares the efficacy of Ovaprim and catfish pituitary gland in induction of breeding in *C. gariepinus,* using appropriate reproductive and cost-benefit indices.

2. Materials and Method

2.1. Collection and Selection of Broodfish

Ten gravid female *C. gariepinus* fish and nine mature male fish with weight ranging from 550g – 1000g were purchased from a fish farm in Port Harcourt, Rivers State. All broodfish were selected by external morphological characteristics, using the method of Ayinla *et al.* (1994). Females were selected on the basis of their bulging abdomen as well as egg colour. The selected fish were kept in an outdoor concrete tank (2.5m x 2.0m x 1.5m) at the University of Port Harcourt Demonstration Farm for seven days prior to the breeding date. They were fed 5% of total biomass with Coppens pelleted fish feed (48% crude protein) twice daily. Feeding was suspended a day prior to the hormonal treatments.

2.2. Water Holding Facilities

Eighteen litres of bore-hole water was introduced into ten plastic containers (20-litre capacity) arranged in rows of fives on an elevated platform in the hatchery section of the Demonstration Farm. The dimension of the plastic containers was 40cm x 30cm x 26cm. The fish holding containers were equipped with a simple flow through system with an overhead reservoir containing water properly conditioned before use. Each of the plastic containers was filled with calcium carbonate treated water (50g of $CaCO_3$ to 1000 liters of water) and kept standing for twelve hours prior to the time of injecting the broodfish.

2.3. Pituitary Extraction and Preparation

Pituitary glands were extracted from five of the male broodfish. They were weighed so as to get a corresponding weight to that of the recipient fish. The head of the male donor was cutoff after stunning the fish, and subsequently the lower jaw was also cut off. The ventral side of the brain was opened to expose the pituitary gland. Glands were collected with a pair of tweezers and placed in a beaker containing 2ml of 0.9% normal saline solution. Each of the glands was crushed in a mortar using a pestle. Two millimeter of 0.9% normal saline solution was added and the suspension decanted and collected into a 2ml syringe. The freshly collected pituitaries were immediately injected into the female spawners. This was done in the evening hours at about 9pm.

2.4. Administration of Hormone

The ten gravid female were divided into two groups of five fish with one set of fish representing the replicates for each treatment. Ovaprim and pituitary extract were administered on

each set of gravid female fish. The mean weight of female brooders used in pituitary and Ovaprim treated fish was 815.5 g. Pituitary suspension was drawn into a 2ml syringe, and then injected intramuscularly above the lateral line of the fish toward the dorsal section and pointed to the ventral side. After withdrawal of the needle, the fish were gently rubbed at the site of injection to avoid back flow of the injected fluid. Each female from the second group was injected with a dose of 0.5ml Ovaprim/kg of body weight (Haniffa and Sridhar, 2002). The injected fish were kept separately in well-labeled containers measuring 40cm x 30cm x 26cm containing water. The containers with the injected fish were covered with heavy boards so as to prevent the fish from leaping out.

2.5. Preparation of Milt

Four male fish were killed, dissected carefully and their milt sac obtained. The weight of each male was obtained and recorded alongside the weight of each of the gonad. A small incision was then made on the lobes with a sharp razor blade and the milt squeezed into a dry Petri dish. Milt was washed into the Petri dish with 0.9% normal saline solution (Ayinla, 1991; Nwadukwe *et al.,* 1993).

2.6. Stripping, Fertilization and Incubation of Eggs

Latency period was recorded for each of the fish in each group and stripping took place within 10 and 12 hrs after injection at a mean temperature of 24.05°C. With slight pressure at the ventral part of the abdomen, ovulated eggs oozed out freely and were collected into a dry Petri dish of known weight. This was done for each treated female fish, and collected eggs were weighed and recorded. Workers fecundity was then determined from the data. A sample of 1g was collected from the stripped eggs from each female and fertilization was done by pouring the prepared milt onto the eggs. The mixture of 1g was incubated separately on the spawning substrate (*kakaban*) placed in water in each of the plastic containers (Szabo *et al.,* 2002).

Basic water quality parameters were determined. Temperature was measured with mercury in-glass thermometer. Dissolved Oxygen and pH were measured using pH meter (WTW pH 330) and DO (Model MW600) meter, respectively. The flow through system was used so as to enhance proper aeration. Post hatching, dead eggs were removed by siphoning. The *kakabans* were removed and percentage hatchability determined by recording the number of dead eggs in each container.

Larvae were reared with constant water supply and by the third day their yolk sacs were fully absorbed and the fry were seen swimming in the containers. Fry Survival rate was determined by records of number of dead fry in each treatment medium. Live larvae were fed with decapsulated Artemia, five times daily.

2.7. Determination of Reproductive Success Parameters

Latency period (time taken from injection of female brood fish to time of stripping) was recorded.

Workers fecundity was estimated by counting the number of eggs stripped from each female broodfish as follows:

F = Total weight of eggs x no. of egg per gram

$$\% \ hatchablity = \frac{Total \ no \ of \ egg \ incubated - no \ of \ unhatched \ egg}{Total \ no \ of \ egg \ incubated} \times 100\%$$

The Survival rate per rearing tank was determined at the end of the experimental period with the formular:

$$\% \ survival = \frac{number \ of \ survived \ fry}{Total \ no \ of \ fry \ stocked} \times 100\%$$

The percentage of deformed fry per rearing tank was determined as follows:

$$\% \ deformity = \frac{number \ of \ deformed \ fry}{Total \ number \ of \ fry} \times 100\%$$

2.8. Statistical Analysis

The data obtained were statistically analyzed using student's t-test and for all the analyses, probability values < 0.05 were considered significant.

3. Results

3.1. Physico–Chemical Parameters

Mean pH in the incubation tanks was 6.92±0.08, while that of the rearing tank was 6.80±0.18. Mean temperature and DO in the incubation tank were 26.0 ±0.70 and 6.9± 0.32 mg/l, respectively, while in the rearing tank the values were 26.8±0.83^{0}C and 6.26 ±0.45mg/l, respectively, as shown in Table 1. There was no significant difference in pH and temperature in both tanks despite the higher values recorded in the incubation tank, but there was significant difference in the DO values recorded (P<0.05).

3.2. Latency Period

The latency period ranged from 10:13 to 12:25 hours (Table 2). Ovaprim recorded a mean latency period of 10:25 hours while pituitary recorded 11:49 hours. There was a significant difference between the latency periods of the treatments (P< 0.05).

3.3. Workers Fecundity /Relative Workers Fecundity

Average number of eggs stripped from pituitary treated fish was 20978.4 ±0.50 while 36086±72 eggs were recorded from Ovaprim treated fish as shown in Table 2. Mean relative workers fecundity was also computed and the values recorded were 453.3±2 124.95 and 781.67± 11 eggs for pituitary and Ovaprim treatments, respectively. There was significant difference in both workers fecundity and relative workers fecundity for both treatments (P< 0.05).

Relative workers fecundity was estimated by dividing the number of eggs stripped (fecundity) by the length per fish.

Percentage hatchability was determined by estimating the number of unhatched egg as follows:

3.4. Egg Hatchability

The period of hatching ranged from 10:20 to 24:06 hours for pituitary treated fish, which recorded a longer hatching period at a temperature of 26^{0}C. A mean percentage hatchability rate of 63.77.±29% was recorded for pituitary treated fish while 83.53 ±2.77% was recorded for Ovaprim treated fish. There was significant difference between the percentage hatchability of the two groups (p <0.05) as shown in Table 2.

3.5. Percentage Mean Fry Survival

Females injected with pituitary recorded 77.7±1.33% while Ovaprim treated fish recorded a mean value of 81.9% fry survival. There was significant difference between the treatments (P< 0.05)

3.6. Percentage Deformity of Fry

The percentage deformity of fry is presented in Table 3. Ovaprim treated fry had the lowest mean percentage of deformed fry (0.1± 0.28 %), while fry from pituitary gland injected females had 0.9±0.50% deformity. Student's t–test showed significant difference (P <0.05) for both treatments.

3.7. Percentage Mean Fry Weight

Ovaprim treated females had higher mean fry weight (0.12±0.023g), whereas fry from female treated with pituitary gland had lower mean fry weight (0.09±0.018g). Student's t–test showed significant difference in this parameter (p<0.05).

3.8. Survival of Spent Brood Stock

Spent fish from both treatments were kept in two separate recovery tanks. Ovaprim treated spent fish were relatively less active and died one after the other after six hours. However, pituitary treated spent fish showed signs of recovery and no death was recorded in this group six hours post stripping.

3.9. Cost Benefit Analysis of Hormonal Treatment

Table 4a shows the comparative mean cost per quantity of Ovaprim and pituitary gland required to induce spawning in C. gariepinus. Two hundred and fourteen naira eight kobo (N214.8±171.84 ~ $1.08) was used to procure 1.74ml of Ovaprim, as against N204.48±163.58 used to procure 19mg of pituitary. Table 4b shows the comparative cost benefit analysis required for a gram of egg when Ovaprim and pituitary are used as inducing agents. Females injected with pituitary gland were stripped of a total of 257g of eggs and the cost of pituitary per gram of egg was ₦19.95. For females injected with Ovaprim a total of 334g of eggs were stripped, implying ₦15.9 per gram of egg.

Table 1. Mean values of Physico-Chemical parameters of water.

Ovaprim/pituitary	Incubation tank	Rearing tank
Temperature(^{0}C)	26.0 ±0.71	26.8±0.84
pH	6.9 ±0.08	6.8± 0.19
Dissolved oxygen (mg/L)	6.9 ± 0.34	6.26 ± 0.45

Table 2. Mean values of reproductive indices of C. gariepnus brood fish induced to breed with Ovaprim and pituitary gland.

	Pituitary	Ovaprim
Workers fecundity	20978.4±6782.15[b]	36086.0± 7215[a]
Relative workers fecundity	453.3±124.95[b]	781.7± 11 [a]
Latency period (minutes)	709.0±34.00 [b]	625.0±08.00 [a]
Percentage hatchability (%)	83.5±2.77[a]	63.8±4.29[b]

[a-b]Values with the same superscript are not significantly different (p<0.05)

Table 3. Survival and rate of deformity of progeny of C. gariepinus brood fish induced to breed with Ovaprim and pituitary gland extract.

	Pituitary gland	Ovaprim
% of deformed fry	0.9 ±0.50[a]	0.1 ±0.28[b]
(%)Mean fry survival	77.7 ±1.33[b]	81.9 ±1.10[a]
Mean fry weight (g)	0.09 ± 0.023[b]	0.12 ±0.023[a]

[a-b]Values with the same superscript are not significantly different (p<0.05)

Table 4a. Comparative cost of Ovaprim and PGE required for inducing spawning in C. gariepinus.

Parameter	Total Quantity	Mean cost (in naira)
Ovaprim	1.74ml	214.8±171.84
Pituitary	19mg	204.48±163.58

Table 4b. Comparative cost benefit analysis of Ovaprim and PGE for inducing breeding in C. gariepinus.

	Mean egg wt (g)	Total cost (naira) per gram
Pituitary gland extract	257.0±41.12	19.95±3.19
Ovaprim	334.0±53.44	15.9±2.54

4. Discussion

Prolonged latency period has indirect implication on the quantity and quality of eggs produced as well as the quality of the fry. Thus, the shorter the latency period, the better the reproductive output. The latency period recorded in this experiment is similar to De Leeuw et al. (1985), who reported a latency period of 12:3hrs when C. gariepinus was injected with Gly[10](D-Ala[6]) LHRH – ethylamide and pimozide. Richter et al. (1987) recorded 16hrs latency period, using the same treatment as De Leeuw et al. (1985) and Kouril et al. (1992). Tan-Fermin and Emata (1993) recorded a latency period of 12 to 16 hrs when C. gariepinus and C. macrocephalus were induced to breed using pituitary and Ovaprim. Mohammed et al. (2000) recorded a latency period of 26hrs for fish injected with 3000 IU HCG. Francis (1992) also recorded 26hrs latency period for Heterobranchus fossilis

and C. batrachus. The difference in the latency period is generally attributable to variable potency of the different hormonal materials used as well as the responses of the different fish species and temperature. Prolonged time of spawning is a drawback of using certain substance in induced breeding (De Leeuw et al., 1985; Kouril et al., 1992; Bruzuska, 2000). The shorter latency period recorded in Ovaprim treated fish than those treated with pituitary gland extract is a reflection of the superiority of the former in induction of breeding of C. gariepinus.

Workers fecundity is an important index in determining the reproductive capacity of fish that are undergoing artificial spawning and is, therefore, a measure of the efficiency of the inducing agent. From the values recorded for this index, it is obvious that Ovaprim is more effective in induction of ovulation in C. gariepinus. The hatching time recorded in this study was similar to that of Haniffa et al. (2000) who recorded hatching duration of 39 – 43 hours for C. gariepinus treated with pituitary gland, 23 hours for Ovaprim induced eggs and 36 – 38 hours for eggs induced with HCG at a mean temperature of 25± 0.35^{0}C. The hatching duration recorded in this study was a function of combined effect of temperature and the potency of the hormonal agent. The percentage hatchability is analogous to the data reported by Olubiyi et al. (2005) on H. longifilis, induced with Ovaprim. Nwadukwe (1993) recorded 63% mean hatchability in H. longifilis induced with pituitary gland. Similarly, Ude et al. (2005) reported 67% hatching success for C. gariepinus induced with LHRHa. Hatching success is also hormonal dose dependent. Haniffa and Sridhar (2002) recorded different hatching rates at variable hormonal doses. The hatching rate was 50.5% for eggs from brooders injected with 0.3ml/kg Ovaprim, while hatching rate was increased to 60% when 0.5ml/kg was administered.

Fry survival rate depends on several factors such as feed availability, pH, temperature, dissolved oxygen, ammonia, nitrite, nitrate, etc (Ajah, 2007). In addition, type of hormonal agent also determines fry survival rate as recorded in this study. Nwokoye (1985) reported 75 – 80% fry survival rate following pituitary gland induced breeding in C. gariepinus. In this study, Ovaprim showed superior fry survival rate over PGE.

Spawners induced with Ovaprim were characterized by aggressive swimming behaviour. The persistent intense swimming activity is energy sapping, which aggravated post stripping weakness in the Ovaprim treated individuals, and may have contributed to their mortality. However, Ovaprim proved to be more cost effective than pituitary gland. Nwokoye et al. (2007) reported similar production cost effectiveness in H. bidorsalis when induced with Ovaprim. It is inferred that both Ovaprim and pituitary can be used effectively for artificial breeding of C. gariepinus. However, Ovaprim has more comparative advantages in terms of reproductive outputs and fry quality. Notwithstanding the above merits, absolute mortality suffered by Ovaprim treated spent fish is indicative of substance toxicity, which may have food safety implications. Also loss of broodfish in Ovaprim treated lot has cost implication, which to some extent could offset the comparative advantage the synthetic hormone has over PGE. Future studies should be directed at validating the toxicity of Ovaprim.

References

[1] Adewolu, M.A., Adeoti, A.J. 2010. Effect of mixed feeding schedules with varying dietary crude protein levels on the growth and feed utilization of *Clarias gariepinus* (Burchell, 1822) fingerlings. Journal of Fisheries and Aquatic Science, 5, 304-310. DOI:10.3923/jfas.2010.304.310.

[2] Ajah .O. P. 2007. Fish Breeding and Hatchery Management. Jerry commercial production, Calabar, Nigeria. Pp52.

[3] Akinwande, A.A., Moody, F.O., Umar, S.O. 2009. Growth performance and survival of *Heterobranchus longifilis, Heterobranchus bidorsalis* and their reciprocal hybrids. African Scientist, 10(1), 15-18.

[4] Ataguba, G.A., Annune, P.A., Ogbe, F. G. 2009. Induced breeding and Early Growth of Progeny from crosses between two African Clariid fishes, *Clarias gariepinus* (Burchell) and *Heterobranchus longifilis* under hatchery conditions. Journal of Applied Biosceiences, 14,755-600.

[5] Ayinla O.A. 1991. Spawning of selected culturable species: In: Ayinla, O.A. (ed.) Proceeding of the fish seed propagation course. African Regional Aquaculture Centre (ARAC), Aluu, Port Harcourt, 14 – 28 August, 1991. Pg 104.

[6] Ayinla O.A, Kayode O, Idoniboye–Obu, Oresegun A, Aidu V.T. 1994. Use of Tadpole meal as substitute for fishmeal in the diet of *Heterobranchus bidorsalis* (Geoffery St Hillarie 1809). Journal of Aquaculture in the Tropics, 9, ,25-33

[7] Bruzuska E. 2000. Artificial spawning of carp *Cyprinus carpio*: difference between the effects on reproduction in females on Polish and Hungarian provenance treated with carp pituitary and D –Ala⁶, GnRHproNET (Kobarelin). Aquaculture Resources, 31, 457 – 465.

[8] De Leeuw, R., Goods, H.J.T., Richter, C.J.J., Edind, E. H. 1985. Pimozide-LHRHa induced breeding in the African catfish, *Clarias gariepinus* (Burchell). Aquaculture, 44, 229 – 302.

[9] Ekunwe, P.A., Emokaro C. O. 2009. Technical Efficiency of Catfish Farmers in Kaduna Nigeria. Journal of Applied Science Research, 5(7), 802-805.

[10] Francis, T. 1992. Induction of oocyte maturation and ovulation in the freshwater Asian catfish, *Clarias macrocephalus* by LHRHa and pimozide. Journal of Applied Ichthyology, 80, 90-98.

[11] Haniffa M.A, Mohamed Shaik, Merlinrose T. 1996. Induction of ovulation in *Channa striatus* (Bloch) by SGnRH. Fishing Chines, 23-24

[12] Haniffa M.A. K., Sridhar, S. 2002. Induced spawning of spotted murrel (*Channa punctatus*) and catfish (*Heteropneustes fossilis*) using human chorionic gonadotropin and synthetic hormone (Ovaprim). Veterinarski Arhiv, 72, 51 – 56.

[13] Kouril, J., Hamackova, J., Barth, T. 1992. Induction of ovulation in African catfish (*Clarias gariepinus*) using GnRH analogues, dopaminergic inhibitor of isophoxythepin and carp pituitary. Zoological section of the Slocac Academy of

Sciences, Bratislava. Proceedings of the Ichthyologic Conference, November 4, 1992. Pg 81-85.

[14] Mohammed, A.H., Thangarose, M., Junaity, S. M. 2000. Induced spawning of the striped murrel *Channa striatus* using pituitary extracts, Human Chorionic Gonadotropin, Luteinizing Hormone Releasing Hormone Analogue, and Ovaprim. Acta Ichthyology, 30, 53 – 60.

[15] Ndimele, P.E, Owodiende, F.B. 2012. Comparative reproduction and growth performance of *Clarias gariepinus* (Burchell, 1822) and its hybrid induced with syntheic hormone and pituitary gland of *Clarias gariepinus*. Turkish Journal of Fisheries and Aquatic Sciences, 12, 619-626

[16] Nwadukwe, F.O. 1993. Inducing oocyte maturation, ovulation and spawning in African catfish, *Heterobranchus longifilis* Valenciennes (Pisces: Claridae), using frog pituitary extract. Aquaculture and Fisheries Management, 24, 625 – 630.

[17] Nwadukwe, F.O., Ayinla, O.A. Abby-Kalio, N. J. 1993. Effect of various doses of Acetone-dried powdered carp pituitary extract and season on hatchery propagation of *Heterobranchus longifilis* (Val. 1840) Pisces: Claridae. Journal of Aquaculture Tropical, 8, 333 - 340

[18] Nwokoye, C.O., Nwuba, L.A., Eyo, J.E. 2007. Induced propagation of African clariid catfish, *Heterobranchus bidorsalis* (Geoffrey Saint Hillarie, 1809) using synthetic and homoplastic hormones. African Journal of Biotechnology, 6, 2687-2693.

[19] Okoro, C.B, Nwadukwe, F.O. Ibemere, I. 2007. The use of Ovaprim in oocyte maturation and ovulation in *Clarias gariepinus* (Burchell, 1822). African Journal of Applied Zoology and Environmental Biology, 9, 83–84.

[20] Oladosu, G.A., Ayinla, O.A., Adeyemo, A.A., Yakubu, A. F. Ajani, A.A. 1993. A comparative study of the reproductive capacity of the African catfish species *Heterobranchus bidorsalis* (Geoffery), *Clarias gariepinus* (Burchell) and their hydrid "Heteroclarias". ARAC Technical Paper 92, 1 -5.

[21] Olubiyi O. A, Ayinla O.A., Adeyemo A. A 2005. The Effect of Various Doses of Ovaprim on Reproductive Performance of the African Catfish *Clarias gariepinus* (Burchell) and *Heterobranchus Iongifilis* (Valenciennes). African Journal of Applied Zoology and Environmental Biology, 7, 101 -105

[22] Osuigwe D.I. Erondu E.S.1997. Reproductive Biology of *Clarias gariepinus* (Burchell, 1822) in the River Ezu (Southeastern Nigeria). Acta Hydrobiologia, 39, 53 – 60

[23] Richter, C.J.J., Eding, E.H., Goos, H.J., De Leeuw, R.M., Scott, A.P., Van Oordt, P.G.W.J. 1987. The effect of pimozyde/LHRH-a and 17a -hydroxyprogesterone on plasma steroid levels and ovulation in the African catfish, *Clarias gariepinus*. Aquaculture, 63, 15 –168.

[24] Rottmann, R.W., Shireman, J., Chapman, F.A 1991. Introduction to Hormone – Induced Spawning of Fish. SRAC Publication, No. 421.

[25] Szabo, T., Medgyasszay, C., Horvath, L. 2002. Ovulation induction in nase (*Chondrostoma nasus*, Cyprinidae) using pituitary extract or GnRH analogue combined with domperidone. Aquaculture, 203, 389 -395,

[26] Tan-Fermin, J.D. and Emata, A.C. (1993). Induced spawning by LHRH-a andpimozide in the catfish *Clarias macrocephalus* (Gunther). Journal of Applied Ichthyology 9, 86–96.

[27] Viveen, W., Richter, C.J.J., Van, O., Janssen, J., Huisman, E.A. 1986. Practical manual for the culture of the African catfish (*Clarias gariepinus*). Section for research and technology, Ministry of Development Cooperation. The Hague, Netherlands, P128.

10

Comparison of different fertilizer management practices on rice growth and yield in the Ashanti region of Ghana

Roland Nuhu Issaka[1, *], Moro Mohammed Buri[1], Satoshi Nakamura[2], Satoshi Tobita[2]

[1]CSIR-Soil Fertility and Plant Nutrition Division, Soil Research Institute, Academy Post Office, Kwadaso, Ghana
[2]Crop Production and Environment Division, Japan International Research Center for Agricultural Sciences. Ohwashi, Tsukuba, 305-8686, Japan

Email address:
rolandissaka@yahoo.com (Issaka R. N.), moro_buri@yahoo.com (Buri M. M.), nsatoshi@affrc.go.jp (Nakamura S.),
bita1mon@jircas.affrc.go.jp (Tobita S.)

Abstract: Nutrient management is critical in increasing and sustaining rice yield. A field experiment was conducted to examine the effects of inorganic fertilizer (IF), poultry manure (PM) and their combinations on rice yield and possible residual effects. A randomized complete block design with three replications was used and the trial was conducted on a *Gleysol*. In 2011 SPAD values for IF and PM/ IF combinations (except 2.0 t/ha PM + 22.5-15-15 kg N: P_2O_5: K_2O/ha) were significantly higher in the sixth week onwards than PM. Number of panicles/plant and number of panicles m^2 were significantly higher for 90-60-60 kg N: P_2O_5: K_2O/ha and 2.0 t/ha PM + 22.5-15-15 kg N: P_2O_5: K_2O/ha than 6.0 and 4.0 t/ha PM resulting in significantly higher grain yield. Grain yield of IF was similar to grain yield of PM/IF combinations. In 2012 the residual effects showed a significantly higher SPAD value for the 6.0 t/ha PM. Also 6.0 t/ha PM, 4.0 t/ha PM and 4.0 t/ha PM + 30 kg N/ha had significantly high number of panicles/plant and number of panicles/m^2 than IF. Residual effect of PM applied at 4.0 t/ha and above gave significantly higher grain yield than IF. Mean grain yield for the three years showed that 4.0 t/ha PM + 30 kg N/ha and 2.0 t/ha PM + 22.5-15-15 kg N: P_2O_5: K_2O/ha gave significantly higher yields than the other treatments. The results indicate that integrating IF and PM is a better option in increasing and sustaining rice production.

Keywords: Grain Yield, Inorganic Fertilizer, Poultry Manure, Rice

1. Introduction

Rice production in Ghana faces several challenges, notably water shortage, low soil fertility and poor soil and water management [1]. Several authors have identified low inherent soil fertility as a major cause for low rice yield [2; 3; 4; 1; 5] in Ghana. The problem is compounded as farmers are not able to purchase adequate amounts of mineral fertilizer (high cost) and rely mostly on natural soil fertility which is low and declining. Rice yield can vary from less than 1.0 t/ha to as high as 6.0 t/ha due to variation in both water and soil fertility management [3 and 1].

The availability of large amounts of organic materials in Ghana [6] can be harnessed to improve rice production in the country. Organic materials available for usage include poultry manure, cow dung, rice straw and human excreta. [4] observed that application of 7.0 t/ha cow dung gave similar yield as sole mineral fertilizer. The authors also reported that application of 3.5 t/ha cow dung + half rate of mineral fertilizer gave similar effect as full rate of mineral fertilizer.

According to [7] cited by [8], organic materials are fundamentally important in that they supply various kinds of plant nutrients including micronutrients, improve soil physical and chemical properties and hence nutrient holding and buffering capacity, and consequently enhance microbial activity. In addition, organic matter slowly but continuously releases N as plant need it. N is the most limiting nutrient in irrigated rice systems, P and K deficiencies also reduce rice yield under continuous cultivation particularly in the inland valleys [4].

Many researchers including [9], reported that the effects of organic matter application can be observed after 3 to 5 years. A good residual effect ensures good soil health and sustains productivity over a longer period of time compared to mineral fertilizer.

To ensure that the fertility of the soil is improved, it is imperative to find cheaper alternatives which can be used solely or in combination with mineral fertilizer. In this study poultry manure, sole or in combination with mineral fertilizer was evaluated for its potential to improve and sustain rice production.

2. Materials and Methods

Experimental site and design: The experiment was conducted at the Central Agricultural Station, Kwadaso (Latitude 6o 40' 59" and Longitude 1o37'0") in 2010 and repeated in 2011 and 2012. The soils are gleysols, developed over granite [10]. A Randomized Complete Block design with 3 replications was used. Jasmine 85 was used as the test crop in 2010 while TOX 3108-56-4-4-2 (Sikamo-local name) was planted in 2011 and 2012. Two seedlings were transplanted per hill at 20 x 20 cm. Treatments and times of application are presented in Table 1. Poultry manure (PM) was broadcast and incorporated into the soil a week before transplanting.

Table 1. Treatment combinations in 2010 and 2011.

Treatment	Basal Application (1WAT)	Top Dressing (5 WAT)
6 t/ha PM	-	
4 t/ha PM	-	
4 t/ha PM + 30 kg N/ha	-	30 kg N/ha
2 t/ha PM + 30 kg N/ha	-	30 kg N/ha
2 t/ha PM + 22.5-15-15 kg N:P_2O_5:K_2O/ha	All P&K + 10 kg N/ha	20 kg N/ha
90-60-60 kg N:P_2O_5:K_2O/ha	All P&K + 30 kg N/ha	60 kg N/ha

WAT: weeks after transplanting

In 2012, the residual effects of the various treatments were evaluated. 30 kg N/ha was applied to all plots. The field was designed according to the "sawah" system (plots were bunded, ploughed, puddled and leveled with inlets and outlets for supplementary irrigation and drainage when necessary).

2.1. Soil Sampling and Analysis

Initial soil samples were taken (0-20 cm) before the field layout was done. Soil samples were brought to Soil Research Institute laboratory and air-dried at room temperature. The air-dried soil samples were ground and passed through 2 mm sieve. Soil pH was measured using a glass electrode (pH meter) in a soil to water ratio of 1:2.5 [11]. Organic carbon was determined by the wet combustion method [12]. Total nitrogen was determined by micro Kjeldahl method [13] and available phosphorus according to [14] Exchangeable cations (Ca, Mg, K) were extracted with 1.0 M ammonium acetate solution and determined by atomic absorption spectrometry [15]. Exchangeable acidity was determined first by extracting it with KCl and titrating the extract with sodium hydroxide [15]. Effective cation exchange capacity was determined by the addition of all the cations.

2.2. Plant Growth Characteristics

Soil and plant analysis development (SPAD) value was measured weekly up to the eighth week. SPAD value was measured using a chlorophyll meter (SPAD-502, Konica Minolta Sensing Inc., Osaka, Japan). Measurement of SPAD value was done on the youngest fully expanded leaf. Number of panicles per plant and plant height were recorded at maturity.

2.3. Dry Matter and Grain Yield

At maturity, an area of 2.0 m^2 per treatment was demarcated and harvested. Yield components (grain and stover) were measured and yield per hectare estimated. The statistical software, Statistics 8, was used for data analysis. Standard error was used as the mean separator.

3. Results and Discussion

3.1. Soil and Poultry Manure Properties

Initial soil properties for the site are presented in Table 2. Exchangeable cations are very low reflecting a low effective cation exchangeable capacity. The soil is strongly acidic. Both organic matter and available phosphorus are also low.

The nutrient content of the poultry manure used is shown in Table 3. The manure was collected from a farm within the district in which the experiment was conducted. The manure was collected from laying birds hence the high values for calcium and magnesium.

Table 3. Nutrient content of poultry manure.

Total N	P	K%	Ca	Mg
2.6	0.61	1.10	8.50	7.00

3.2. 2010

3.2.1. Yield Components

In 2010 plant height, number of panicles/plant and grain yield for all the treatments were similar (Table 4). Jasmine 85 normally gives up to 5.0 t/ha grain yield or more under good environment and recommended fertilization. However, the experimental field experienced a severe attack of blast and the disease was finally brought under control after 2 weeks of spraying with fungicide. This probably masked the effect of the various treatments and also resulted in grain yield reduction.

3.3. 2011

3.3.1. Effect of Treatments on SPAD Values

Figure 1 shows changes in SPAD values in 2011. SPAD values increased significantly for all the treatments up to the second week. SPAD value for the 90-60-60 kg N: P_2O_5:

K_2O/ha treatment increased significantly over all the other treatments in the third week but declined steadily during the fourth and fifth week. SPAD value for this treatment again increased significantly to above 45 after 60 kg N/ha was applied in the fifth week. 4.0 t/ha PM + 30 kg N/ha and 2.0 t/ha PM + 30 kg N/ha treatments showed similar trend as 90-60-60 kg N: P_2O_5: K_2O/ha. SPAD values for sole PM at 6.0 t/ha and 4.0 t/ha showed a decline from above 40 in the second week to below 40 in the eighth week. The amount of nitrogen and rate of mineralization when PM was applied at 6.0 t/ha was probably not enough to provide adequate amounts of nitrogen for the plants in the eighth week hence the decline in SPAD value. According to [8], SPAD value is the most important growth character. The SPAD value reflects nutrient (especially nitrogen) availability in the soil and its influence on plant growth. Thus high nitrogen content in the 90-60-60 kg N: P2O5: K2O/ha treatment largely explains why SPAD values are generally higher for the 90-60-60 kg N: P2O5: K2O/ha treatment followed by 4.0 t/ha PM + 30 kg N/ha and 2.0 t/ha PM+ 30 kg N/ha treatments. Lower SPAD values for 6.0 t/ha PM, 4.0 t/ha PM and 2.0 t/ha PM+ 22.5-15-15 kg N: P2O5: K2O/ha is partially due to lower amount of N available for the growing plants.

3.3.2. Yield Components and Grain Yield

In 2011 plant height was similar for all the treatments (Table 5). Number of panicles/plant however varied significantly between treatments. Sole poultry manure applied at 6.0 t/ha and 4.0 t/ha gave plants with lower number of tillers/plant than sole mineral fertilizer and 2 t/ha PM + 22.5-15-15 kg N: P2O5: K2O/ha (Table 5) probably due to initial slow release of nutrients during tillering. Number of panicles/m^2, stover and grain yields are also presented in Table 5. Number of panicles/m^2 were significantly higher under 90-60-60 and 2.0 t/ha PM + 22.5-15-15 kg N: P_2O_5: K_2O/ha than PM applied at 2.0 t/ha + 30 kg N/ha and 4.0 t/ha + 30 kg N/ha. PM applied 6.0 t/ha and 4.0 t/ha gave significantly lower number of panicles/m^2 than all the other treatments. Stover yield was similar for all the treatments but grain yield was significantly lower under sole PM than 90-60-60 and 2.0 t/ha PM + 22.5-15-15 kg N: P_2O_5: K_2O/ha. As photosynthesis is enhanced the greener the leaves become. Hence higher SPAD values mean high production of photosynthates. Higher SPAD values for sole mineral fertilizer and mineral/organic fertilizer combinations partly explains higher number of panicles/plant and number of panicles/m^2 than sole poultry manure treatments. Availability of photosynthates may enhance and sustain tiller production which may result in differences in grain yield.

Table 2. Initial soil properties.

Depth (cm)	Soil H	TC	OM	TN	Avail. P	Ca	Mg	K	EA	ECEC
			(g/kg)			(mgkg⁻¹)			cmol(+)/kg)	
0-20	4.9	7.5	11.3	0.8	8.5	2.30	1.40	0.30	0.75	4.85

EA:-exchangeable acidity

Table 4. Effect of treatment on selected yield parameters in 2010.

Treatments	Height (cm)	Number of panicles/plant	Grain yield (t/ha)
6 t/ha PM	59	2.7	3.3
4 t/ha PM	58	3.0	3.3
4 t/ha PM + 30 kg N/ha	58	2.7	3.7
2 t/ha PM + 30 kg N/ha	60	2.7	4.3
2 t/ha PM + 22.5-15-15 kg N:P₂O₅:K₂O/ha	59	2.7	4.2
90-60-60 kg N:P₂O₅:K₂O/ha	60	3.0	3.3
LSD (0.05)	4.6	1.1	1.9

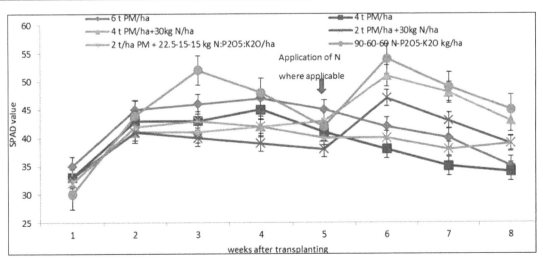

Figure 1. Effect of treatments on changes in SPAD values with time.

Table 5. Effect of treatments on rice growth and yield in 2011.

Treatment	Plant height (cm)	No. of stands/m²	Number of panicles/plant	No. of panicles/m2	Stover yield (t/ha)	Grain yield (t/ha)
6 t/ha PM	131a	25.0a	2.7b	185c	5.7a	4.7b
4 t/ha PM	133a	25.0a	3.0b	200c	4.5a	4.8b
4 t/ha PM + 30 kg N/ha	133a	25.0a	3.7ab	235b	5.2a	6.1ab
2 t/ha PM + 30 kg N/ha	132a	25.0a	3.7ab	235b	5.3a	5.3ab
2 t/ha PM + 22.5-15-15 kg N:P_2O_5:K_2O/ha	134a	25.0a	4.3a	265a	5.6a	6.4a
90-60-60 kg N:P_2O5:$_{K2O}$/ha	132a	25.0a	4.7a	285a	5.9a	6.5a

Within a column figures followed by the same letters are not significantly different by a margin of the standard error.

3.4. 2012: Residual Effect of Treatments

3.4.1. Residual Effect of Treatments on SPAD Values

In 2012 changes in SPAD value during the growth of the plants are presented in Figure 2. SPAD values were similar in the second week due to application of 10 kg N/ha to all plots. After the second week SPAD values started declining but at different rates. SPAD values fell to the lowest level in the fifth week during which additional 20 kg N/ha was applied to all plots. This raised the SPAD values of plants growing on plots that earlier received 4.0 t/ha or 6.0 t/ha PM to over 45 in the sixth week. SPAD values again started declining after the sixth week showing similar trend as observed between the second and fifth weeks. Plants growing on plots that received 6 t/ha PM had SPAD values above 40 up to the eighth week of growth. This was followed by plants on plots that received 4 t/ha PM showing values above 35. The residual effect of mineral fertilizer was less effective in maintaining high levels of SPAD values. SPAD values fell to below 25 for plants growing on plots that received only mineral fertilizer (Figure 2). Gradual release of nutrient over a longer period, especially

nitrogen, by the manure explains the observed trend. Higher SPAD values

3.4.2. Residual Effect of Treatments on Yield Components and Grain Yield

The residual effects of each treatment on yield components are presented in Table 6. Number of panicles/plant and number of panicles/m² were significantly higher for 6.0 t/ha PM, 4.0 t/ha PM and 4.0 t/ha PM + 30 kg N/ha than 2 t/ha PM + 22.5-15-15 kg N: P_2O_5: K_2O/ha and 90-60-60 kg N: P_2O_5:K_2O/ha but similar to 2.0 t/ha PM + 30 kg N/ha. Poultry manure at 6.0 t/ha, 4.0 t/ha and 4.0 t/ha + 30 kg N/ha gave significantly higher grain yield than the other treatments (2 t/ha PM + 30 kg N/ha, 2 t/ha PM + 22.5-15-15 kg N: P_2O_5: K_2O/ha and 90-60-60 kg N: P_2O_5: K_2O/ha). This implies that at 4.0 t/ha PM and above, optimum grain yield can be obtained. Organic fertilizer (poultry manure) showed a better residual effect than mineral fertilizer. This observation supports the findings of several authors [4; 8; 9]. PM is therefore an important source of plant nutrient for effective rice production for poor resource farmers.

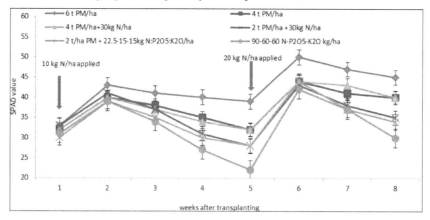

Figure 2. Residual effects on changes in SPAD values with time.

Table 6. Residual effect of treatments on rice yield.

Treatment	Plant height (cm)	No. of tand/m2	Number of panicles/plant	No. of panicles/m2	Stover yield (t/ha)	Grain yield (t/ha)
6 t/ha PM	125a	25.0a	3.0a	200a	6.8a	5.8a
4 t/ha PM	122a	24.3a	3.0a	194a	6.4ab	5.3ab
4 t/ha PM + 30 kg N/ha	121a	24.3a	3.0a	194a	6.7a	5.6a
2 t/ha PM + 30 kg N/ha	121a	24.7a	2.7ab	183ab	5.8bc	4.6c
2 t/ha PM + 22.5-15-15 kg N:P_2O_5:K_2O/ha	122a	24.7a	2.3b	163b	5.7bc	4.7bc
90-60-60 kg N:P_2O_5:K_2O/ha	125a	24.0a	2.3b	158b	5.4c	4.4c

Within a column figures followed by the same letters are not significantly different by a margin of the standard error.

3.5. Mean Grain Yield

Mean grain yield for the three years is presented in Figure 3. The mean grain yield for the treatments 4 t/ha PM + 30 kg N/ha and 2 t/ha PM + 22.5-15-15 kg N: P2O5: K2O/ha were significantly higher than all the other treatments. Particularly the 4 t/ha PM + 30 kg N/ha treatment gave consistent grain yield of above 5.0 t/ha for 2011 and 2012. Mineral fertilizer and poultry manure combinations was more efficient in increasing rice yield.

4. Conclusion

Poultry manure is a cheaper and an alternative fertilizing material that can enhance rice production. Integrating poultry manure and mineral fertilizer provides a more efficient means of improving and sustaining rice yield. 4.0 t/ha PM+ 30 kg N/ha and 2 t/ha PM + 22.5-15-15 kg N: P_2O_5: K_2O/ha gave a significantly higher mean yield for the 3 years. Generally all the combinations gave similar grain yield as 90-60-60 kg N: P_2O_5: K_2O/ha.

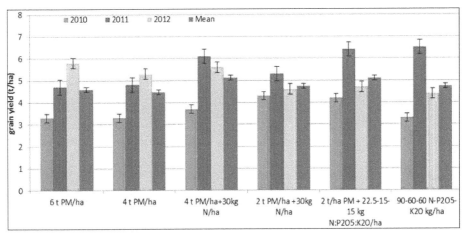

Figure 3. Mean yearly grain yield for the various treatments.

Acknowledgments

The authors are grateful for the funds received from Japan International Research Center for Agricultural Sciences as an out source contract for the Study on "Improvement of Soil Fertility with Use of Indigenous Resources in Rice Systems of sub-Sahara Africa" with Dr. Satoshi Tobita as Project Leader . We also wish to thank technicians and all staff at Soil Research Institute laboratory for their assistance.

References

[1]   Susumu S. Abe, M. Moro Buri, Roland N. Issaka, Paul Kiepe and Toshiyuki Wakatsuki. 2010. Soil Fertility Potential for Rice Production in West African Lowlands. JARQ. 44 (4), 343 – 355

[2]   Buri, M. M., Iassaka, R. N., Fujii , H. and Wa-katsuki, T. 2009 Comparison of Soil Nutrient status of some Rice growing Environments in the major Agro-ecological zones of Ghana. International Journal of Food, Agriculture & Environment Vol. 8 (1): 384-388

[3]   Buri, M. M., Issaka, R. N., Wakatsuki, T. and Otoo, E. (2004) Soil organic amendments and mineral fertilizers: Options for sustainable lowland rice production in the Forest agro-ecology of Ghana. Agriculture and Food Science Journal of Ghana. Vol., 3, 237-248.

[4]   Senayah J. K., Issaka R. N. and Dedzoe, C. D. 2008. Characteristics of Major Lowland Rice-growing Soils in the Guinea Savanna Voltaian Basin of Ghana. Agriculture and Food Science Journal of Ghana. Vol. 6. P 445-458

[5]   Issaka, R. N., Buri, M. M., Tobita, S., Nakamura, S. and Owusu-Adjei, E. 2011. Indigenous Fertilizing Materials to Enhance Soil Productivity in Ghana. In Soil Fertility (editor: Joann K. Whalen) Improvement and Integrated Nutrient Management: A Global Perspective. Published by InTech, Janeza Trrdine 9, 51000 Rijeka, Croatia. ISBN 978-953-307-945-5

[6]   Suzuki, A. (1997). *Fertilization of rice in Japan.* Japan FAO Association, Tokyo, Japan

[7]   Myint, A. K., Yamakawa, T., Kajihara, Y. and Zenmyo, T. 2010. Application of different organic and mineral fertilizers on the growth, yield and nutrient accumulation of rice in a Japanese ordinary paddy field. *Science World Journal Vol 5 (No 2)* Pp 47-54

[8]   Ohyama, N., Katono, M. & Hasegawa, T. (1998). Effects of long term application of organic materials to the paddy field originated form Aso volcanic ash on the soil fertility and rice growth. I. Effects on the rice growth and nutrient uptake for the initial three years. *Proceeding of Faculty of Agriculture, Kyushu Tokai University*, 17:9-24. (In Japanese with English summary)

[9]   Dedzoe C. D., Senayah J. K., Asubonteng K. O. and Otoo E. 1997. Characterization of inland valleys in Ghana: A semi-detailed study of the Mankran system in Ashanti Region. In Efficient Soil and Water Management: A Prerequisite for Sustainable Agriculture. Proc. 14th and 15th Annual General Meeting of the Soil Science Society of Ghana.

[10] Mclean, E. O. 1982. Soil pH and lime requirement. In Page, A.L., Miller, R.H. and Keaney, D. R. (eds). Methods of Soil Analysis. Number 9 Part 2, Am. Soc. of Agron.

[11] Walkley, A. and Black, I.A. 1934. An examination of the method for determining soil organic matter and proposed modification of the chromic acid titration method. Soil Sci. 37:29-38.

[12] Bremner, D.C. and Mulvaney, J.M. 1982. Total nitrogen. In Page, A.L., Miller, R. H. and Keaney, D.R. (eds). Methods of Soil Analysis. Number 9 Part 2, Am. Soc. of Agron.

[13] Bray, R. H. and Kurtz, L.T. 1945. Determination of total, organic and available forms of phosphorus in soils. Soil Sci. 59:39-45.

[14] Thomas, G.W. 1982. Exchangeable cations. In Page, A.L., Miller, R.H. and Keaney, D.R. (eds). Methods of Soil Analysis. Number 9, Part 2. Am. Soc. of Agron.

11

Virtual water and food security in Tunisian semi-arid region

Lamia Lajili-Ghezal, Talel Stambouli[*], Marwa Weslati, Asma Souissi

ESA Mograne, 1121 Mograne, Tunisia

Email address:

tstambouli@cita-aragon.es (T. Stambouli)

Abstract: To confront water scarcity and support food security, the concept of virtual water is used. As defined by Allan (1997) virtual water is "the water embedded in key water-intensive commodities such as wheat" or "the water required for the production of commodities". The importance of this concept is related to its potential contribution for saving water, especially in water short regions like Tunisia. This research study tries to evaluate the strategic importance of polluted or gray water, which is a component of virtual water. Reduction of virtual water for strategic agricultural products can be obtained by the gray water reduction. The latter is defined as "water required diluting polluted water to reach the normalized quality, different with countries". Water pollution is especially related to use of chemical products (fertilizers, pesticides, etc.) for some crops like vegetables. Besides having a lower opportunity cost, the use of green water for crop production has generally less negative environmental externalities than the use of blue water (irrigation with water abstracted from ground or surface water systems). Tunisia exports some crops and gray water volumes in exports have rarely been estimated. Thus, estimation of gray water plays a role in ensuring water and water-dependent food security and avoiding further potential damage to the water environments in both importing and exporting countries. In this context, Tunisian semi-arid region is chosen because the presence of a long period of dry and shiny, occurring after a cold and rainy one, useful for vegetables crops and family food security. The aim of this study is to present: Methodologies which can be used to reduce virtual water for some strategic vegetables crops in Tunisian semi-arid region, based on irrigation techniques improvements and the control of runoff and leaching water; Resources management practices that can be used to improve family income, especially women and children and target food security.

Keywords: Virtual Water, Food Security, Water Quality Conservation, Family Income, Gender

1. Introduction

Agriculture has always occupied an important place in the socio-economic development in Tunisia. The agriculture expansion is heavily relied on available natural resources, especially on water. Because of its geographical location, Tunisia undergoes the influence of two climate types: the Mediterranean type in the north and the Saharan type in the south which are at the origin of space and time variability in water resources. Therefore, the annual rainfall average varies from less than 100 mm in the extreme south to more than 1500 mm in the extreme northern parts of the country. Water resources are evaluated in 2000 to 4825 million m^3, with 2700 million m^3 of surface water and 2,125 million m^3 of ground water. Tunisia is, then, a country with relatively limited renewable water resources (SEMIDE, 2002).

That's why; taking into consideration the scarcity of water resources in the planning of agricultural policies is necessary

to improve the trade balance, to ensure a level of food security, and to enhance the countryside and the environment. In this context, many countries attempt to reach new alternatives for the management and the sustainable use of water resources in particular. These new approaches have brought about a relatively new idea called virtual water (VW) which tries to give an explanation to water use management (Velázquez, 2007).

The virtual water concept, defined by Allan (1997), as the amount of water needed to generate a product of both natural and artificial origin, this concept establish a similarity between product marketing and water trade. Virtual Water trade can alleviate, in arid countries, the problem of water scarcity, increasing imports of products with high virtual water content and so, it can allocate scarce resources to higher priority uses.

Given the influence of water in food production, virtual water studies focus generally on food products. At a global scale, the influence of these product's markets with water management was not seen. Influence has appreciated only by analyzing water-scarce countries at global scale, but at the detail level, should be increased, as most studies consider a country as a single geographical point, leading to considerable inaccuracies. For this reason, we consider the value of exploring virtual water strategy at smaller scales such as an irrigated area. Besides having a lower opportunity cost, the use of green water for the production of crops has generally less negative environmental externalities than the use of blue water (irrigation with water abstracted from ground or surface water systems). Also shinny regions are generally useful for Mediterranean crops and have less negative impact on environment.

In this context, Tunisian semi-arid region is chosen because of presence of dry and shiny period, occurring after a cold and rainy one, useful for vegetables crops and family food security.

Objectives of this work are:

a) Estimation of virtual water balance of strategic irrigated crops (vegetables) in semi-arid areas of Tunisia to determine their influence on the water resources management and to establish patterns for improving it ;

b) Study of resources management practices that can be used to improve family income, especially women and children and target food security.

Based on farmer's surveys, crop and meteorological data, irrigation management and regional statistics.

2. Material and Methods

To estimate the virtual water for different crops, several models were used with the objective to determine the water consumed by the plant. In this study, net irrigation requirements for studied crops and regions were computed following the FAO56 method (Allen et al, 1998) from meteorological data available.

Crop evapotranspiration (ETM, equation 1) was estimated from reference evapotranspiration (ET_0) and the appropriate crop coefficients (K_c).

$$ETM = K_c \; ET_0 \qquad (1)$$

Reference evapotranspiration (ET_0) was computed using the Penman-Monteith method (Smith 1993). The crop coefficients values at the initial, medium and end of the crop stages ($K_{c \; ini}$, $K_{c \; med}$ and $K_{c \; end}$), the general lengths (L) for the different growth stages (L_{ini}, L_{dev}, L_{mid} and L_{late}) and the total growing period for the main crops. Net Irrigation requirements (NIR, equation 2) were calculated using the standard FAO procedures, as described by Allen et al. (1998). Effective precipitation (EP) was calculated using the empirical USDA method (Cuenca, 1989). Following these procedures, reference evapotranspiration (ET_0), crop

coefficients (K_c), crop evapotranspiration (ET_c), effective precipitation (EP) and net irrigation requirements (NIR) were estimated for the main crops in the AID in 2011.

$$NIR = \left(K_c \; ET_0\right) - EP \qquad (2)$$

Net irrigation requirements calculations are based on the soil moisture regime and the phenological stage of the crop, while keeping the other variables at the optimal production level. On this basis, we can calculate crop coefficients for a given location. It is also possible to construct the mathematical function that connects the crop water consumption to the desired crop yield.

The choice of a model depends on the objectives of the study. When the most important is the relationship between water and crop production, which is the case, FAO models (AQUACROP and CROPWAT) are frequently used. CROPWAT is the simplest, based on empirical relationships between water availability and production.

In this study, virtual water consumed by crops was calculated as green (water provided by rain) and blue (water provided by irrigation) water. The present study estimates the green and blue water footprint of 1 kilogram of vegetables produced in semi-arid area in Tunisia following the method described by Hoekstra et al. (2009).

In the study, vegetable production in the different Tunisian semi-arid regions was considered, distinguishing production throughout the year as well as between growing systems. The study focuses on the production stage, that is, the cultivation of the product, from sowing to harvest. The crop virtual water was calculated for each year distinguishing the green and blue water components.

The virtual water of vegetables (rainfed or irrigated) has been calculated distinguishing the green and blue water components Within the CROPWAT model (FAO, 2009)., the 'irrigation schedule option' was applied, which includes a dynamic soil water balance and keeps track of the soil moisture content over time. The calculations have been done using climate data from representative meteorological stations located in the major crop-producing regions, selected depending on data availability.

Low virtual water values can be obtained by use of green water and reduction of blue water, based on improve of irrigation techniques and control of runoff and leaching water. For Tunisian semi-arid region, the best seasons for this are spring and autumn.

Vegetable crops generally need a large amount of workers, which can bring agricultural income, especially for women and children and then target food security for them. To perform this, statistical and field analysis of workers (ONAGRI, 2010), for Tunisian semi-arid region, were used.

3. Results and Discussion

Based on the results of the research project "Virtual Water and Food Security in Tunisia: from observation to support development" (VWFST), related to virtual water in food security strategy in Tunisia and its implications on the economy of water resources, the following data were used:

- Spatial bioclimatic distribution (governorate), to select governorates that are located on Tunisian semi-arid region ;
- Estimation of gross margin of water, obtained from collected data sheets elaborated by the regional research team of the project.

3.1. Semi-Arid Governorates

Based on the results of the research project (VWFST), 5 governorates which are mainly located in the Tunisian semi-arid region (> 50% of the total area of the governorate), were identified: Ariana, Ben Arous, Manouba, Tunis and Zaghouan. Field investigation results, related to the latter, were used in this paper.

3.2. Virtual Water

Figure 1 presents the net irrigation requirements (NIR) for the studied crops and irrigated zones in semi arid area. NIR slightly increase for all studied crops from the north to the center of Tunisia, this increase is due to that central regions of the country receive less rain. Some vegetables such as

pepper, pea and bean do not need irrigation in some regions and have a negative NIR, which means that virtual water consumed by these crops was only provided by the rain or green water. Cultivating these crops in these areas of the country helps minimize irrigation water consumption and thus the virtual water reduction. Summer crops such as tomato and pepper have the highest NIR from 430 mm in Bizerte to 600 mm in Monastir.

Awareness of the farmer to managing water resources is related to knowing the real crop water requirements at different stages of crop development although he manages well the irrigation scheduling. Over irrigation as infra-irrigation, have a negative impact on crop productivity. Finally, the virtual water concept, contextualized in space and time can provide useful information for benchmarking, indentifying best practices and achieving a more integrated water resource management. Nevertheless, to obtain a comprehensive picture, not only the (eco) efficiency in terms of m^3/ton should be considered, but also the context-specific total cumulative virtual water.

Analysis of field investigation results for Zaghouan governorate showed that the most cultivated vegetables crops on 2013 are: Tomato, Watermelon, Potato and Artichoke.

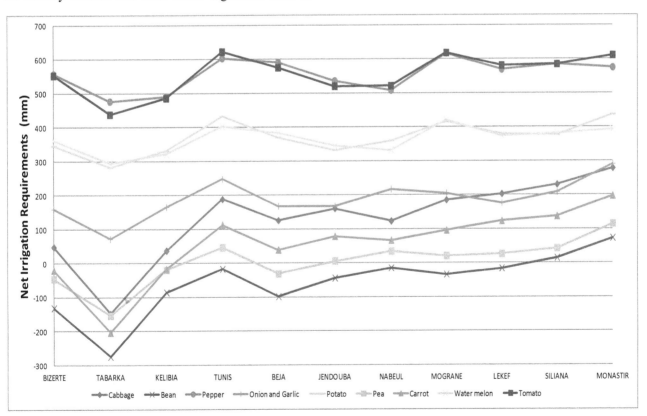

Figure 1. *Net irrigation Requirements (mm) for studied vegetables in Tunisian semi arid areas.*

Figure 2, represents estimation of Virtual Water (m^3/kg) and Gross Margin of Water (DT/m^3), for the mentioned crops.

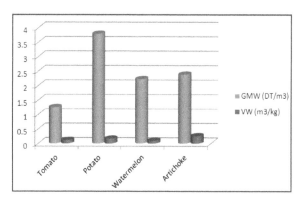

Figure 2. *Variation of virtual water (m³/kg) and gross margin of water (DT/m³) with vegetable crops for Zaghouan governorate.*

3.3. Labor

Figure 3 showed the evolution of agricultural worker's structure.

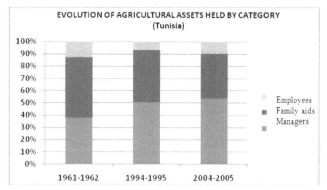

Figure 3. *Evolution of Tunisian agricultural workers*

Figure 3 shows increases in farm manager's percentage compared to family aids, which proves that the Tunisian agriculture has evolved into modern and productive agriculture.

Statistical data related to Tunisian agricultural and gender workers, are presented on the figures below.

Variations of gender permanent and family workers for Tunisian semi-arid governorates are showed by figures 4 and 5 respectively.

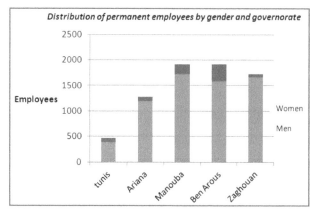

Figure 4. *Variation of gender permanent workers for Tunisian semi-arid governorates*

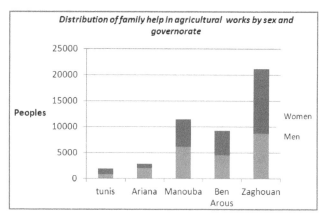

Figure 5. *Variation of gender family workers for Tunisian semi-arid governorates*

Variation of Tunisian permanent workers with area's farm is presented by figure 6, while the variation of Tunisian family workers with area's farm and gender is showed by figure 7.

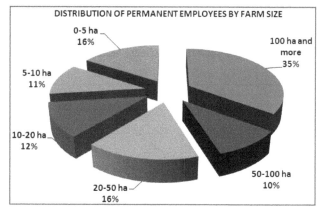

Figure 6. *Variation of Tunisian permanent workers with area's farm*

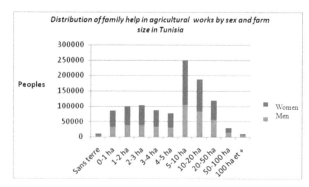

Figure 7. *Variation of Tunisian family workers with area's farm*

Agricultural family workers are the most important for Zaghouan governorate (Figure 5). For this governorate, field investigation for a sample of farmers showed that agricultural women workers represent 75% of the total agricultural workers. The tasks performed by women are: seedlings, planting, weeding, replacement, fertilizing, mulching and crop harvesting. Mechanical and chemical tasks are typically performed by men: plugging, mechanical and chemical weeding, fertilizer application and marketing.

4. Conclusion

The use of the virtual water concept to confront water scarcity and support food security, in Tunisian semi-arid region showed that:

- Spring and autumn vegetable crops present low virtual water and are thus recommended for this region;
- vegetables are one of the most important agricultural activities in the country contributing to food security needs in water are relatively high compared to other agricultural products;
- Citrus present the high virtual water in the studied regions;
- Reduction of virtual water for these vegetables crops in Tunisian semi-arid region, can be obtained by improve of irrigation techniques and control of runoff and leaching water, using drip, localized and underground irrigations ;
- The virtual water estimation by using a model that gives enough information to perform the value of consumed water each crop and the water wasted by the farmer can help to guide agricultural policy for better water management;
- The concept of virtual water should be treated with caution in trying to both manage water resources according to speculation and ensure food security;
- It seems clear that integrated water allocation, planning and management is needed in the Tunisian semi-arid regions, considering the environmental water requirements together with the blue (surface and ground) and green virtual water, to achieve a more compatible agricultural production.

Acknowledgments

This research was funded by the IRDC of the Government of Canada through grant "Virtual water and food security in Tunisia: from observation to support development. Authors would like to thank the Regional Commission for Agricultural Development managers and farmers for their support. We are also thankful for the comments by reviewers.

References

[1] Allan, J.A., 1997. "Virtual water": A Long Term Solution for Water Short Middle Eastern Economies? : Occasional paper, no. 3. Water Issues Study Group, School of Oriental and African Studies, University of London.

[2] Allen, R.G., Pereira, L.S., Raes, D., Smith, M., 1998. Crop evapotranspiration: guidelines for computing crop water requirements, FAO Irrigation and Drainage Paper 56. United Nations Food and Agriculture Organization, Rome.

[3] Cuenca, R.H., 1989. Irrigation System Design: An Engineering Approach. Prentice-Hall Inc., Englewood Cliffs, NJ, USA, 552 pp.

[4] FAO, 2009. CROPWAT 8.0 model, Food and Agriculture Organization, Rome, Italy. [online] www.fao.org/nr/water/infores_databases_cropwat.html. [Accessed on March 2012].

[5] Hoekstra, A.Y., Chapagain, A.K., Aldaya, M.M. and Mekonnen, M.M., 2009. Water footprint manual: State of the art 2009, Water Footprint Network, Enschede, the Netherlands.

[6] Loiseau, E. 2010. Environmental impacts evaluations methods of water use. Bibliotheque synthesis. AgroParisTech–ENGREF centre de Montpellier. 18p

[7] ONAGRI, 2010: http://www.onagri.tn/STATISTIQUES/ENQUTES%20STRU CTURES/ESEA%202004-2005.htm#_Toc125361774. [Accessed on May 2012].

[8] Point Focal Tunisien du SEMIDE, (2002). http://www.semide.tn/english/contressource.htm Accessed on 20/07/2013

[9] Smith, M., 1993. CLIMWAT for CROPWAT, a climatic database for irrigation planning and management. FAO Irrigation and Drainage Paper 49, Rome, 113 pp.

[10] Velázquez, E., (2007).Water trade in Andalusia. Virtual water: An alternative way to manage water use. Ecological economics Vol. (63), (201-208).

12

Factors Influencing Pesticide Use in Smallholder Rice Production in Northern Ghana

Benjamin Tetteh Anang[1, *], Joseph Amikuzuno[2]

[1]Department of Agricultural and Resource Economics, FACS, University for Development Studies, Tamale, Ghana
[2]Department of Climate Change and Food Security, FACS, University for Development Studies, Tamale, Ghana

Email address:

abtben@yahoo.com (B. T. Anang), amikj26@yahoo.com (J. Amikuzuno)

Abstract: Rice production is an important economic activity among smallholder farmers in northern Ghana serving as source of income and household food security. The production of rice is often associated with the use of pesticides to control harmful pests of rice, a practice which also poses environmental and human health risks. The study sought to investigate the factors which influence smallholder rice farmers' use of pesticides in rice farming in northern Ghana. Rice farmers were selected from three irrigation schemes in northern Ghana, namely the Botanga, Tono and Vea Irrigation Schemes. A multi-stage stratified random sampling technique was used to identify 300 rice farmers who were interviewed with a semi-structured questionnaire. The data was analyzed both descriptively and inferentially. A probit model was used to study the determinants of pesticide use. The study showed that farm size, farm income, mechanization, extension contact, distance to source of pesticide and production system were the influencial factors in rice producers' choice to use pesticide in rice farming. The study recommends extension education to farmers on pesticide use in order to avoid misuse and the risks factors associated with improper application.

Keywords: Northern Ghana, Pesticide Use, Probit Model, Rice Production, Smallholder Farmers

1. Introduction

Pesticides are chemical substances containing a very diverse array of chemical structures. Pesticides can be classified under agrochemicals which are crop protection products or agents used to control plants or weeds, diseases, insects or animals that are undesirable or harmful to man, and/or also to promote the growth and development of crops. According to [1] and [2], the commonly used agrochemicals in Ghana are insecticides, herbicides, fungicides, fumigants, fertilizers and growth regulators.The complex nature of many pesticides does not make the classification of pesticides simple. A convenient classification of pesticides is based on the targeted pest. For example, insecticides target insects, herbicides target weeds, fungicides target fungi, nematicides target nematodes while rodenticides target rodents. There is a further subdivision of each class into smaller subclasses based on chemical structure. In terms of total use, herbicidesaccount for the largest proportion of pesticide use.

Pesticides are widely used in agricultural production to protect crops and animals from pest such as insects, mites, birds, ticks, rodents, nematodes, weeds, fungi, and other organisms that cause losses. They are also used to maintain product quality. Some pesticides are prepared locally by farmers and these are called natural pesticides such as neem extract and wood ash. Other pesticides are manufactured in industries through advanced procedures and are known as synthetic or artificial pesticides. This study focused on the use of synthetic or artificial pesticides by smallholder rice farmers.

It has been shown that the pesticides used in agricultural, commercial, home and garden applications are associated with great health risk including cancer [3]. According to]4], acute poisoning with pesticides is a global public health problem accounting for as many as 300,000 deaths worldwide every year.

As noted by [5], low yields and productivity are compounded in the long-run by production shocks arising from environmental stresses such as drought, pests, and diseases. Despite chemical control measures to increase agricultural production, 20 – 40 % of potential food production is still lost every year to pests and diseases [6].

The use of pesticides is thus important to farmers to control pests and diseases. From an agricultural industry perspective, pesticides are an important component of economic and effective pest control and their continued use is considered essential [7]. It therefore comes as little surprise that pesticide use has increased over time in Ghana and is particularly elevated in the production of high-value cash crops and vegetables [8]. Chemical pesticides are also used improperly or in dangerous combinations as noted by [9]. The promotion of safe use of pesticides on vegetables has been put on the agenda of Ghana's Food and Agriculture Sector Development Policy in recognition of the misuse of chemical pesticides in the country [10].

2. Methodology

The following section presents the methodology employed in gathering and analyzing the data for the study together with the analytical and empirical frameworks.

2.1. Analytical Framework and Empirical Model

Probit analysis permits the estimation of dichotomous dependent variables within the regression framework. Many response variables are binary by nature, requiring either a yes or no (or 1/0) response. Ordinary least squares (OLS) regression is inadequate when we have dependent variables that are discrete [11],[12]. Probit and logit analyses become more convenient when dealing with such situations.

The probit model constrains the estimated probabilities to lie between 0 and 1, relaxing the constraint that the effect of the independent variable is constant across different predicted values of the dependent variable. The probit model makes the assumption that while we only observe the values of 0 and 1 for the dependent variable Y, there is a latent, unobserved continuous variable Y^* that determines the value of Y[13]. This study uses the probit model to determine the factors influencing pesticide use in smallholder rice production at three irrigation schemes in northern Ghana.

Suppose the response variable Y is binary; with only two possible outcomes denoted as 1 and 0. Consider also a vector of regressors X, which are assumed to influence Y. Specifically we assume that the model takes the form:

$$Pr(Y = 1 \mid X) = \Phi(X'\beta) \qquad (1)$$

Where Pr stands for probability and Φ is the Cumulative Distribution Function (CDF) of the standard normal distribution. The parameters (β) are estimated by the method of maximum likelihood. The Probit model for the study can be written as:

$$Y = F(\alpha + \beta x_i) = F(z_i) \qquad (2)$$

Where Y = the discrete choice variable, that is, pesticide use; F = the cumulative probability distribution function; β = a vector of parameters; x = the vector of explanatory variables, and z = the Z-score of βx for the area under the normal curve.

It is assumed that Y can be specified as follows:

$$Y = \beta_0 + \beta_1 X_{1i} + \beta_2 X_{2i} + \ldots + \beta_{11} X_{11i} + V_i \qquad (3)$$

And that:

$$Y = 1_{if Y^* > 0} \text{ and } Y = 0_{Otherwise} \qquad (4)$$

Where $X_1, X_2, \ldots, X_{11}$ represent vector of random variables, β is a vector of unknown parameters, and V is a random disturbance term.

2.2. The Empirical Model

The probit model for the study was specified as follows:

$$Y_i = \beta_0 + \sum_{i=1}^{11} \beta_i X_i + v \qquad (5)$$

where Y_i = pesticide use (=1 if the farmer used pesticide, 0 otherwise), x_1 = sex of farmer, x_2 = educational attainment, x_3 = years of experience in rice production, x_4 = farm size; x_5 = farm income squared, x_6 = farm income squared; x_7 = mechanization using tractor; x_8 = number of extension contact; x_9 = distance to pesticide selling point; x_{10} = production system (binary): 1 if farm is irrigated, 0 otherwise; x_{11} = regional dummy (binary): 1 if in Northern Region, 0 otherwise. The βs are unknown parameters to be estimated.

2.3. Study Area

The study was carried out in the three regions of Northern Ghana, namely the Upper East, Upper West and Northern Regions. The Northern and Upper East Regions were purposively selected for the study because of their irrigation potential for rice production. The three largest irrigation schemes notable for rice cultivation are located in these Regions. Ghana has a total land area of 238,540 km^2 and has a warm humid climate. There are six ecological zones in Ghana namely the rain forest, the deciduous forest, the transitional zone, the Guinea Savannah, the Sudan Savannah, and the Coastal savannah. Rainfall distribution is bimodal for the forest, transitional and coastal zones resulting in two cropping seasons, the major and the minor seasons. On the other hand, the savannah zone which comprises the three Regions of northern Ghana has a mono-modal rainfall distribution which gives rise to only one cropping season. The rainy season permits a growing season of 150–160 days in the Upper East Region and 180–200 days in the Northern Region. Mean total annual rainfall ranges from 1,000 mm in the Upper East Region to 1,200 in the south-eastern part of the Northern Region. Northern Ghana has a total area of 98,000 km^2, of which 16,000 km^2 is intensely farmed and about 8,000 km^2 is less intensely farmed [14].

2.4. Sampling and Data Collection

The data was obtained from a farm household survey conducted during the 2013/2014 farming season. A total of

300 rice farmers were selected from fifteen (15) communities and interviewed using semi-structured questionnaire which was pre-tested. A stratified multi-stage sampling technique was used. The three largest irrigation schemes in northern Ghana were purposively sampled. These are the Tono Irrigation Scheme in the Kassena-Nankana District of the Upper East Region, the Vea Irrigation Scheme in the Bolgatanga District also in the Upper East Region, and the Botanga Irrigation Scheme in the Tolon-Kumbungu District of the Northern Region. Five communities were randomly selected from each catchment area of the three irrigation schemes to form the target farming communities in which the questionnaire were administered. Rice farmers in each community were stratified into irrigators and non-irrigators with equal samples of irrigators and non-irrigators being randomly selected. Pesticide use in rice cultivated was solicited from farmers as well as other household and production data.

The data was subjected to both descriptive and inferential statistical analysis. The descriptive statistics included measures such as frequencies, percentages, means, and standard deviations. The probit model was used to determine the factors influencing pesticide use in rice production by smallholder farmers in northern Ghana.

3. Results and Discussions

This section provides a description of the data used for the study followed by a presentation of the results of the probit model which estimates the role of the factors influencing pesticide use bysmallholder rice farmers in Northern Ghana.

3.1. Description of Variables Used in the Study

Table 1 shows the explanatory variables used in the study and their description. The hypothesized signs of the variables are also presented.

Table 1. Description of the explanatory variables used in the study.

Variable Description	Measurement	Expected sign
Sex of respondent	Dummy: 1 for male; 0 for otherwise	+/-
Educational attainment in years	Continuous	-
Experience in rice production (years)	Continuous	+/-
Farm size (in acres)	Continuous	+
Farm income (Ghana Cedi)	Continuous	+
Mechanization indicated by tractor use	Dummy: 1 for tractor use; 0 otherwise	+/-
Number of extension contacts made	Continuous	-
Distance to pesticide selling point	Continuous	-
Production system (access to irrigation)	Dummy: 1 if farm isirrigated; 0 for otherwise	+/-
Regional dummy	Dummy: 1 if in Northern Region; 0 for otherwise	+/-

The effect of gender, farming experience, mechanization use and access to irrigation on the choice to use pesticides are hypothesized to be indeterminate. This means that the effect can be either positive or negative. However, educational attainment, number of extension contacts and distance to pesticide selling point are hypothesized to have a negative effect on pesticide use. For example, an increase in the distance to the pesticide selling point is associated with additional cost of production to the farmer which is likely to reduce pesticide adoption. Similarly, farmers who receive extension contacts as well as farmers who are educated are expected to acquire knowledge of various methods of controlling pests and diseases and may be less disposed to use pesticides in rice production. An increase in both farm size and farm income is hypothesized to increase pesticide use because as farm size increases, pest control becomes more difficult while an increase in income makes pesticide acquisition more affordable to the farmer.

Table 2 gives the descriptive statistics of the relevant variables used in the model.

Table 2. Descriptive statistics of relevant variables.

Variable	Pesticide Users N=218	Non-users N=82	Overall N=300
Farm size (in acres)	2.39	2.02	2.29
Experience in rice production (years)	15.2	15.9	15.4
Farm income (Ghana Cedis)	1879	1696	1829
Mechanization (%)	72.5	45.1	65.0
Number of extension contacts made	2.58	5.23	3.31
Distance to pesticide selling point (km)	7.40	9.31	7.92
Production system (irrigation use) (%)	48.0	55.0	50.0
Northern Region (%)	14.6	40.4	33.3
Upper East Region (%)	85.4	59.6	66.7
Male (%)	82.6	67.1	78.3
Female (%)	17.4	32.9	21.7

Table 3. Probit regression results.

Variables	Coefficient	P>z	Marginal Effect
Sex	0.283 (0.222)	0.202	0.090
Education	-0.019 (0.017)	0.262	-0.006
Experience	-0.007 (0.008)	0.413	-0.002
Farm size	0.159 (0.077)	0.040**	0.048
Farm income	0.716 (0.318)	0.024**	0.215
Farm income squared	-0.268 (0.084)	0.001***	-0.080
Mechanization	0.734 (0.189)	0.000***	0.234
Extension contact	-0.049 (0.017)	0.005***	-0.015
Distance	-0.083 (0.023)	0.000***	-0.025
Production system	-0.470 (0.217)	0.030**	-0.140
Region	0.173 (0.247)	0.483	0.051
Constant	0.626 (0.376)	0.096*	-

***, **, and * indicate statistical significance at the 1, 5 and 10 percent level, respectively. R2=0.20, Percentage correctly predicted = 77.0. Figures in parentheses are standard errors.

Pesticide users did not differ much from non-users in terms of farm size and years of farming experience. Pesticide users had slightly higher farm income compared to non-users. Hence an increase in income is likely to positively influence pesticide use. Pesticide users also had less extension contact and shorter distance to pesticide selling point compared to the non-users. Thus nearness to the pesticide selling point enhances the use of pesticides. Irrigation access was found to be lower for pesticide users compared to the non-users while mechanization (tractor use) was higher among pesticide users compare to the non-users. Thus irrigation access is likely to have a negative effect on pesticide use while mechanization is likely to have a positive effect. In addition, majority of pesticide users were found in the Upper East Region while men exceeded women in the use of pesticide in rice cultivation. Pesticide use is therefore likely to be positively related to being a male farmer as well as being located in the Upper East Region.

3.2. Factors Influencing Pesticide Useby Smallholder Rice Farmers

Table 3 presents the result of the probit analysis and the marginal effects of the probit regression.

Farm size was positively related to pesticide use and was significant at the 5 percent level. Thus an increase in farm size increases the probability of adoption of pesticide in rice production in the study area. The marginal effect of 0.048 indicates that a unit increase in farm size increases the probability of pesticide use by 4.8 percent. The result is at variance with [15] who found a negative correlation betweeen farm size and pesticide use. The study is however consistent with [16] who found a positive relation between agrochemical use and farm size. From our study, we deduce that as farm size increases, farmers are unable to effectively control pests and diseases manually without resorting to pesticide use. Hence an increase in farm size increases the likelihood of pesticide use in rice production by farmers.

Farm income also exhibited a positive and significant relationship with pesticide use which is consistent with [17] who found income of farmers to be positively related to the level of pesticide use in Nigeria. This shows that an increase in farm income increases the predicted probability of pesticide use in rice farming by 0.22, as indicated by the marginal effect. The squared value of farm income was significant but negative, indicating that as farm income increases beyond a critical point, the probability of pesticide use decreases by 0.08. Thus an initial increase in farm income increases pesticide use but futher increase in income leads to a decrease in the use of the chemical input. The result is to be expected because as farm income increases, farmers are expected to be able to afford to purchase the quantity of pesticides they need. However, with further increase in income, the farmer is less likely to continue to increase pesticide use because beyond a certain point, pesticide use is no longer beneficial. Diminishing marginal returns sets in with the application of the chemical and less amount is used.

Mechanization or tractor use had a positive and significant relationship with pesticide use. Hence farmers who used mechanization were more likely to use pesticide in rice farming compared to those who did not use mechanization. The probability of pesticide use by farmers who use mechanized farm implements was 0.23 higher than those who did not use mechanization.

Contact with extension agents exhibited a significantly negative relationship with pesticide use, implying that pesticide use decreases with contact with extension agents. Hence an additional increase in extension contact decreases the probability of pesticide use in rice farming by 0.02. It is generally expected that in order to reduce yield losses due to pests and diseases farmers will use more pesticides, all things being equal. The study however shows that extension contacts tend to reduce the use of pesticide in rice farming. One possible reason is the introduction of integrated pest management (IPM) to farmers to reduce pesticide use particularly in rice farming. For example, IPM Farmer Field Schools which started in Indonesia in 1989 was designed to reduce pesticied use in rice production [18]. The IPM farmer field schools have been replicated in over 30 countries worldwide. In Ghana, similar activities by the Ministry of Food and Agriculture through extension education with farmers have been ongoing and this could account for the reduction in pesticide use in rice farming by farmers with more extension contacts.

Distance to the source of pesticide had a significantly negative relationship with pesticide use, indicating that pesticide use decreases with distance and hence the transaction cost needed to travel to acquire the input. An increase in the distance to the pesticide selling point from the farm decreases the probability of pesticide use by 0.03. The result is consistent with *a priori* expectation since an increase in transaction cost due to long travel distance is expected to decrease input use, all things being equal.

Finally, the production system used by the farmers exhibited a negatively significant relationship with pesticide use. This implies that irrigation users were less likely to use pesticide in rice production compared to non-irrigation users. The probability of pesticide use by users of irrigators was 0.14 lower than those who did not use irrigation.

4. Conclusion

Pesticide use in rice production is influenced by certain household characteristics such as farm size, farm income, the adoption of farm mechanization, extension contact, distance to the source of pesticide and whether the farmer uses irrigation or not. In addition, rain-fed farmers were more disposed to use pesticides in rice cultivation compared to irrigation users. However regional variation, sex, education and farming experience were less likely to influence pesticide use in the study area.

The study recommends the intensification of extension education to farmers and particularly, the IPM Farmer Field School system since it has been found to influence pesticide use by farmers. As noted by [18], production losses due to pests and diseases compel farmer to excessively use pesticides. Thus extension service to farmers is one major channel to educate farmers on pesticide use to avoid misuse and the risks factors associated with improper application.

Acknowledgement

The authors would like to thank Wienco Ghana Ltd and Wienco Chair Research Committee of the Faculty of Agriculture of the University for Development Studies for providing funding for the research.

References

[1] Ntow, W. J., 2004. Organochlorine Pesticides in Water, Sediments, Crops and Human Fluids in a Farming Community in Ghana. Journal of Archives of Environmental Contamination and Toxicology. 40(4): 557-563.

[2] Pan African Regulation (PAR), 2000. Pan Africa Regulation of Dangerous Pesticides in Ghana. *Pan* African Monitoring and Briefing Series No. 5, Dakar, Senegal, 16Pp.

[3] Alavanja M. C. R., Ross M. K., and Bonner M. R,.2013. Increased Cancer Burden Among Pesticide Applicators and Others Due to Pesticide Exposure. CA Cancer J Clin 2013 (63):120–142. doi:10.3322/caac.21170. Available online at cacancerjournal.com.

[4] Goel A. and Aggarwal P., 2007. Pesticide poisoning. The National Medical Journal of India, Vol. 20, No. 4. Review Article 182. July/August 2007.

[5] Horna D, Smale M., Al-Hassan R., Falck-Zepeda J., and Timpo S. E.,2008. Insecticide Use on Vegetables in Ghana: Would GM Seed Benefit Farmers? IFPRI Discussion Paper 00785, August 2008.

[6] Obeng-Ofori, D., 1998. Post Harvest Science. Crop Science Department. University of Ghana, Legon, October, 1998. 71 Pp.

[7] Kent, J., 1991. Education and Training in Farm Chemical Management. Proceedings of Conference on Agriculture, Education and Information Transfer. Murrumbidge College of Agriculture, 1991.

[8] Gerken, A., Suglo, J.-V. and Braun, M., 2001. Crop Protection Policy in Ghana. Pokuase-Accra, Ghana: Integrated Crop Protection Project, PPRSD/GTZ.

[9] Obeng-Ofori, D., Owousu, E. O. and Kaiwa E. T., 2002. Variation in the level of carboxylesterase activity as an indicator of insecticide resistance in populations of the diamondback moth Plutella xylostella (L.) attacking cabbage in Ghana. Journal of the Ghana Science Association 4 (2): 52–62.

[10] Ministry of Food and Agriculture, 2002. Food and agriculture sector development policy. Accra: Government of Ghana.

[11] Collett D., 1991. Modelling Binary Data, London: Chapman and Hall.

[12] Agresti A., 1990. Categorical Data Analysis, New York: John Wiley & Sons.

[13] Sebopetji TO, Belete A., 2009.An Application of Probit Analysis to Factors Affecting Small-Scale Farmers' Decision to take Credit: a Case Study of Greater Letabo Local Municipality in South Africa. Afri. J. Agric. Res. 4(8):718-723.

[14] Al-hassan, S., 2008. Technical Efficiency of Rice Farmers in Northern Ghana. AERC Research Paper 178, African Economic Research Consortium, Nairobi. April 2008.

[15] Yasin G, Aslam M, Parvez I. and Naz S., 2003. Socio-economic Correlates of Pesticide Usage: The Case of Citrus Farmers. Journal of Research (Science), Bahauddin Zakariya University, Multan, Pakistan 14 (1):43-48, June 2003.

[16] Alabi O. O., Lawal A. F., Coker A. A. and Awoyinka Y. A., 2014. Probit Model Analysis of Smallholder's Farmers Decision to Use Agrochemical Inputs in Gwagwalada and Kuje Area Councils of Federal Capital Territory, Abuja, Nigeria. International Journal of Food and Agricultural Economics 2(1):85-93.

[17] Idris A, Rasaki K., Folake T., and Hakeem B., 2013. Analysis of Pesticide Use in Cocoa Production in Obafemi Owode Local Government Area of Ogun State, Nigeria. Journal of Biology, Agriculture and Healthcare 3(6):1-9.

[18] Rahman, A. M. A. and Hamid, M E., 2013. Impact of FFS on Farmer's Adoption of IPM Options for Onion: A Case Study from Gezira State, Sudan. World Journal of Agricultural Sciences 9 (1): 38-44. DOI: 10.5829/idosi.wjas.2013.9.1.1720.

13

Investigating the role of apiculture in watershed management and income improvement in Galessa protected area, Ethiopia

Tura Bareke Kifle, Kibebew Wakjira Hora, Admassu Addi Merti

Holeta Bee Research Centre, Oromia Agriculture Research Institute, Holeta, Ethiopia

Email address:

trbareke@gmail.com (T. B. Kifle)

Abstract: Beekeeping gives local people economic incentive for the preservation of natural habitats and is an ideal activity in watershed conservation program. The study was designed to assess and demonstrate the contribution of improved beekeeping for income generation and sustainable watershed management in Galessa protected area. For this purpose households were purposively selected based on their interest in beekeeping, experience in traditional beekeeping and proximity of residence to watershed areas. Training on beekeeping and integrations of beekeeping with watershed management were provided. Data of honey yield, bee plants, and annual income obtained from honey and field crops before and after improved beekeeping intervention were collected. Accordingly, the mean annual honey yield, income obtained from honey sales, bee forage planting practice and number of transitional hives owned by the beekeepers are significantly different between the sample households ($P<0.05$) before and after intermediate beekeeping intervention but the number of traditional hives owned was not significantly different between the household . The total honey yield has increased almost by two fold and the annual revenue increased by 6.5 folds. Therefore integration of intermediate beekeeping technology with conservation of watershed can enhance the income of household and encourages planting of bee forages which directly contributes for sustainable watershed managements. Thus demonstration and scaling up improved beekeeping technology should be promoted for sustainable watershed rehabilitation and to diversify the household income.

Keywords: Watershed, Rehabilitation, Beekeeping, Honey, Bee Forages

1. Introduction

Plants are primary producers in terrestrial ecosystems and direct providers of many ecosystem services such as carbon sequestration, prevention of soil erosion, nitrogen fixation, maintenance of water tables, greenhouse gas absorption, and food and habitat providers for most other terrestrial and many aquatic life forms (FAO, 2001). In Ethiopia, unwise extractions of the available forest resources for various purposes have resulted in the rapid depletion of natural forest resources. Because of that, there is a rapid decrease in the number of springs and pronounced decline in ground water table in highlands of Ethiopia (Kindu Mekonnen and Zenebe Admassu, 2008). The Holeta Agricultural Research Center identified 18 watershed constraints of which the losses of indigenous trees, the decline of spring discharge and land shortage due to high

population pressure. These problems were also confirmed by most farmers at Galessa watershed to be the top priority problems. This calls for coordinated effort by policy makers, development experts and researchers to look for alternative development options to enhance forest conservation and which in turn improves watershed system in general.

Beekeeping is an incentive for planting trees and protecting existing trees, because trees are very important for bees and therefore for beekeepers as well (Crane 1999; FAO, 2003). On top of this honeybees serve as pollinating agents for numerous species of plants and contribute to their survival, genetic prosperity and play a crucial role in the maintenance of ecosystem services (FAO, 2004). In past few years, few attempts have been made by Farm Africa to implement beekeeping technologies around Chillimo state forest as incentive for conservation of existing natural forest by providing beekeeping accessories and training to farmers living in or close to the forests to practice beekeeping for honey production (Regassa Ensermu *et al.*, 1998.). However, due to lack of proper follow-up, research intervention and also the pilot project did not encompass all farmers in and around the forest area; the attempts did not hit the desired target. Therefore, in this research activity an attempt was made to assess the role of beekeeping in watershed conservation around the Galessa watershed area with active involvement of the community to improve their livelihood and to enhance the watershed rehabilitation.

2. Methodology

2.1. Study Site

The study was undertaken at Galessa watershed which is delineated by Holeta Agriculture Research Centre and located in Dendi Woreda, West Shewa Zone of Oromia Regional State of Ethiopia. The altitude of watershed area ranges from 2900–3200 m.a.s.l with bimodal rainfall patterns (Mekonnen, 2007) and located in grid points of *09°06'54" N to 09°07'52"N and 37°07'16"E to 37°08' 54"E.*

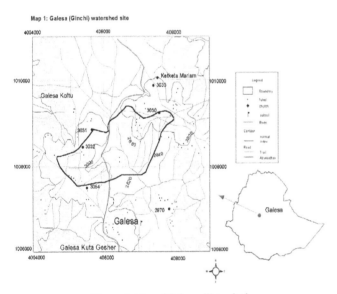

Map 1: Galesa (Ginchi) watershed site

Figure 1. *Map of Galessa Watershed*

2.2. Methods

2.2.1. Household Selection and Farmer Research Group (FRG) Establishments

Proximity to watershed areas is a factor that was accounted when selecting the demonstration site. Galessa Koftu kebele which is close to the watershed was selected for demonstration of the technology. People who practice beekeeping were purposively selected based on their interest in beekeeping,

experience in traditional beekeeping and proximity of residence to watershed areas and 2 FRG groups consisting a total of 20 households were selected to conduct the demonstration of beekeeping technology. Only 105 traditional hives were found in the watershed. Of these 60 were included in our demonstration and transferred to transitional hives. After the establishment of FRG initial sensitization workshop was held and memorandum of agreement was signed by the member of the FRG, Holeta Bee Research Center Researchers and Ministry of Agriculture and Rural Development. Practical training on honeybee transferring from traditional to improved hives, seasonal colony manipulation and intermediate beehive construction were carried out. The participants were also introduced about the integration of beekeeping with watershed management. Before starting, watershed preliminary assessment was done to get present information concerning the households living in the delineated watershed areas. Socioeconomic data was collected from FRG members using semi structured questionnaire before and after improved beekeeping technology involvement. All the necessary beekeeping management practices such as inspection, feeding, supering, super reducing etc were undertaken. Honey yield and income were obtained and compared to income obtained from sale of major crops (barley, wheat and potatoes) grown in the area. On top of this the attitude of farmers towards watershed rehabilitation was recorded. A number of plant species conserved and planted was recorded around home garden

2.2.2. Data Management and Statistical Analysis

Data of honey yield, income obtained from honey, the watershed management practice in relation to beekeeping, the contribution of beekeeping in the livelihoods of the households, the attitude of the farmers toward planting bee forage planting and constraints of beekeeping were collected. The data analysis was undertaken using index (Falconer and Mackay, 1996), descriptive statistics and t-test by *Statistical Package for the Social Sciences* (Spss) software.

3. Result and Discussion

According to the beekeepers socioeconomic assessment, the mean annual honey yield, income obtained from honey sales, bee forage planting practice and number of transitional hives owned by the beekeepers are significantly different between the sampled households ($P<0.05$) before and after beekeeping intervention but the number of traditional hives owned was not significantly different between the household (Table1).

Table 1. *Mean ± Standard error of honey yield, number of hives owned and bee forage planted before and after beekeeping intervention*

Intervention	Annual average honey yield in kg	Mean number of traditional hives owned	Mean number of transitional hives	Bee forages planted
Before beekeeping	7± 0.94a	1.7±0.193a	0±0a	35.2 ±5.6a
After beekeeping	22 ± 2.3b	0.85 ±0.29a	2 ± 0.25b	70 ±10.4b
	Different letters show significance difference			

Figure 2. Home garden beekeeping apiary of the sample household

3.1. Honey Yield

Amount of honey harvested was 7kg in traditional hives before project intervention and increased to 22 kgs per hive after changed to intermediate hives. The total honey yield has increased almost by two fold and the annual revenue increased by 6.5 folds after beekeeping projects. This indicated that improved beekeeping practices contributes to income generation of the households and reduce deforestation of the trees in the watershed. This was supported by Hussein`s (2000) findings which confirm that beekeeping enhances the income generation potential of small holders and promotes the conservation and utilization of natural resources that are being rapidly depleting. Beekeeping is a practical tool for raising an awareness of the communities to manage watersheds and could favor watershed conservation (Alemtsehay, 2011; Albersand and Robinson, 2011). The products of the beehives (honey, beeswax, pollen and Propolis) are a rich source of nutrients and can be of world quality, and for which there are significant local and international markets (Lietaer, 2009). These activities are not only generating income from sale of honey for watershed user group but also able to sustain the resources through tree plantation, access to improved beekeeping technologies and expensive bee equipment.

Farmers did not increase the number of traditional hives after improved beekeeping technology demonstration because of its low yield and unsuitability to manage honeybee colonies (GRM International, 2007).

Figure 3. Honey harvested from intermediate hives in the watersheds

3.2. Contribution of Beekeeping to Household Income Compared with Major Crops Grown

The results indicated that potatoes *contribute* 68.7%, barley 17.5, wheat 9.6% to households' incomes while honey contributes only 4.2%. Even though the income obtained from beekeeping was low compared to those from major crops, (Figure 4).

proportion(%)

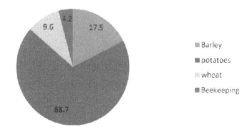

Figure 4. Proportional contributions of beekeeping and major crops to household income

3.3. Bee Forage Planting

The mean number of bee forage planted annually by the beekeepers before beekeeping intervention was 42 plants per sampled farmers and 124 plants per beekeepers after beekeeping intervention. The bee forage development has increased by 2 folds (Table 1). Attitude of beekeepers towards watershed integrated beekeeping technology is a very important phenomenon to take into consideration for multipurpose bee forages planting. Many countries introduced improved beekeeping as reforestation incentives, paying special attention to plant flowering trees that provide nectar and pollen whilst generating income for local communities from bee products (FAO, 2003; Steffan and Kuhn, 2003; Decourtye, et al, 2010). Diversification of cropping systems team such as vegetables, legumes, oilseeds, and forage crops in watershed improved the rainwater harvesting capacity and the impacts on environmental resources (Adugna, 2002).Crop varieties planted in watershed observed as major honeybee forage and important to maximize honey yield and spread the farmer's economic risk. Moreover, the crop growers benefited from the pollination services of the honeybees indirectly but not yet quantified. A mixture of different weedy species maintained between crop boarders and uncultivated land of watershed contributed as major honeybee forage, rain water harvesting, watershed biodiversity conservation and climate adaption as well (Tolera Kumsa, 2014).

The plant species planted by the beekeepers before beekeeping intervention were *mainly* planted for fuel wood requirement, cash income and watershed conservation. Some of them are not visited by honeybees. (Table 2)

Table 2. *List of plant species before the project*

Plant species	family	growth habit
Juniperus procera	cupressaceae	tree
Arundinaria alpine	poaceae	shrub
Eucalyptus globulus	Mrytaceae	tree
Acacia decurrens	Fabaceae	tree
Dombeya torrida	Sterculiaceae	tree
Chamecytisus proliferus	Fabaceae	shrub
Buddleja polystachya	Buddlejaceae	shrub
Hagenia abyssinica	Rosaceae	tree

Table 3. *List of bee forage plant species newly adopted after the project*

Plant species	family	growth habit
Callistemon *citrinus*	*Mrytaceae*	*shrub*
Albizia gummifera	*Fabaceae*	*tree*
Dovyalis caffra	Flacourtiaceae	shrub
Olea africana	Oleaceae	tree
Prunus africana	Oleaceae	tree

The number of bee forage trees planted varied from household to household. Most of households planted *Eucalyptus globulus* more than the other species. Because the timber of *Eucalyptus globulus* also provide income through selling for different construction.

3.4. Types of Plant Species Preferred for Watershed Management

Researchers together with farmers identified more than 16 trees, shrubs and herbaceous plant species around homesteads, farm land and other niches (Table 4). Of these 88.2% were bee forage plants. Therefore, integration of improved beekeeping technology with watershed management is very crucial to diversify the annual income of the household. It is an alternative income generating activities which can be an appropriate solution for sustainable watershed development and encourage the farmers toward tree planting. Past conservation efforts in Ethiopia have only concentrated in developing watershed conservation programs without addressing the socioeconomic of watershed communities. Community ownership and participation in conservative initiatives is critical to sustainable conservation of watersheds (Lietaer, 2009). Therefore, integrating improved beekeeping

technologies and natural resources development offers a pathway that guarantees sustainable watershed management. It is common knowledge that beekeeping is dependent on natural resources and therefore any effort to improve beekeeping in watershed areas, should be hand in hand with the natural resources development (Tolera Kumsa, 2014). Legesse Negash in (2002) stated that nowadays it has become clear that the land degradation issue could be solved through a holistic approach that addresses land reclaiming, sustainable utilization and diversifying livelihood options so as to reduce the pressure on the biodiversity through beekeeping technology.

Figure 5. *Galessa Watershed and beekeeping practice*

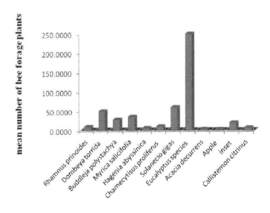

Figure 6. *Bee tree planting status of beekeepers around apiary (Home garden)*

Table 4. *Types of plant species preferred for Galessa watershed management*

No	plant species	Family	local name	Habit	other uses
1	*Acacia decurrens*	Fabaceae	Kacha	tree	fuel wood
2	*Arundinaria alpine*	Poaceae	Bamboo	shrub	construction
3	*Buddleja polystachya*	Buddlejaceae	Anfara	shrub	fuel wood
4	*Chamaecytisus proliferus*	Fabaceae	Treelucern	shrub	animal feed and fuel wood
5	*Dombeya torrida*	Sterculiaceae	Dannisa	tree	fuel wood and fences
6	*Eucalyptus globulus*	Mrytaceae	Barzaf	tree	construction, fuel wood and
7	*Hagenia abyssinica*	Rosaceae	Heto/Koso	tree	timber, medicine and
8	*Juniperus procera*	Pupressaceae	Gatira	tree	timber, fence and house construction
9	*Kalanchoe deficiens*	Crassulaceae	Bosoke	herb	fuel wood
10	*Maesa lanceolata*	Myrsinaceae	Abayyi	shrub	fence and fuel wood
11	*Myrica salicifolia*	Sapindaceae	chongi	shrub	fence and fuel wood

No	plant species	Family	local name	Habit	other uses
12	*Phytolacca dodecandra*	Phytolaccaceae	Andode	herb	cloth washing and traditional medicine
13	*Solanecio gigas*	Asteraceae	Osole	shrub	fences
14	*Urtica simensis*	Uritaceae	Dobi	herb	food
15	*Vernonia amygdalina*	Asteraceae	Ebicha /Girawa	shrub	animal feed, medicine and to scour pots used for making Tela, local beer & Tej and fence
16	*Vernonia auriculifera*	Asteraceae	Reji	shrub	fence, fuel wood and animal feed

3.5. Beekeeping Constraints of Galessa Watershed

Based on the response of the respondents among the seven constraints, herbicide use was the first serious problem for the decline of bee colonies. One of the mismanagement practicing is that farmers extensively using various types of pesticides near apiary sites without considering the damages caused on bee colonies, so that a number of bee colonies either die or abscond. The study conducted by Arse G. et al, (2010) also revealed that shortage of honeybee colonies due to poisoning from agro-chemical is the current obstacle to beekeeping.

Table 3. Ranked beekeeping constraints of Galessa Watershed

Major beekeeping constraint of the area	1st rank	2cd rank	3rd rank	4th rank	5th rank	6th rank	Index
Lack of bee forage	14.3	21.3	17.9	0	0	14.3	0.31
Decline of bee colony	0	10.7	10.7	39.3	0	0	0.24
Herbicides and pesticides problem	46.4	42.9	21.4	3.6	0	3.6	0.65
Pests and predators	0	14.3	25	3.6	3.6	0	0.21
lack of bee colony	28.6	3.6	14.3	21.4	75	46.4	0.56
Cost of beekeeping equipment	10.7	7.1	10.7	21.4	21.4	21.4	0.3
Lack of knowledge	0	0	0	10.7	0	14.3	0.05

Index=sum of [6 for rank 1 + 5 for rank 2 + 4 for rank 3 +3 for rank 4+2 for rank 5+1 for rank 6] for particular beekeeping constraints divided by sum of [6 for rank 1 +5 for rank2 +4 for rank 3+3for rank 4+2 for rank 5 +1 for rank 6] for all beekeeping constraints of the area

3.6. Attitude toward Improved Beekeeping

Attitude of the beekeepers towards watershed integrated beekeeping technology is a very important phenomenon to take into consideration for sustainable adoption of improved beekeeping in watershed conservation. They have developed awareness on the value of beekeeping for conservation and income generation as the result they have brought relatively better attitudinal change towards improved beekeeping technology and planting of different bee forage plants after this demonstration. It indicated that the majority of the watershed respondents (95%) had positive attitude towards watershed integrated beekeeping and honey production. However, 5 % of the respondents had neutral attitude and none of the respondents had negative attitude towards the technology in the study area. This showed how much the beekeeper farmers are understood the economical and ecological importance of beekeeping.

4. Conclusion and Recommendations

As a result of this demonstration, the average annual income of the beekeeper household from honey sale has increased, indicating that the integration of beekeeping with conservation and rehabilitation of natural resources would be an important incentive to mobilize communities to participate in rehabilitation programs for both economic and environmental reasons. The bee forage growing practice of the beekeepers have increased and thus beekeepers have due regard for watershed management and planting bee forages for bees.

Based on this study, Galessa watershed is a suitable area to initiate bee farming. However, attention must be given to maintain the existing bee flora and multiplication of multipurpose bee plant species in order to make it sustainable. Cost of beekeeping equipment and lack of bee colony is also the other main bottle neck problem to expand beekeeping technology and therefore queen rearing technology is recommended. As well as the demonstration and scaling up this technology should be promoted for sustainable watershed rehabilitation and to diversify the household income

Acknowledgements

I am thankful to Holeta Bee research Center and Oromia Agricultural Research Institute for providing required facilities and logistics. My sincere thanks also extended to Konjit Asfaw and Tesfaye Abera, for their inspiration and support in the implementation and follow-up of the research. I am thankful to all the watershed user groups and partners for their contribution towards this successful of the project.

References

[1]   Adugna, W. 2002. Genetic diversity analysis of linseed under different environments. PhD thesis, Department of Plant Breeding, Faculty of Agric. Univ. of the Free State, Bloemfontein, South Africa.

[2]   Albers, H.J. and Robinson, E.J. 2011. The trees and the bees: using enforcement and income projects to protect forests and rural livelihoods through spatial joint production. *Agricultural and resource economics review* 40/3: 424–438.

[3] Alemtsehay, T. 2011. Seasonal availability of common bee flora in relation to land use and colony performance in Gergera watershed Atsbi Wembwrta District Eastern Zone of Tigray, M.Sc Thesis, Wondo Genet College of Forestry, School Of Graduate Studies, Hawassa University, Hawassa, Ethiopia

[4] Arse Gebeyehu, Tesfaye Kebede, Sebsibe Zuber, Tekalign Gutu, Gurmessa Umeta, Tesfaye Lemma and Feyisa Hundessa, 2010. Participatory rural appraisal investigation on beekeeping in Arsi Negelle and Shashemene districts of West Arsi zone of Oromia, Ethiopia

[5] Debissa, L. 2006. *The Roles of Apiculture in Vegetation Characterization and Household Livelihoods in Walmara District, Central Ethiopia.* M.sc. Thesis, Wondo Genet College of Forestry, School Of Graduate Studies, Hawassa University, Hawassa, Ethiopia

[6] Decourtye, A; Mader, E. & Desneux, N. 2010. Landscape enhancement of floral resources for honey bees in agroecosystems.*Apidologie*, Vol.41, pp. 264-277.

[7] Falconer D.S and Mackay T.F.C.1996 Introduction to Quantitative Genetics.4th ed. Longman Group Ltd., Malaysia, 464 pp.

[8] FAO (2001). *Global Forest Resources Assessment 2000–main report.* FAO Forestry Paper No. 140, Rome.

[9] FAO, 2003. *State of the World's Forests.* Food and Agriculture Organization of the United Nations, Rome.

[10] FAO, (2004). Conservation and management of pollinators for sustainable agriculture–The International response. A contribution to the International Workshop on solitary bees and their role in pollination held in Berberibe, Cerara, Brazil, PP 19-25.

[11] GRM International (2007) .Livestock Development master plan study phase I report-Data collection analysis Volume - Apiculture

[12] Hussein, M. 2000. Beekeeping in Africa. In: *Apiacta* 1/2000: 32 – 48.

[13] Legesse Negash. (2002). Review of research advances in some selected African trees with special reference to Ethiopia. [Review article]. *Ethiop. J. Biol. Sci.* 1: 81-126.

[14] Lietaer, C. 2009. Impact of beekeeping on forest conservation, preservation of forest ecosystems and poverty reduction. XIII World Forestry Congress, 18 – 23, Argentina.

[15] Mekonnen, K. (2007),"Evaluation of selected indigenous and exotic tree and shrub species for soil fertility improvement and fodder production in the highland areas of western Shewa, Ethiopia", doctoral thesis, Department for Forest and Soil Sciences, University of Natural Resources and Applied Life Sciences, Vienna.

[16] Paterson, P.D. (1999), Constraints in transforming traditional to modern beekeeping in Kenya. In the conservation and utilization of commercial insects, Proceedings of the first international workshop, (Raina, S.K, Kioko, E.N. and Mwanycky, eds.). Nairobi, 18-21 August, 1997. pp 95-102.

[17] Regassa Ensermu, W.Mwangi, Hugo Verkuijle and Mohammed Hussen. 1998. Farmers' Seed Sources and Seed Management in ChilaloAwuraja, Ethiopia. Mexico, D. F: IAR/CIMMYT

[18] Steffan-Dewenter, I. and Kuhn, A. 2003. Honeybee foraging in differentially structured landscapes. *Proceedings of the* Royal Society of London Series B-Biological Sciences, Vol.270 pp. 569-575.

[19] Tolera Kumsa Gemeda. Integrating Improved Beekeeping as Economic Incentive to Community Watershed Management: The Case of Sasiga and Sagure Districts in Oromia Region, Ethiopia. *Agriculture, Forestry and Fisheries.* Vol. 3, No. 1, 2014, pp. 52-57.doi: 10.11648/j.aff.20140301.19

14

Human-wildlife conflicts: Case study in Wondo Genet District, Southern Ethiopia

Muluken Mekuyie Fenta

Animal nutritionist, Hawassa University, Wondo Genet College of Forestry and Natural Resources, Wondo Genet, Ethiopia

Email address:

mulukenzeru@gmail.com

Abstract: The purpose of this research was to identify the type of human- wildlife conflicts and wild mammals that cause the conflict, determine the extent of damage and to provide a better understanding of the causes of human-wildlife conflict in Wondo Genet district. The study was carried out from December 2013 to June 2013. Four sample areas were selected to collect data on human-wildlife conflict; Gotu, Wosha Soyoma and Wethera Kechema villages and Wondo Genet College of Forestry and Natural Resources. Data were collected using questionnaires, one to one interviews, observations and cross checking of crop loss using quadrants in selected crop lands, reviewing of literature, and was later analyzed using statistical package for social scientists (SPSS). In order to achieve the objectives of the study, the target population comprised the households living in sampled areas, local administrators and staffs who lived within Wondo Genet College of Forestry and Natural Resources. The study established that crop damages, livestock killing, human disruption and property destruction were some of the mostly reported damages. The results of the study further indicated that animal species most involved in HWC were warthog (*Phacochoerus africanus*), bush pig (*Potamochoerus larvatus*), vervet monkey (*Chlorocebus pygerythrus*), Olive baboon (*Papio anubis*), porcupine (*Hystrix cristata*), Giant mole rat (*Tachyoryctes macrocephalus*) and African civet (*Civettictis civetta*). They were involved mostly in crop raiding/ damage. Most raided crops were maize (*Zea mays*), sugar cane (*Saccharum africanum*) and Enset (*Ensete ventricosum*). Over 75% of the population of Wondo Genet district was affected by crop raiders. Therefore, conservation education is paramount, coherent land use plans should be emphasized to determine where certain crops can be grown.

Keywords: Human-Wildlife Conflict, Crop Loss, Raiding Pests

1. Introduction

1.1. Background to the Study

Human-Wildlife Conflict (HWC) or negative interaction between people and wildlife has recently become one of the fundamental aspects of wildlife management as it represents the most widespread and complex challenge currently being faced by the conservationist around the world. HWC arises mainly because of the loss, degradation and fragmentation of habitats through human activities such as, logging, animal husbandry, agricultural expansion, and developmental projects [2]. As habitat gets fragmented, the boundary for the interface between humans and wildlife increases, while the animal populations become compressed in insular refuges.

Consequently, it leads to greater contact and conflict with humans as wild animals seek to fulfill their nutritional, ecological and behavioral needs [7]. The damage to human interests caused by contact with such animals can include loss of life or injury, threats to economic security, reduced food security and livelihood opportunities. The rural communities with limited livelihood opportunities are often hardest hit by conflicts with wildlife. Without mitigating HWC the results are further impoverishment of the poor, reduced local support for conservation, and increased retaliatory killings of wildlife causing increased vulnerability of wildlife populations. Understanding the ecological and socio-economical context of the HWC is a prerequisite to bring about an efficient and long-term management of wildlife and its habitats [7]. Human wildlife conflict is one of the major threats to house hold food security and rural incomes. In Africa, the great dependence of a large proportion of the human population for their survival on land, coupled with the presence of many species of large mammals leads to many sources of conflict between people and wildlife [3].

1.2. Problem Statement

Wondo Genet college of Forestry has a national significance regarding conservation of biodiversity and endemic species. Currently, Human encroachment in terms of the density and distribution of the human population around this area is increasing and the wildlife and their habitats have faced challenging because of deforestation and agricultural expansion. According to [1], the varied natural resources of Wondo Genet district are degrading at fastest rate. Land once covered by natural forest is now converted into agricultural land and settlement. For instance in 1977, 13% of the Wondo Genet Catchment was under natural forest but in 2000 it was reduced to 2 percent. During the same period land under vegetation cover (forests, woodland and forest remnant) decreased from 36 percent to 24 percent. On the other hand, area for cultivation and settlement had increased from 55 percent in 1977 to 65 percent in 2000 [1].

The people within and around the reserve are small scale farmers who entirely depend on subsistence agriculture for their livelihoods. Preliminary survey showed that some farmers around Wondo Genet College of Forestry complain that the wild mammals such as baboon, pigs etc are damaging their crops like sugar cane, maize and other vegetable crops. Therefore, the present study is initiated to assess the various dimensions of human-wildlife conflict.

Objectives of the Study:
- To assess nature and extent of human-wildlife conflict in the study area
- To identify animal species most involved in human-wildlife conflict
- To explore major causes giving rise to HWC

1.3. Significance of the Study

The specific significance of the study was to come up with recommendations that will help prevent future HWC while ensuring sustainable conservation. The outcome of the recommendation could be used to review the current wildlife conservation policies in order to enhance its effectiveness and to formulate new policies. National's parks could also benefit by adopting measures suggested in the study. The findings are also important for decision and policy makers in providing them with greater insight on the problems that are usually associated with wildlife conservation. The area community developers can use the findings as a tool of awareness creation to the local community. Finally, the report contributes to the pool of wildlife conservation knowledge and hence is useful to the academic fraternity and those interested in wildlife conservation.

1.4. Research Questions

- What are the animals species most involved in human-wildlife conflict?
- What type of conflict is the community encounter by wild animals and to what extent?
- What are the underlying causes of the problem of HWC in the study area?

Definition of significant term

For the purpose of this study, the following terms had the attached meaning:

Human – This was taken to imply anthropogenic activities (relating to people) which include Agricultural activities, and settlement

Pest - means wildlife mammals which has caused crop raiding

Wildlife – This was constructed to imply the presence of wild animals within the context of their natural environment.

Human wildlife conflict (HWC) - was taken to imply negative results from the interaction between human and wildlife.

2. Materials and Methods

2.1. Description of the Study Area

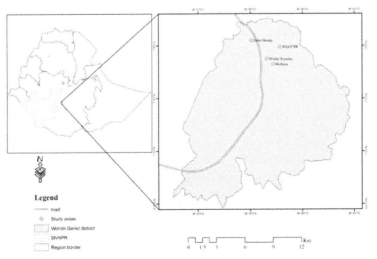

Figure 1. *Map of the study area*

Wondo Genet is located in the southeastern escarpment of the Ethiopian Great Rift Valley (7°06–07′N, 38°37′–42′E), approximately 260 km south of Addis Ababa(Figure 1). The altitude ranges from 1,800 to 2,580 m a.s.l. The average yearly rainfall is 1,210 mm, with a rainy season during March to September, and a relatively dry period from December to February. The average annual temperature is 20°C. The study area comprised 897 ha of natural and plantation forests, farmland and human settlements. The remnant forest vegetation is dry Afro-montane and is dominated by Cordia africana, Albizia gummifera, Croton macrostachyus, Ficus, Celtis africana and Milletia ferruginea [8]. Several cash crops are grown in the plantation areas, such as Saccharum, Coffea arabica and Catha edulis. Exotic plant species such as Grevillea robusta, Pinus patula, Eucalyptus and Cupressus lusitanica occupy the plantation forest.

2.2. Data Collection

2.2.1. Questionnaire Survey

Prior to data collection, extensive discussions with the key informants was undertaken to locate the sites with the highest incidences of HWC in the study area. Data was collected by employing combination of social survey methods involving participatory techniques (focal group discussions and key informant interview), structured questionnaire survey of households. Our queries was designed to solicit information such as the issues of HWC faced in the study area such as the type and number of incidences, extent of damage to wildlife and humans, attitude of humans in relation to HWC, and mitigation measures the community took towards damages by wild mammals. The questionnaire survey was carried out between December, 2013 and June, 2013 among local community in three villages of Wondo genet district; Wosha Soyama, Gotonoma and Wethera Kechema. Interviews were held with households, administrators at the local level and agriculturalists to establish in depth information about human wildlife conflict. We took approximately 10% sampling intensity in each village. These households were randomly selected by following a pattern of skipping two households, and the third house hold interviewed. The interview was conducted together with three research assistants; all were well-versed in local languages. Before initiating the fieldwork, the research assistants were trained to administer the survey and the questionnaire. Interview was in the form of a conversation, structured around a written questionnaire consisting of both fixed-response and open-ended questions. Respondents were asked whether conserving wildlife is important or not. Then their answer was considered as positive if they replied as conserving wildlife is important. However, if they responded as conserving wildlife is not important; their answer was considered as negative response. The positive attitude represents the respondents' good will to protect and utilize the local wildlife wisely whereas the respondents' negative attitude towards wildlife represent unwilling to utilize their natural resource in a wise way rather wish to destroy it. This approach mainly collected primary data.

2.2.2. Secondary Data

Secondary data on the other hand was sought from previous studies carried out on human-wildlife conflicts at global, regional and local levels. Such information was obtained from published reports such as journals, thesis and relevant documentation and the internet.

2.2.3. Observation / Use of Quadrants

For crop damage assessments, field visits and observations were mainly used to confirm the respondents" responses so that accurate and reliable information would be collected since most farmers have a tendency of exaggerating the problem. Observation was also important in identifying the particular problem animal species responsible for the damage through assessing the teeth marks left on the damaged plants and foot marks of the animals. Three Quadrants of 4m by 4m were placed randomly within the crop stand of 10 farmers in each village and observed two times a week to check if there was crop damage and to identify the type of animal that caused the damage. The proportion of damaged crops was derived from calculating the number of damaged or missing plants or plant parts, divided by the total crop population planted in the farm land. The mean of the three quadrant values for each damaged stand is a measure of the proportion of crop damage sustained in any one sample. The mean percentage crop losses for each farm, taking into account the number of stands planted was estimated, of each crop and the proportion of stands that sustained crop damage.

2.3. Data Analysis Techniques

First data were organized into different topics by following the objectives of the study and coded according to the topics already described.

Various techniques were used for the analyses and presentation of data. These include both quantitative and qualitative techniques. In quantitative technique, the analyses are to be characterized by the use of statistical package for social science, proportions, percentages and averages to arrive at a general picture for the generation of conclusion. Qualitative data from questionnaires as well as interviews were analyzed thematically. Qualitative techniques on the other hand were employed in the computation of statistical tables and bar graphs.

3. Results and Discussions

3.1. Demographic Information

The researcher begun by a general analysis on the demographic data got from the respondents which included; - the gender, age and educational level.

3.1.1. Gender of Respondents

Table 1: indicated the gender of the respondents. From the findings, it was indicated that 66.7% were male and33.3% of the respondents were female.

Table 1. Gender of respondents

Gender	Frequency	Percent
female	40	33.3
male	80	66.7
Total	120	100

3.1.2. Age of Respondents

The study also sought to establish the respondent's age group. From the findings, the majority of the respondents in the age bracket of 15 – 35years were shown by 46.7%, 30.8% of the respondents were between 36 - 50 years, 12.5% of the respondents were between 51 - 65years and 10% of the respondents were above 65 years.

Table 2. Age of the respondents

Age	Frequency	Percent
15-35 years	56	46.7
36-50 years	37	30.8
51-65 years	15	12.5
above 65 years	12	10
Total	120	100

3.1.3. Education Level

The study sought to know the level of education of the respondents. From the findings, 73.3% of the respondents were able to write and read, 10% of the respondents indicated that they had reached secondary level, 16.7% of the respondents were uneducated.

Table 3. Education Level of respondents

	Frequency	Percentage
Reading & writing	88	73.3
Secondary	12	10
Uneducated	20	16.7
Total	120	100

3.2. Wildlife Conflicts

Table 4. Percentage of respondents encountered by wildlife conflicts

	Frequency	Percent
Yes	120	100
No	0	0
Total	120	100

All data regarding crop and livestock loss, animal behavior (preference, frequency, activity etc), raiding intensity, human threat and property destruction represents perceived data; this information was provided by farmers from each site, though actual crop losses were observed on farms through quadrant sampling, as a means of cross-checking. All crop loss reported refers to loss in the presence of guarding. Without guarding, farmers reported 100% loss to raiding pests.

The study sought to establish whether the locals encountered any conflicts with wild animals.

From the findings, all the respondents indicated that they encountered conflicts with wild animals as shown by 100%.

3.2.1. Type of Conflict

The study sought to establish the type of conflicts that the locals encounter with wild animals.

Table 5. Type of conflict the community encountered by wild animals

	Frequency	Percentage
Crop damage	90	75
Destruction of property	15	12.5
Human threat	5	4.2
Livestock killing	10	8.3
Total	120	100

From the findings, 75% of the respondents indicated that they encountered crop damage, 12.5% of the respondents indicated that they encountered destruction of property, 4.2% of the respondents indicated that they encountered human threat, 10% of the respondents indicated that they encountered livestock killing

3.2.2. Wild Animals Involved in Crop Damage

Wild animals that caused greater damage to crops damage were warthog (Phacochoerus africanus), bush pig (Potamochoerus larvatus), vervet monkey (Chlorocebus pygerythrus), Olive baboon (Papio anubis), porcupine (Hystrix cristata), African civet (Civetictis civetta) and Giant mole rat (Tachyoryctes macrocephalus).

3.2.3. Crop Loss Assessment

Table 6. Sugarcane damaged/0.1ha by village and species

Species	Average Sugarcane stalk/0.1ha			Average Sugarcane stalk damage/0.1ha		
	Gotu	W/Soyama	Wethera	Gotu	W/Soyama	Wethera
baboon	4375.00	6200.00	5000.00	600	500	600
Warthog	4375.00	6200.00	5000.00	400	NR	500
Bush pig	4375.00	6200.00	5000.00	300	NR	400
Porcupine	4375.00	6200.00	5000.00	250	650	250
Vervet Monkey	4375.00	6200.00	5000.00	100	400	150
Total	4375.00	6200.00	5000.00	1650	1550	1900
Total loss (%)				37.67	25.0	38

NR- not reported, W/Soyama-Wosha Soyama

The extent of crop damage varied depending upon the villages and the type of animal that actually caused the crop damage. Wild mammals damage to sugarcane (Saccharum africanum) plantation accounted 1650(37.67%), 1550(25%) and 1900(38%) in Gotu, Wosha Soyoma and Wethera kechema villages respectively (Table 6). Warthog damage to sugarcane plantation was not only through consuming sugarcane but also through digging burrow to get a place to hide itself from predators, and the cane guards. Warthogs damage sugarcane plantation at the roots and base of the sugarcane stalk which was supposed to give more yields per unit area.

Wild mammals damage to maize(Zea mays) accounted 200kg, 110kg and 140 kg for Gotu, Wosha Soyoma and Wethera Kechema villages respectively (Table 7) while Enset(Ensete ventricosum) loss accounted 160kg, 250kg and 208kg for Gotu, Wosha Soyoma and Wethera Kechema villages respectively (Table 8). The present study also indicated that there was no significant difference between villages for sugar cane and Enset loss (p>0.05) while significant difference was observed between villages for maize loss.

The competitions for resources cause conflict between wild animals and people. Wild animal population is increasing and at the same time human populations expand year after year, which resulted in competitions for resources

between wild animals and human populations [4]. In the study area, the natural habitats of the animals were modified into crop cultivation like sugar cane, maize and Enset and settlement. As a result, animals in the surrounding area enter the sugarcane, maize, Enset plantation and caused damage. Furthermore, [6] reported that agricultural crops such as maize, sugarcane and sorghum which grow over two meters conceal larger animals such as Bush pig, warthog and primates.

Similarly, the current finding showed that the sugarcane plant was observed concealing baboon, vervet monkey and other large mammals that enter the plantation fields. As a result, the sugarcane was exposed to damage because the animals were not easily observed by the cane guards. Especially this enables warthog to dig burrows in the sugarcane plantation fields and hide itself against the cane guards.

Table 7. Maize loss (kg)/0.1ha by village and species

Species	Average maize yield (kg)/0.1ha			Average maize loss(kg)/0.1ha		
	Gotu	W/Soyama	Wethera	Gotu	W/Soyama	Wethera
baboon	700	400	375	60	40	40
Warthog	700	400	375	50	NR	30
Bush pig	700	400	375	35	NR	10
Porcupine	700	400	375	30	35	30
Vervet Monkey	700	400	375	25	30	25
African civet	700	400	375	NR	5	5
Total(kg/0.1ha)	700	400	375	200	110	140
Total loss (%)				28.5	27.5	37.3

NR- not reported, W/Soyama-Wosha Soyama

The crop damage by baboon and Vervet monkey was due to the social organization and intelligence of the animal to recognize the absence of cane guards and then immediately rushes into the plantation fields forming different groups in different directions. This kind of social organization of vervet makes the damage incidence high because it is difficult to chase them away since they come to the plantation fields in

different directions in large numbers.

Similar result by [6] reported that in Uganda primates are dominant pests and responsible for over 70% of the damage events and 50% of the area damaged due to their intelligence, adaptability, wide dietary range, complex social organization and manipulative abilities.

Table 8. Enset loss (kg)/0.1ha by village and species

Species	Average Enset yield(kg)/0.1ha			Average Enset damage(kg)/0.1ha		
	Gotu	W/Soyama	Wethera	Gotu	W/Soyama	Wethera
baboon	400	600	450	60	90	60
Warthog	400	NR	450	40	NR	52
Bush pig	400	NR	450	30	NR	41
Porcupine	400	600	450	20	85	30
Vervet Monkey	400	600	450	10	75	25
Total	400	600	450	160	250	208
Total loss (%)				40.0	41.7	46.2

NR- not reported, W/Soyama-Wosha Soyama

According to [4], farms located within 300 meter of a forested boundary probably are exposed more for crop-raiding by Vervet monkeys. In the study area, Vervet monkey and Baboon were the top worst pests in Wosha Soyoma and Wethera villages this is because the plantation fields were in areas where there are plenty of trees which support Vervet

monkeys by providing shelter to escape from the crop guards. Further more in the study area, the people in the area do not have a trend to hunt vervet monkeys except chasing them away from their crops. This is because the people in Wondo Genet area neither eat Vervet monkey meat nor use their hide for different purposes. This has increased the crop damage in the area. The presence of large trees has helped Vervet monkeys as an escaping site from guards. The respondents during interview said that Vervet monkeys cut the sugarcane stalk around the middle and chew it like humans. They also damage the stalk while they jump from one cane to another. According [3], crop losses by wild animals can be enormous both in the direct economic terms and through indirect costs on time and energy devoted to protection of crop damage. In the present study area, the farmers were observed spending considerable amount of time to prevent the crops damage both during the day and night time.

Table 9 summarizes the profile given by each village for all pests ranked in the top six. Ranking considered the combined affect of each animal on crop loss in both villages. Ranking identifies pest severity with (1) indicating the worst pest. Gotu ranked the baboon, Warthog, bush pig and porcupine in the same order as the top 4 worst pests in its site while Wosha Soyama ranked Mole rat, Vervet monkey, baboon and porcupine as the top 4 worst pests. Furthermore, in Wethera kechema village the top four worst wild pests in descending order were Warthog, porcupine & baboon, Vervet monkey and bush pig, Mole rat.

3.2.4. Pest Profiles

Table 9. Animal profiles listed in order of pest severity

Animal	Site	Site rank	Crops-age/part	Time of day	Frequency
Bush pig	Gotu	3	maize- all parts, sugar cane	Night	1X/weak
	Wethera	3	maize- all parts root crops like Enset false banana, sugar cane	Night*2	Every 3 days
Vervet monkey	Gotu	5			
	W/soyama	2			
	Wethera	3	maize- cob, new shoots, planted seeds, sugar cane, fruit- banana, mango, papaya	5am(before humans wake) to 6pm; Present all day.	Every day
Olive baboon	Gotu	1			
	W/soyama	3			
	Wethera	2			
Warthog	Gotu	2	maize- all parts, sugar cane	Night	2X/weak
	Wethera	1	maize- all parts, sugar cane	Night*2	Every 2 days
Porcupine	Gotu	4			
	W/soyama	2	maize- cob only, all tubers, roots of all crops, bananas		
	Wethera	2		Night	Every night
Mole rat	Gotu	6			
	W/soyama	1	root and tuber crops		
	Wethera	4			

*2 depends- may start to come as early as 3pm or 5pm, W/Soyama-Wosha Soyama

3.2.5. Dominant Foragers by Crop and Village

Dominant foragers on each crop can be identified in Figure 2, 3 and 4. It summarizes the loss of sugar cane, maize and Enset crops for each village by crop and village types. In Gotu village sugar cane, maize and Enset were primarily dominated by the baboon followed by Warthog and bush pig while in Wethera Kechema and Wosha Soyoma villages baboon followed by Porcupine and warthog, vervet monkey, bush pig and African civet were dominant foragers for maize crop. For Wosha Soyama village the Mole rat was the worst pest for both sugar cane and Enset crops followed by Vervet monkey and porcupine.

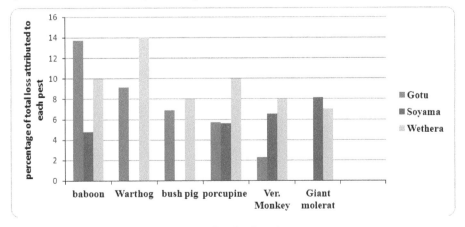

Figure 2. Loss Attributed to Pests for sugar cane

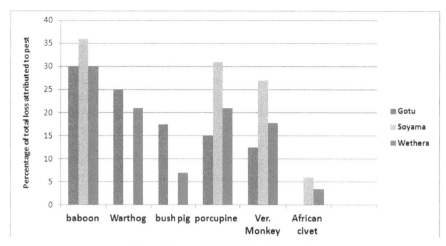

Figure 3. Loss Attributed to Pests for maize

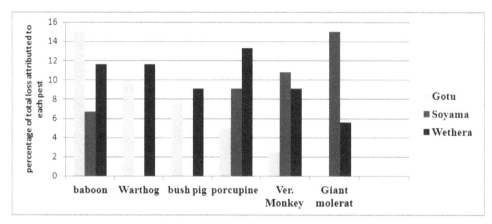

Figure 4. Loss Attributed to Pests for Enset

3.2.6. Livestock Depredation

In the study area, when asked about perception of people on livestock depredation, a very large number (90 %) of respondents said that livestock loss by carnivores was not as such a big problem. Hyena seemed to be a threat to livestock in the study area. White tailed Mongoose considered the leading predators to poultry production. 20% of the households were experiencing conflict with White tailed Mongoose and 6 chicken losses per house hold were observed in the study area.

3.2.7. Household's Opinion on Causes and Workable Solutions to Human-Wildlife Conflicts

Selected respondents were also asked about their views on the causes of human-wildlife conflict, possible solutions to this problem, and reasons why wildlife should be protected. Their responses are tabulated in Table 10 and 11 below, which showed a need to educate farmers about the possible solutions to the problem of human-wildlife conflict, as well as potential benefits of effective wildlife conservation measures in the study area.

Table 10. Surveyed Household's opinions on the causes of human-wildlife conflict

In the HH's opinion, what are the causes of human-wildlife conflict?	Number of Responses	Percentage of Respondents
Don't know	50	41.7
Human moved into animal habitat	15	12.5
Lack of food	25	20.8
Too many wild animals	30	25
Total	120	100

From the Table 10, above, it can be seen that the majority of surveyed respondents, 41.7 percent, responded that they did not know the basic causes of human-wildlife conflict, while 20.8 percent felt the cause was a lack of food for wild animals, and a further 25 percent felt the problem was simply that there were too many wild mammals, both basic misconceptions. Thus, in total, 87.5 percent of households

were unaware of actual root causes of human-wildlife conflict, such as humans occupying wildlife habitat, only 12.5 percent of survey respondents were able to cite accurate.

When asked to comment on a series of possible solutions to human-wildlife conflict, the following responses were given (Table 11):

Table 11. *Surveyed Household's opinions on workable solutions to human-wildlife conflict*

In the HH's opinion, which of the following are workable solutions to human-wildlife conflict?	Number of Responses	Percentage of Respondents
Don't know	40	33.3
Kill wild animals	45	37.5
Moving to a different place	5	4.2
Guarding	30	25
Total	120	100

As with the previous question concerning causes of wildlife conflict, the surveyed HHS most common response when asked about possible solutions to the human wildlife conflict was simply to kill problem wild animals, accounting for 37.5 percent of the total which is the worst solution and against wildlife conservation. The next popular solution expressed by 33.3 percent of survey respondents was that I don't know and the next 25% of the respondents felt that guarding was the appropriate solution against wild pests. Just 4.2 percent of survey respondents felt moving wildlife to a different place indicating their negative attitude towards wildlife conservation. Thus from the largely uninformed responses to the above two survey questions it can easily be concluded that there is an immediate need for a sweeping wildlife conservation education program to educate farmers living in Wondo Genet district about the purpose and benefits of wildlife conservation, the causes of human-wildlife conflict, and methods for reducing or eliminating various forms of this conflict.

3.2.8. Underlying Causes of Crop Raiding

Research findings showed that proximity to the forest reserve around Wondo Genet College of Forestry and Natural Resources is the main reason leading to frequent crop raiding by wild animals, followed by increased habitat destruction, high population, lack of grazing land and inadequate help from government institutions are the most underlying causes of crop raiding in Wondo Genet district. In Wondo Genet district the rapid population growth has changed the life style resulting into destruction of habitats through encroachment for agriculture and human settlement. Habitats destruction is also through fragmentation of natural habitats, killing of animal species, cultivation and settlement near the forest reserve around Wondo Genet College of Forestry and Natural Resources. This has resulted in human-wildlife conflicts around Wondo Genet. This was in harmony with [5] who indicated that most natural wildlife buffer zones have led to competition for food, water, habitats, and space for both humans and wildlife hence resulting in a conflict for survival.

3.2.8.1. Guarding Strategies

Table 12 summarizes descriptions of guarding strategies used in the study areas. Efficacy within site is a ranking value, with (1) being the most effective strategy. Guarding in field was indicated to be the primary and most effective means of guarding against pests.

Table 12. *Profile of Guarding Strategies*

Preventive method	Description	Target animal	Efficiency
Guarding in field	at least one person is in the field guarding for 24 hours (day/night)	any	1
Perfume/ soaps	Place perfume and soaps in a plastic bag and hang from a stick. Mix soaps with water and spray around crops	Porcupine, Pig, night pests	Supplementary
Smoking	Burn wood and smoke around the crop	Any	Supplementary
Goat urine/ water	Goat urine /water is inserted in to the hole of pest	Mongoose	1

3.2.8.2. Guarding in Field

Scare tactic

In all study areas, a daytime pest is scared away from the plot by silently approaching the foraging animal and, when in close proximity, throwing objects, such as stones, sticks and spears at the pest. Women and children scare away an invading pest by shouting at the animal.

Evening in-field guarding is only carried out by men in all sites. Evening raiding pests are detected by sounds made while foraging. Shouting is mainly used to deter evening raiders due to poor visibility.

Problems encountered and difficulties with in-field guarding

In the study areas, guarding against baboons was listed as a problem encountered while guarding. Each site reported that guarding against baboons was unmanageable for women and children because the baboons are not afraid of them and therefore continue to raid. Farmers described accounts of baboons dispersing themselves around plots followed by one or two baboons darting through the field to attract the farmers' attention. While the farmers are busy chasing the invading baboon, the other baboons dispersed around the boarder raid the crops.

Other difficulties listed include: 1) difficulty to guard during heavy rainfall and during illness; 2) pests adapt to new strategies in guarding within one week; 3) almost all of the strategies used (supplemental and primary) are ineffective in deterring pests; 4) When an animal is scared away with the "approach and scar" method, it merely moves to another spot in the same plot and continues to forage.

Perfume/ soaps

Small Perfume/ soaps were placed in a plastic bag and hang from a stick. The smell originating from the Perfume/ soaps gave the false feeling of the presence of human beings in the field which acted as a deterrent to the wild pig and porcupine. This method was only effective for few days.

3.2.9. Human Attacks

The present study indicated that spotted hyenas (Crocuta crocuta) was considered to threaten human beings although human deaths and injuries were less common in the study area. The respondents also indicated that people were fearful to spotted hyenas and other carnivores when they slept in makeshift huts in the field to prevent animals from damaging standing crops, and walking at night.

3.2.10. Bark Stripping

From the findings olive baboons were responsible for stripping bark from trees. Baboons raided for the inner bark of Avocado, Eucalyptus, Acacia and Cupressus. They bit into the bark, lifting and pulling it from the tree.

3.3. Human-Wildlife Conflict in WGCFNR

The study revealed that wild mammals caused human threat, destruction of property and taking food from kitchens both in home and staff lunch at Wondo Genet College (Table, 14). A majority of the residents of the Wondo Genet College of Forestry and Natural resources have expressed strong dissatisfaction over the presence of the baboons and monkey population in the campus. Due to their intolerable activities, people now view them as a vermin species rather than a species of conservation importance. The residents were even more apprehensive of increased degree of the conflict in the coming years as they speculate that the baboons and monkeys' population will not decrease. The study sought to know the waiting time of respondents within the campus.

Table 13. Number of years of residents within the College

Waiting in the campus	Frequency	Percent
1-2 years	5	16.7
3-5 years	10	33.3
>5 years	15	50
Total	30	100

3.3.1. Type of Conflict

The study sought to establish the type of conflicts that the residents encountered with wild animals.

Table 14. Type of conflict the residents of Wondo Genet College encountered by wild mammals

Type of conflict	Frequency	Percentage
Snatch food items from kitchens, tables &people's hands	20	63.3
Destruction of property	6	20
Human disruption	3	10
Human threat when walking night time	2	6.7
Total	30	100

From the findings, 63.3% of the residents indicated that they encountered with snatching of food items by baboons and monkeys, 20% of the respondents indicated that they encountered destruction of property, 10% of the respondents indicated that they encountered human disruption especially at the morning when they were in sleeping, 10% of the respondents indicated that they encountered threats with nocturnal carnivores such as spotted hyenas when walking night time from their office to home.

3.3.2. Causes of Human-Wildlife Conflict in Wondo Genet College

1) Habitat destruction

Extensive cutting of forest trees and plantation of exotic tree species in place of natural food plants forced the monkeys and baboons to invade the staff lunch & home of residents in Wondo Genet College which provided a wide range of food items for them.

2) Food provisioning by the residents

Beside habitat destruction, food provisioning by local residents and students of the campus was also another cause of conflict. Offering of eatables whenever the monkeys/baboons visit the area attracts most of the monkeys in the area thus increasing the conflict.

3) Improper waste disposal

The improper disposal of wastes also account for the prevailing human-monkey conflict in the study area. Careless dumping of kitchen wastes and garbage in the open areas provides easy food for the monkeys, which resulted in their frequent visits to the campus premises.

3.3.3. Damages at Stuff Lunch

Field observation and survey of selected stuff responses showed that Vervet monkey and Olive baboons did damage in stuff lunch. Most type of damages included destruction of property and taking food from kitchens, tables and out of people's hands. Development has greatly reduced the number of food tree species in the area. This is coupled with an increase in the birth rate compared to truly wild troops as human food and rubbish is readily available which requires considerably less expenditure of energy to access and the foods have a higher caloric value than wild forage. Vervet monkey behavior is also enhanced through the enticement of monkeys to approach people for photos through the offering of food.

3.3.4. Damages at Stuff Home

The present study showed that the stuffs, who lived in Wondo Genet College, suffered from vervet monkeys and baboon because of their damage to properties, food from kitchen and disruption when they jump on roof of the house. Especially early in the morning they run here and there on the roof of house causing disturbance for sleeping. They also climbed on wire/rope where washed clothes were mounted for drying and made it to drop on the ground. In homes, typically food availability from open doors leading to the kitchen was the leading issue encouraging vervet monkey and baboon pest behaviors. Hence, windows and doors were closed the whole day which leads to insufficient fresh air to the family inside the house. Fungus had got opportunity to multiply on the roof of the house because of insufficient ventilation and this was a real health problem for many stuffs living there

4. Conclusion and Recommendation

This report provides strong evidence that in Wondo Genet district where the study was conducted, farmers perceived crop damage by wild animals as a great hindrance to their agricultural development, and crop losses varied from farmer to farmer depending on the amount of time invested to guard the fields. Guarding in the field was indicated to be the primary and most effective means of guarding against pests. During guarding the aim was to kill the animal using stone or other harmful instruments. This indicated that there is an immediate need for a sweeping wildlife conservation education program to educate farmers living in Wondo Genet district about the purpose and benefits of wildlife conservation, the causes of human-wildlife conflict, and methods for reducing or eliminating various forms of this conflict. The top 6 animals responsible for the most damage to crops were determined to be baboons, Warthog, bush pig, Vervet monkeys, Porcupine and Mole rat. The impact of the baboon followed by Warthog, bush pig and porcupine was strongest in Gotu and Wethera villages, while crop loss in Wosha Soyama was largely due to Mole rat followed by Vervet monkey and baboon. The study further revealed that Vervet monkeys and Olive baboons caused human disruption, destruction of property and taking food from kitchens both in home and staff lunch at Wondo Genet College of Forestry and Natural Resources.

Therefore, human-Wild life Conflict issues must be treated with concern, and placed in the context of local community and individual needs, as well as conservation objectives and those of the government and industry involved. Measures which might seem to be appropriate strategy to researchers might not necessarily be acceptable and practical to community or individual farmers. To establish measures which are sustainable and efficient may not be an overnight event, requiring adoption of a series of strategies. Interventions that can solve one type of conflict might not be applicable to others.

Intervention methods are therefore likely to be more successful if they are financially and technologically within the capacity of the people, organizations, institutions or bodies who will implement them. Farmers need to take responsibility for protecting their own crops, which requires assisting them to develop locally-appropriate schemes to successfully reduce loss.

Based on the obtained results of the present study and reviews of previous works, the following points are recommended for the rural villages of Wondo Genet district:

- Planting unpalatable plants such as sisal or hot pepper spray on the boarder line of the crops plantation will help minimize the animals that visit the plantation fields.
- Wondo Genet College of Forestry along with the community should work against deforestation because with increase in deforestation, vegetation in croplands will presumably become a major food resource for foraging animals.

- School of wildlife and Ecotourism in collaboration with other concerned sectors should educate the community around Wondo Genet regarding:
  - the purpose and benefits of wildlife conservation,
  - the causes of human-wildlife conflict, and
  - Methods for reducing or eliminating various forms of this conflict.

For Vervet monkey and baboon management in Wondo Genet College of Forestry and Natural resources, both short-term and long-term measures can be adopted to control the man-monkey/baboon conflict in the campus.

Short-term measures
- All catering staff should be aware that it is also their duty to deter by approved humane methods any monkey seen in the food preparation, serving and eating areas
- Waste from kitchens awaiting removal to external rubbish sites must be stored in monkey proof bins. This is also desirable to maintain proper kitchen hygiene.
- Meals, drinks and snacks served outside must be cleared away immediately after the guest has finished.
- Guests should not, under any circumstance, feed monkeys either directly by hand or by throwing food in their direction.
- Staff and guests must be discouraged from approaching monkeys.

Long-term measures
Long- term measures aim at removing the factors responsible for the monkey/baboon depredation and at creating ideal living conditions for the monkeys/baboons within the forests viz:
  1) Ban on illegal encroachment of the forest lands.
  2) Extensive cutting of trees and plantation of exotic tree species must be minimized.

Precautionary Measures
Besides these recommendations some precautions can also be adopted to minimize direct encounter with the monkeys/baboons.
- People must not let their children go out to play in the open if the monkeys are nearby.
- It is advisable to keep their doors and windows shut properly whenever the monkeys visit their localities.
- People should be warned not to cause any injury to an infant monkey, as mother monkeys are very aggressive.

Acknowledgments

I am very grateful to Research and Development office of Wondo Genet College of Forestry and Natural Resources for the financial support offered for this study. I would also like to acknowledge individual respondents from around Wondo Genet district for their patience and time they sacrificed, without them this research would have been short of what it contains. Last but not least, I am grateful to my research assistants, Zewuditu Zeleke and Fanaye Werara for the commitment shown during data collection for this study. Thank you all.

Acronyms

Ha	Hectare
HWC	Human-Wildlife conflict
SNNPR	Southern Nation Nationality Peoples Region
WGCFNR	Wondo Genet College of Forestry and Natural Resources

References

[1]   Belaynesh Zewdie. 2002. Perceptions of Forest Resource Changes in and around Wondo Genet Catchment and Its near Future Impacts, unpublished MSc Thesis, Wondo Genet College of Forestry. Stakeholder attitudes in Virginia. Wildlife Society Bulletin 30:139-147.

[2]   Fernando, P., E. Wikramanayake, D. Weerakoon, L. K. A. Jayasinghe, M. Gunawardene, and H. K. Janaka. 2005. Perceptions and patterns of human-elephant conflict in old and new settlements in Sri Lanka: insights for mitigation and management. Biodiversity and Conservation 14:2465-2481.

[3]   Hill, C. M. 1998. Conflicting Attitudes Towards Elephants Around the Budungo Forest Reserve. Environ. Conser. 25: 244-250.

[4]   Hill, C.M. 2000. Conflict of interest between people and baboons: Crop raiding in

[5]   Kagiri, J.W (2000) —Human –Wildlife conflicts in Kenya: A conflict Resolution conceptl. Farmers Perspective 43-45 6. Lahm, S. 1996. A Nation Wide Survey of Crop Raiding by Elephants and other Species in Gabon. Pachyderm, 21: 69-77.

[6]   Naughton-Treves, L. 1998. Predicting Patterns of Crop Damage by Wildlife around Kibale National Park, Uganda. Conserv. Biol. 12: 156-168.

[7]   Sukumar, R. 1990. Ecology of the Asian elephant in southern India. II. Feeding habits and crop raiding patterns. Journal of Tropical Ecology 6:33-53

[8]   Yirdaw, E. 2002. Restoration of the native wood species diversity, using plantation species as faster trees, in the degraded highlands of Ethiopia. University of Helsinki (Ph.D. thesis), Helsinki, Finland.

15

Characteristics of indigenous mycorrhiza of weeds on marginal dry land in south Konawe, Indonesia

Halim[1], Fransiscus S. Rembon[2], Aminuddin Mane Kandari[3], Resman[4], Asrul Sani[5]

[1]Specifications Weed Science, Department of Agrotechnology, Faculty of Agriculture, Halu Oleo University, Southeast Sulawesi, Indonesia
[2]Specifications Soil Nutrition, Department of Agrotechnology, Faculty of Agriculture, Halu Oleo University, Southeast Sulawesi, Indonesia
[3]Specifications Agroclimatology, Department of Agrotechnology, Faculty of Agriculture, Halu Oleo University, Southeast Sulawesi Indonesia
[4]Specifications Soil Science, Department of Agrotechnology, Faculty of Agriculture, Halu Oleo University, Southeast Sulawesi, Indonesia
[5]Specifications Biomathematics, Department of Mathematics, Faculty of Sciences, Halu Oleo University, Southeast Sulawesi, Indonesia

Email address:
haliwu_lim73@yahoo.co.id (Halim), fransrembon@yahoo.com (F. S. Rembon), manekandaria@yahoo.com (A .M. Kandari),
resman_pedologi@yahoo.com (Resman), saniasrul2001@yahoo.com (A. Sani)

Abstract: South Konawe is one of the areas that have the potential for the development of marginal farming dry land, which is wide enough, with a predominance of Ultisol type. In such area, more than 80% of farming communities who are dependent on the farming activities are still conventional to characterize the shifting cultivation. In many cases, most weeds that grow in their land are always considered to be destructing and disturbing the human interests, both during the land clearing and after the fields abandoned. On the other hand, the presence of weeds can be useful for the growth of plant as it provides benefits against microorganisms. One of the microorganisms which is associated with roots of weed is mycorrhiza. This study aims to determine the characteristics of indigenous mycorrhiza being present on dry weeds from marginal land. This study was conducted from May to November 2013 in South Konawe, Indonesia. The result shows that two types of indigenous mycorrhiza were present on the marginal dry land; *Glomus* sp and *Gigaspora* sp. The highest percentage of indigenous mycorrhiza infection was found in the roots of weeds *Amaranthus gracilis* and *Sida rhombifolia*, each of which by 90%. The presence of the vesicles and internal hyphae on the roots of weeds indicate the indigenous mycorrhiza infection.

Keywords: Marginal Dry Land, Ultisols, Indigenous Mycorrhiza, Weeds

1. Introduction

The presence of weeds in crop acreage greatly affects all aspects of growth and yield. This happens because the weeds have a high ability to compete for water, nutrients, sunlight and CO2. Thus, the weeds that grow in the area of the plant must be controlled before incurring losses. On the other hand, the presence of weeds can provide benefits for the life of the soil microorganisms. One of microorganisms associated with roots of weeds is mycorrhiza fungi, which are generally found associated with weed species about 80% -90% [3]; [15]; [19]), and even 90% - 95%, spread across the Arctic to the tropics and from the desert to the forest area [24]; [8]). Mycorrhiza spread almost all over the earth's surface and can be associated with most of the weeds. Approximately 83%

dikotiledon, 79% monokotiledon and all gymnosperms studied was infected by mycorrhiza [25].

The types of weed that are found in association with mycorrhizal include *Imperata cylindrica, Cyperus rotundus, Eupatorium odorata, Ageratum conyzoides, Amaranthus spinosus, Cleome rutidosperma, Euphorbia hirta, Dactyloctenium aegyptium, Digitaria ciliaris, Heliotropium indicum,* and *Scoparia dulcis* [8]; [11]. The type of weeds that are infected by mycorrhiza has a very rapid growth, making it possible to be used as a propagation medium mycorrhiza [10]. Weed growth is very rapid in nature even though the marginal lands. Some research shows that the growth of weeds were allegedly due to mutualism between mycorrhiza associated with the roots of weeds. The relationship between mycorrhiza with weed roots lasted from weed seeds form sprouts. This is in accordance with [26], the relationship causes the weeds easily absorb nutrients, while mycorrhiza is able to take advantage of root exudates of weeds as a source of carbon and energy.

The use of weed as a propagation medium mycorrhiza, because weeds have high adaptability to marginal lands [10]. With the mycorrhiza, weeds can absorb water and nutrients (especially P) optimally. Moreover, mycorrhizae can improve the formation and spread of the roots of weeds through external hyphae which resulted in an increased uptake of other nutrients by plants and weeds [14]. [12] reported that the *Eupatorium* odorata L. and *Imperata cylindrica* (L.) Beauv found the types of mycorrhiza like *Acaulospora* sp, sp *Gigaspora* sp and *Glomus* sp with spore number density of each 883 spores, 667 spores and 994 spores in the 250g of soil sample. However, if the weeds were inoculated on maize plant roots, the weeds around the maize plant were infected by mycorrhiza [11]. The types of weeds infected by mycorrhiza are *Ageratum conyzoides* (L.), *Amaranthus gracilis* Desf, Borreria alata (Aubl.) DC, *Centrosema plumieri* (Pers) Beath, *Mimosa invisa* Mart.ex. Colla, and *Digitaria adscendes* (HBK) Henr [11]. This shows that the weed has the potential to be used as a medium for the growth of mycorrhizae, especially for long-term goals that will support sustainable agriculture by utilizing local resources. This local resource, if used optimally, is able to restore the health of the soil and increase the strength of soil biological power, all of which could potentially improve the welfare of farmers. In this study, the characteristics of indigenous mycorrhiza being present on weeds from marginal dry land were examined.

2. Methodology

2.1. Exploration Mycorrhiza

Exploration of indigenous mycorrhiza of dry weed was carried out on marginal land, which is dominated by weeds classified as a secondary vegetation. The sampling method used is a nested plot technique as a minimum model for taking samples of weed species which have been determined [20]. The identification process of the weeds of indigenous mycorrhiza was conducted in the Laboratory of Forestry, Faculty of Agriculture, Halu Oleo University, Kendari, Indonesia. Propagation of indigenous mycorrhiza propagules was held in a plastic house Sindang Kasih village, district of West Ranomeeto, Southeast Sulawesi Province. The materials used in this study were *raffia* ropes, water, soil, the roots of weed, corn seed, polybag (size 10cm x 20cm), 30% sucrose, Acero Formalin Alcohol (FAA), 10% KOH solution, a solution of hydrogen 10% alkaline peroxide (H_2O_2), a solution of HCl 1%, dyes carbol fuchin 0.05%, laktogliserol, filter paper and paper labels. The tools used were tillage tools, machetes, meter, digital cameras, filters to see mycorrhizal spore size (mesh size of 500lm, 250lm, 90lm, 60lm, and

50lm), analytical balance, autoclave, microscope, glass measuring, petridish, pipettes, scissors, and stationery.

2.2. Identification of Indigenous Mycorrhiza

Identification of indigenous mycorrhiza was performed by using wet sieving method on soil taken from around the roots of weeds to observe the types of indigenous mycorrhiza spores. The process of getting indigenous mycorrhiza spores weeds by wet screening was as follows (1) 250gram of soil taken from the field was mixed with 500ml of water, (2) pour the liquid portion passes through a sieve with a mesh size of 500lm and then collect suspension that passes through the filter, (3) The suspension obtained in step 2 was filtered with a sieve mesh size of 250 lm, 90 lm and 60 lm, (4) suspend the pellets are retained on the filter in 30% sucrose, and then centrifuged for 1 min at 2000 rpm. Spores in sucrose supernatant were poured into a sieve with a mesh size of 50 lm and washed with water to remove sucrose, (5) the spore was observed with a microscope, and (6) other similar spores collected to make a pot culture for propagation of mycorrhiza [5].

2.3. Propagation of Indigenous Mycorrhiza on Weeds

The types of weed that produce seeds were planted with two seeds per polybag. In detail the propagation of mycorrhiza were as follows: (1) taking soil samples from the field on weed rhizosphere region as growing media, (2) sterilizing weed's seeds with a solution of FAA, (3) putting mycorrhiza propagules into a polybag with the planting hole of around 5cm in depth, (4) planting the weeds, (5) watering the plant, (6) allowing the plant to grow until the preferred age, (7) cutting the top of the weeds, (8) storing the rest of the weed roots in plastic bags that have been labeled, (9) staining roots, and (10) observing them by using a microscope [11].

2.4 Staining Roots

The steps in staining roots are as follows: (1) washing the roots with water, (2) saving the FAA for fixation prior to painting, (3) soaking in 10% KOH and heat with an autoclave for 15 -20 minutes at 121^0C, (4) washing with distilled water 3 times, (5) soaking in hydrogen peroxide outsmart 10% (H_2O_2), (6) washing with distilled water 3 times, (7) soaking with HCl 1%, (8) wasting HCl without washed with distilled water, (9) soaking in carbon fuchin with concentration of 0.05% w/v in laktogliserol and heat at 900C for several hours or in an autoclave at 1210C for 15 minutes, (10) removing the paint and soak the roots in laktogliserol, and (11) observing the roots sample using a microscope [5].

2.5. Observed Variables

The variables measured in this study are as follows: the types of mycorrhiza, the percentage of mycorrhizal infection on the weed's roots, and the characteristics of mycorrhiza infection on the weed's roots.

3. Result and Discussion

3.1. Identification of Indigenous Mycorrhiza of Weeds

Based on the identification of indigenous mycorrhiza of weeds among the sites, two types of mycorrhiza were obtained, namely *Glomus* sp and *Gigaspora* sp. The characteristics of the indigenous mycorrhiza found in this study were listed in Table 1.

Based on the observation of the form of mycorrhiza spores originating from weeds, two types of mycorrhiza were obtained mycorrhiza; *Glomus* sp and *Gigaspora* sp. The Mycorrhizae found were distinguished by spore surface shape, decoration spores, spore size and color changes due to the reaction spores dye [6]; [18]; [27]; [23]. Each type of mycorrhiza found has different characteristics; the ability to adapt to the environment and also the different host plants. The results of soil analysis at a sampling marginal land were pH 4.9, organic matter 4.80%, Nitrogen 0.15% Phosphorous

0.29 me/100 g and Potassium 17.88 ppm me/100 g. Based on the results of the soil analysis, the mycorrhiza can thrive. This is in accordance with [28] that the differences in the nature of mycorrhiza adaptation are influenced by the chemical properties of the soil. However, indigenous mycorrhiza in general has higher adaptability if compared with mycorrhiza in the form of fertilized mycorrhiza [11]. The observation of Mycorrhiza spores was made after one week from the time period of the roots and the soil in which the mycorrhiza spores are still alive. If the host plants are not present, mycorrhiza is able to survive for 20-30 days [22]. In unfavorable conditions, the presence of mycorrhiza can be observed in the form of spores, either individually or in the form of sporokarp, before it interacts with the roots of host plant [1]; [4]. *Gigaspora* sp was more tolerant on acid soils and soil-high aluminum [28] while the type of mycorrhiza *Glomus* sp was more common in alkaline soils and less in soils sour [7].

Table 1. *Characteristics of Indigenous Mycorrhiza of Weed*

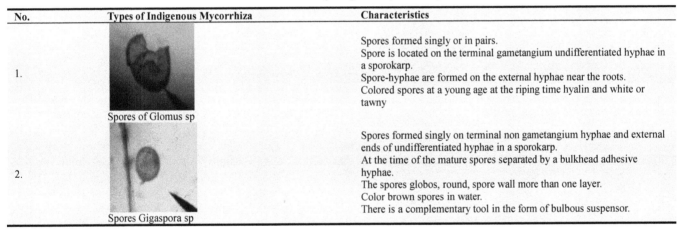

No.	Types of Indigenous Mycorrhiza	Characteristics
1.	Spores of Glomus sp	Spores formed singly or in pairs. Spore is located on the terminal gametangium undifferentiated hyphae in a sporokarp. Spore-hyphae are formed on the external hyphae near the roots. Colored spores at a young age at the riping time hyalin and white or tawny
2.	Spores Gigaspora sp	Spores formed singly on terminal non gametangium hyphae and external ends of undifferentiated hyphae in a sporokarp. At the time of the mature spores separated by a bulkhead adhesive hyphae. The spores globos, round, spore wall more than one layer. Color brown spores in water. There is a complementary tool in the form of bulbous suspensor.

3.2. Percentage of Mycorrhiza Infection on Root Weeds

The percentage of mycorrhiza infection on the roots of weeds is listed in Table 2. The shape of mycorrhiza infection on the roots of weeds is shown in Figures 1 and 2.

Table 2. *Percentage of indigenous mycorrhiza Infection indigenous on weeds rooting*

No.	Types of Weeds	Root sample										Percentage of Mycorrhiza (%) Infection
		1	2	3	4	5	6	7	8	9	10	
1.	*Ageratum conyzoides*	-	-	-	+	-	+	+	+	+	+	60
2.	*Ageratum haustianum*	+	+	-	-	-	+	+	+	+	+	70
3.	*Amaranthus gracilis*	-	+	+	+	+	+	+	+	+	+	90
4.	*Alternanthera sessilis*	-	-	+	+	+	-	-	+	+	+	60
5	*Alternanthera philoxeroides*	-	-	-	-	+	+	+	+	+	+	60
6.	*Croton hirtus*	+	+	+	+	+	-	-	+	+	+	80
7.	*Cleome rutidosperma*	-	-	+	+	+	+	+	+	+	+	80
8.	*Cyperus killingya*	-	-	+	+	+	+	+	+	+	+	80
9.	*Eleusina indica*	-	-	-	+	+	+	+	+	+	+	70
10.	*Fimbristylis aestivalis*	-	-	+	+	+	+	+	+	+	+	80
11.	*Ludwigia hyssopifolia*	-	-	+	+	+	+	+	+	+	+	80
12.	*Mimosa pudica*	-	-	+	+	+	+	+	+	+	+	80
13.	*Mimosa pigra*	-	-	-	+	+	+	+	+	+	+	70
14.	*Nefia spirata*	-	-	-	+	+	+	+	+	+	+	70
15.	*Sida rhombifolia*	-	+	+	+	+	+	+	+	+	+	90

Note: - = not infected, + = infected

The percentage of mycorrhiza infection on the roots of weeds varies between 60% - 90%. This occurs because of differences in weed species, morphology and structure of the root [17]; [21]), the content of nutrients in the root [29] as well as the conformity between mycorrhizae with host plants [16]. The highest percentage of mycorrhiza infection occurred in the weeds of *Amaranthus gracilis* and *Sida rhombifolia* by 90%, respectively. This happens because the content of the root exudates of these weeds is very suitable for the growth of mycorrhiza [13]. In [21], it suggests that the ability of mycorrhiza to infect the roots was greatly influenced by the characteristics of the host plant. The Characteristics of mycorrhiza infection on the roots of weeds are shown in Figures 1 and 2.

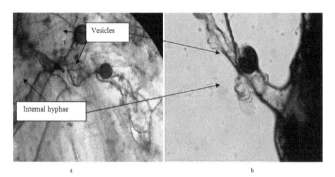

Figure 1. *Characteristics of mycorrhiza infection on the roots of weeds Ageratum conyzoides (a) and Amaranthus gracilis (b)*

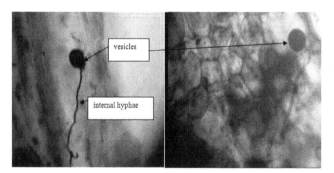

Figure 2. *Characteristics of mycorrhiza infection on the roots of weeds Sida rhombifolia*

Indicators used as a sign of mycorrhiza infection on the roots of weeds are spores, vesicles and arbuscular [9]. From these indicators, spores and vesicles were found in all types of weeds while arbuscular was not found in all types of observed weeds. The presence of mycorrhizal in the root zone of weeds allegedly is so rapid. When the mycorrhizae population on rhizosphere of weeds is very abundant, the competition of mycorrhizae in obtaining root exudates as a source of energy occurs. As a result the failed competitive mycorrhizal forms resting spores [11]. While vesicles are formed as an indication, mycorrhizae can help weed in absorbing nutrients which are then kept as a reserve food in the network weeds. According to [6], the vesicles function are the storage organs of food reserves. Fungi are not found on the network trunk weeds due to long observed age so that arbuscular was not formed again. According to [2],

arbuscular in general began to form about two to three days after the infected roots. Based on the percentage of mycorrhizal infection on the roots of weeds, there is an indication that the dependence of weeds on mycorrhizal was very high. The weed's dependence on mycorrhiza was identical to the percentage of dry weight increase of weeds that were inoculated with mycorrhiza [30]. This means that the higher the value of the mycorrhiza dependence weeds, the higher the percentage of dry weight of weeds to decrease. Thus, there is a positive correlation between the dry weight of weeds decrease with the dependence value of the mycorrhiza weeds.

4. Conclusion

The results of this study are summarized as follows: (1) two types of indigenous mycorrhiza found at the site are Glomus sp and Gigaspora sp, (2) the highest percentage of indigenous mycorrhiza infection was found in the roots of weeds *Amaranthus gracilis* and *Sida rhombifolia* respectively by 90%, (3) Indicators of indigenous mycorrhiza infection on the roots of weeds was the vesicles and internal hyphae.

Acknowledgements

The author would like to thank to the Ministry of Education and Culture of the Republic of Indonesia cq Directorate General of Higher Education for the financial assistance through the scheme of National Priorities Research Grant Master Plan for the Acceleration and Expansion of Indonesian Economic Development 2011-2025 in 2013. The author also thank to the Rector of Halu Oleo University and the Chairman of the Research Institute of Halu Oleo University for the administrative support.

References

[1] Anas I., 1992. Bioteknologi Pertanian 2. Pusat Antar Universitas Bioteknologi. Institut Pertanian Bogor. Bogor.

[2] Barea. J. M., 1991. Vesicular-Arbuscular Mycorrhizas as Modifiers of Soil Fertility. Adv. Soil Science. Vol.15. No.112:399-404.

[3] Brundrett. M., 1999a. Introduction to Mycorrhizas. CSIRO Forestry and Forest Product.Melalui <http://www.ffp.csiro.au/research/mycorrhiza/intro.html>.

[4] Brundrett. M., 2006. Mycorrhizae Mutualistic Plant Fungus Symbioses. Melalui <http://www.mycorrhiza.ag.utk.edu>.

[5] Brundrett. M., N.Bougher, B.Dell, T.Grove and M.Malajczuk, 1996. Working with Mycorrhizas in Forestry and Agriculture. Australian Centre for International Agricultural Research Mohograph.32.374+xp.

[6] Brundrett. M., 1999b. Arbuscular Mycorrhizas. CSIRO Forestry and Forest Products.Melalui http://www.invam.edu/methods/sporas/extraction.html>.

[7] Corryanti, F.Maryadi and Irmawati, 2001. Arbuscular Mycorrhizas under Teak Seed Orchard. Poster Presented on the Third International Conference on Mycorrhizas. Diversity and Integration in Mycorrhizas. Adelaide. South Australia.

[8] Gupta.N. and R.Shubhashree, 2004. Arbuscular Mycorrhizal Association of Weed Found with Different Plantation Crops and Nursery Plants. Regional Plant Resource Centre. Nayapalli. Bhubaneswar. Orissa. India. Melalui <http://www.cababstractsplus.org./google/abstract.asp?>.

[9] Halim dan M.K. Aminuddin, 2012. Perbaikan Pertumbuhan dan Produksi Tanaman Jagung (Zea mays L.) pada Kondisi Kekeringan Melalui Aplikasi Mikoriza Indigenous Gulma. Laporan Penelitian BOPTN Lembaga Penelitian Universitas Halu Oleo Kendari.

[10] Halim dan Resman, 2013. Domestikasi Gulma Penghasil Biji sebagai Media Perbanyakan Mikoriza Indigen Gulma pada Tanah Marginal Masam. Laporan Penelitian Hibah Bersaing Lembaga Penelitian Universitas Halu Oleo, Kendari. Indonesia.

[11] Halim, 2009. Peran Mikoriza Indigenous Gulma *Imperata cylindrica* (L.) Beauv dan *Eupatorium odorata* (L.) terhadap Kompetisi Gulma dan Tanaman Jagung. Disertasi Program Doktor Universitas Padjadjaran Bandung. 45-40 p. (Tidak dipublikasikan).

[12] Halim, 2010. Kelimpahan Populasi Mikoriza Indigen Gulma pada Lahan Sekunder. Majalah Ilmiah Agriplus.Vol.20.No.03.

[13] Halim, 2013. Identification of Indigenous Mycorrhiiza Fungi of Weed in The Biosciences Park Area of Halu Oleo University. Proceedings The 8 TH International Konfrence on Innovation and Collaboration Towards ASEAN Community 2015. Halu Oleo University, Indonesia. ISBN 978-602-8161-57-2.

[14] Harran.S dan N.Ansori, 1993. Bioteknologi Pertanian 2. Pusat Antar Universitas Bioteknologi. IPB.Bogor.

[15] Harrier. L. A., 2003. The Arbuscular Mycorrhizal Symbiosis. A Molecular Review of the Fungal Dimension. J. of Expt. Bot. Vol.52.469-478.

[16] Hasbi. R., 2005. Studi Diversitas Cendawan Mikoriza Arbuskula (CMA) pada Berbagai Tanaman Budidaya di Lahan Gambut Pontianak. Jurnal Agrosains. Jurnal Ilmiah Fakultas Pertanian Universitas Panca Bhakti Pontianak. Melalui <http://www.upb.ac.id/jurnal/vol-1-No.1.pdf>.

[17] Hetrick. B. A. D., G.W.T.Wilson and J.F.Leslie, 1991. Root Architecture of Warm and Cool Season Grasses. Relationship to Mycorrhizal Dependency. Can. J. of Bot. 69:112-118.

[18] Invam, 2006. International Culture Collection of VAM Fungi. Melalui http://www.Invam.cap.wvu.edu/classification.htm.

[19] Miyasaka. S. S., M.Habte, J.B.Friday and E.V.Johnson, 2003. Manual on Arbuscular Mycorrhizal Fungus Production and Inoculation Techniques. Cooperative Extension Service. College of Tropical Agriculture and Human Resource. University of Hawaii. Manoa. Melalui <http://www.ctahr.hawaii.edu>.

[20] Mueller.D. and Ellenberg, 1974. Aims and Method of Vegetation Ecology. John Wiley and Sons.Inc. New York. Chichester Brisbone.Toronto.

[21] Newsham.K.K., A.H.Fitter and A.R.Watkinson, 1995. Multifunctionality and Biodiversity in Arbuscular Mycorrhizas. J. of trends in Ecol. and Evol.. No.10:407- 412.

[22] Nurita.T.M., S.Chalimah, Muhadiono, A.Latifah dan S.Haran, 2007. Kultur Akar Rambut in Vitro serta Pemanfaatan Kultur Ganda untuk Pertumbuhan dan Perkembangan Endomikoriza *Gigaspora* sp. dan *Acaulospora* sp. Jurnal Menara Perkebunan. No.75.Vol.1. 20-31.

[23] Prihastuti, 2008. Isolasi dan Karakterisasi Mikoriza Vesikular Arbuskula di Lahan Kering Masam Lampung Tengah. Balai Penelitian Tanaman Kacang-Kacangan dan Umbi-Umbian. Kendalpayak. Malang.

[24] Setiadi.Y., 1998. Fungi Mikoriza Arbuskula dan Prospeknya sebagai Pupuk Biologis. Makalah Disampaikan pada Workshop Aplikasi Cendawan Mikoriza Arbuskula pada Tanaman Pertanian, Perkebunan dan Kehutanan. Bogor.

[25] Smith.S.E. and D.J.Read, 1997. Mycorrhizal Symbiosis. Second Edition. Academiz Press. Harcourt Brace & Company Publisher. London.

[26] Smith. S. E., E. S. Dickon, F. A. Smith and V. P. Gianiazzi, 1993. Nutrient Transport between Fungus and Plant in Vesicular Arbuscular Mycorrhizal. Proceeding of Second Asian Conference on Mycorrhiza. Chiang Mai. Thailand. Biotrop Special Publication No.42 Seameo Biotrop. Bogor.

[27] Supriatun. T., L.Ulfiah, N.Rosita dan G.Abdullah, 2006. Jenis-Jenis Cendawan Mikoriza Arbuskula (CMA) sebagai Pupuk Hayati di Lahan Aboretum Jatinangor. Fakultas Matematika dan Ilmu Pengetahuan Alam. Universitas Padjadjaran. Bandung.

[28] Tommerup. I.C., 1994. Methods for Study of the Population Biology of Vesicular Arbuscular Mycorrhizal Fungi. Academic Press. London.

[29] Wright. D.P., D.J.Read and J.D.Scholes, 1998. Mycorrhizal Sink Strength Influences whole Plant Carbon Balance of *Trifolium repens* L. Plant, Cell and Environ. 21:881-891.

[30] Yudhy.H.B., 2002. Ketergantungan terhadap MVA dan Serapan Hara Fosfor Tiga Galur Tanaman Kedelai (*Glycine max* L.) pada Tanah Ultisol Bengkulu. Jurnal Ilmu-Ilmu Pertanian Indonesia. Vol.4.No.1. Hal. 49-55.

16

Recent increased incidences of potato late blight on the Jos Plateau: A case for intercropping

Chuwang Pam Zang

Department of Crop Science, Faculty of Agriculture , University of Abuja, Nigeria

Email address:

pamleechuwang@yahoo.com, pzchuwang@gmail.com

Abstract: Potato cultivation on the Jos Plateau is a multi-Billion Naira enterprise which is on the very brink of collapse due to upsurge in the incidences and severity of late blight a disease caused by *Phytophtora infestans* (Mont) DeBary. This paper highlighted the scope of the spread of this scourge by assessing the magnitude of loses due to the disease in four zones of the potato growing region of the Jos Plateau- Bokkos, Ampang, Heipang and Vwang. The production parameters studied were the land area under potato, average yield, proportion of potato produced through sole/mono cropping, severity of the late blight epidemic and the level of adaptation of the new varieties imported from Europe the Americas and Australia. The results revealed that Bokkos was the most important potato growing area in terms of total land area, adoption of new planting materials, and sole/mono cropping system of production. The incidences and severity of the potato late blight was most serious in Bokkos, followed by Ampang, Heipang and Vwang in that order. The average yield of potato tubers (kg/ha) was highest in Ampang and least in Heipang. A brief view of the weather reports from these areas shows erratic patterns of rainfall and rise in temperature which may be attributed to the general climate change. A major trend observed in the weather report is the increase in early rainfall (March-April) which farmers tend to explore for early planting with severe consequences. The increasing tendency to adopt mono cropping by out growers for the multinational seed and other Agro-based companies was highlighted and the attendant risks involved while making a case for mixed/inter cropping. Other benefits suggested for inter cropping were higher resource use efficiencies, security against total crop lost, reduction in the use of pesticides to control diseases and pests as well as favorable environmental effects like shading, erosion control and suppressing weeds.

Keywords: Intercropping, Sole Cropping, Mono Cropping, Resource Use Efficiencies, Climate Change, Potato Late Blight, Incidence, Severity

1. Introduction

The colonial tin miners introduced Potato (*Solanum tuberosum*.L) on the Jos Plateau in the early 19th Century to provide food for their European expatriate population (WPA, 2006). Since then the crop had become an integral part of the farming system of the local farmers who are responsible for producing over 1,500,000 metric tons and 92% of Nigeria's annual output (Okwonkwo *et al*, 1995; *FAO,* 2012, NRCRI annual Reports, 2010). The rainy season crop which accounts for 82% of Nigeria's total annual production is prone to various diseases, the most economically important being late and early blights caused majorly by *Phytophtora infestans* (Mont) DeBary and *Alterania solani* respectively. These, by no means, are the only pathogens but they are the most important.

The incidences of blight have always been observed but the severity and frequency of the cases had never really exceeded the economic threshold to the extent which warranted the drastic control measures that were taken in the past few rainy seasons (2012, 2013 and 2014). This disturbing development predisposes the entire population of the Jos Plateau to food insecurity, economic deterioration and environmental pollution. The assertion is justified when we consider the fact that the potato crop, which is usually harvested early, holds the key to bridging the hunger gap during the critical period between July and August (Chuwang *et al*, 2007). Similarly the Potato business in Nigeria is conservatively estimated to be worth 300 Billion Naira (FAO, 2012, NRCRI, 2012) and the demand for fresh tubers had never been fully met. This potato demand deficit, in Nigeria, can only be effectively supplied from the Jos Plateau given the favorable climatic conditions like cool temperature due to high altitude.

Most of the potato produced (70%) on the Jos Plateau is through intercropping with tropical cereals like maize, sorghum and millet by the small scaled farmers (Okonkwo, et al, 1995). However the most predominant crop mixture seems to be potato/maize considering the wide spread cultivation. The advantages of maize/potato intercropping was highlighted to include security against total crop yield, diversity of nutrition options, environmental management, higher nitrogen use efficiency, higher land/ other resources use efficiencies, creating artificial/physical barriers for pests and diseases (Allard, 1961, Chuwang, 2006, Bouws and Finckh, 2007, Chuwang and Odion, 2008).

However in recent times there appears to be a concerted effort by some key players in the potato industry of Nigeria most especially multinational seed companies, agro based/potato processing companies and some scientists to encourage mono cropping to the detriment of intercropping. The reasons usually advanced for this paradigm shift is that mono cropping provides the platform for easy mechanization and use of agro chemical (Gregor and Author, 1994). Other reasons include the production of clean and attractive tubers for 'seed' and for the market as well as reported apparent high levels of returns on investment with sole cropping (Chuwang, 2010). Plausible as this argument may seem, but the fact remains, that the threat of total crop lost and insecurity arising from it is as real as ever. This was demonstrated by the disaster of the 2014 Blight epidemic where thousands of hectares of potato farms were devastated by the scourge.

The object of this paper is to make a case for intercropping potato with a non- host crop species like maize or other cereal in order to forestall or contain the emerging threat posed by potato late blight and other diseases.

2. Materials and Methods

2.1. The Study Area

The potato producing areas of the Jos Plateau ($9^0 6^1$-10.5^0 N; $8^0 35^1$W-$9^0 45^1$W) were divided into four segments for the purpose of this investigation. These sectors or areas are

- Bokkos/Nbar/Kuba with all the adorning areas
- Ampang/ Kerang with all the adorning areas
- Vwang/Kuru/Miango with all the adorning areas
- Heipang/Kassa with all the adorning areas

2.2. Weather Reports

The weather report for the entire Jos Plateau was collated for analysis from the Potato Program of the National Root Crop Research Institute in Kuru. The information mostly accessed was rainfall distribution and duration as well as maximum and minimum temperature. Where it is available the record of relative humidity is analyzed for the past 5-10 years.

A summary of these elements of the weather is made to

assess the trends of changes as may be observed in the years under consideration along with data from global climate change websites and other similar sources. These summaries and analyzed data are presented in figures.

2.3. Farm Visits, Interviews and Discussions with Potato Farmers and other Stake Holders

Five farms were visited from each of the potato growing areas and the farmers were interviewed. Farmers from contiguous farms were engaged in discussions to obtain information on the increase in incidences of late blight and other problems of production as well as farming system options open to them. These interactions and visits yielded data on average potato tuber yield, total land area and the types and sources of planting materials as well as farming/production system adopted. The late blight incidence and severity levels as well as their economic implications were dully evaluated on the farmers' fields.

2.4. Review of Reports of Potato Late Blight on the Jos Plateau

The reports available were analyzed and compared with what really took place on the field. These reports included among others the field reports of the Plateau Agricultural Development Programme (PADP), the NRCRI- Potato programme Kuru, the Plateau state Ministry of Agriculture and the Agric units of the affected local governments, namely Barkin-Ladi, Bokkos, Mangu, Jos South, Riyom and Bassa.

2.5. Data Analysis

Simple percentages were calculated and the chart were prepared with the help of Micro-Soft Excel 2010 Version

3. Results

The result is presented area by area and all the parameters studied are displayed in figures 1 to 4.

3.1. Bokkos Area

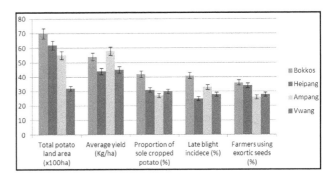

Figure 1. *Potato Production Information for the four farming centers on the Jos Plateau (2014).*

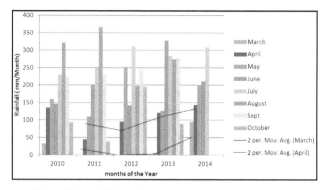

Figure 2. Rainfall Distribution on the Jos Plateau (2010-2014).

The result of the survey presented in figure 1 shows that the Bokkos area had the largest land area committed to potato and the second highest unit yield per hectare when compared to the other areas. However the potato farmers in Bokkos seemed to have accepted mono cropping much more readily than the other areas because the proportion of potato produced through monocropping there was highest. Most of the farmers that used newly introduced potato varieties from Europe America and Australia were found in the Bokkos area. The earliest and most serious cases of late blight were also reported from this potato growing area.

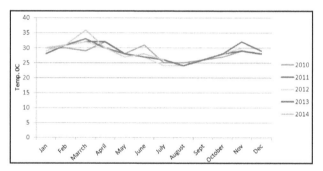

Figure 3. Average Maximum Temperature on the Jos Plateau (2010-2014).

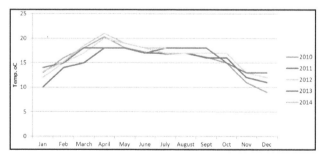

Figure 4. Average Minimum Temperature (°C) On the Jos Plateau (2010-2014).

3.2. Ampang/Kerang

In this area produced the highest yield of potato per unit area of land in the entire state and also has the lowest proportion of potato farmers using new varieties introduced from the western world as well as the smallest quantity of tubers produced through sole cropping. The reported incidences of late blight were serious but not as serious as in the Bokkos area.

3.3. Heipang/ Kassa Area

The total land area dedicated to potato in the Heipang/Kassa area is the second largest in the state but the yield per unit area is less than in both Bokkos and Ampang. The newly introduced varieties had been as fairly well accepted as in Bokkos. Intercropping here as in Vwang and Ampang are still very heavily practiced in view of the low proportion of potato produced as sole crops and the reported incidences of late blight were not as serious as in Bokkos and Ampang.

3.4. Vwang/Miango

All the parameters studied in this area had the least values. However the differences between these values and those obtained at Heipang were not very appreciable.

3.5. Weather Report

The weather data assessed were total rainfall, average minimum and maximum temperature from 2010 to 2014. The records revealed unusual rainfall patterns such as the heavy rainfall early in the year which coincided with unusually high temperatures. (figs 2and3). The average minimum temperature which is usually a fare reflection of night temperatures have remained within the range which allows for effective tuberization by the crop plants but in 2014 there was an upsurge which tends to exceed the 20°C mark (fig 4).

4. Discussion

Most of the potato produced in Ampang/Kerang comes from the foothill of (and the area around) the fertile Kerang volcanic hill which is rich in volcanic ash hence the relatively high yields. The area, like Bokkos, is also located on the very edge of the Jos Plateau characterized by deep gorges and misty weather. This misty weather may have been responsible for the relatively high incidences and equally high severity of the reported late blight of potato. Another probable factor that was responsible for these severe incidences of blight in these two areas was the heavy monsoon-like rainfall which came very early in the season most especially between late March and April of 2014. Heavy rainfall had been reported to be a major factor in accelerating the incidence and severity of potato late blight most especially in the tropical highlands (Hijmans *et al*, 2000). In Bokkos, like in some parts of Ampang, the farmers usually cultivate the potato crop with the earliest rainfall usually between early and middle of April). This year (2014) very heavy rainfall was recorded between March and May which also coincided with a period of unusually high temperature (fig 2 and 3). This situation resulted in the potato blight outbreak early in the growing season before the commencement of full tuberization by the crop plants, hence the outcome was very devastating for the farmers in these areas.

Recently (within the last two or three years), new varieties of potato planting materials have been imported from Europe, the Americas, and Australia by multinational and indigenous companies of Nigeria. These new planting materials may not have adapted very well to the tropical conditions and most probably do not possess any degree of tolerance to the blight pathogens available on the Jos Plateau. More over with global rise in temperature and unpredictable rainfall patterns due to climate change, it is very likely that the blight tolerance capacity of the indigenous potato varieties may have been greatly compromised as there exist reports of this from other places (Hijmans, *et al*, 2000, Forbes and Simons, 2007, Shankar, 2014).The most serious incidences of potato late blight reported are from areas that have a relatively higher level of adoption of these new varieties. Bokkos and areas surrounding it have the highest adoption rate as a result the potato plants grown may have been very susceptible to the pathogens. In the Bokkos area, where the epidermic was most severe (DailyTrust,2014), a comparatively large proportion of the potato farmers practice sole cropping than in other areas, where the disease was not as severe. Growing of potato along with a non-host crop plant species had been reported to reduce the spread of the Potato late blight pathogens significantly (Bouws and Finckh, 2008). Intercropping potato with maize or other cereals like sorghum will not only present the potential to check the spread of blight on the potato plants but also to guard against total yield lost as was the case in some farms where the disease started very early in the season prior to commencement of tuberization. This pest and disease control advantage of intercropping potato with maize or other non-blight host crop species makes it environment friendly due to reduced usage of environment harmful pesticides.

Previously over 90% of all the potato produced in Nigeria was from intercropping (Okwonko *et al.,* 1995). However, in recent years, seed companies, who engage most of the prominent potato farmers as out growers, have tended to discourage them from practicing their traditional farming system which is based predominantly on intercropping potato with cereals like maize and some legumes such as *Phaseolus spp.* With the current Nigerian government transformation of the Agricultural sector of the economy, which encourages private sector investment, more seed companies and other Agro based/potato processing outfits will emerge to engage more out growers and consequently more potato farms lands will be committed to sole/mono cropping in the coming years. This emerging trend will diminish the natural benefits derivable from intercropping such as better pest/disease management, food security and wider dietary options as well as more efficient use of resources like land, labour, fertilizers and other agrochemicals (Gregor and Author, 1994, Bouws and Finckh, 2008). Apart from the pest/disease control benefits and enhanced resource use efficiencies, intercropping is more in tone with nature as the intercropped field seeks to mimic the natural undisturbed or uncultivated fields where variety of life forms coexist in harmony. These harmonious interrelationships could have beneficial effects for the components of the ecosystem as in legume/grass intercrops where the grass benefits from the N-fixing

activities of the legumes. These ideas and the role of the taller plant species (maize/sorghum) in shading the shorter potato to lower the soil temperature for more effective tuberization had been reported (Harris, 1990, Chuwang, 2006,). The environmental impacts of these and other observed benefits opens room for further research.

The consequences of climate change like the erratic rainfall patterns experienced early in 2014 and general increase in air temperature are most likely to continue for some time (Climate Change Reports, 2007, Bindi, 2007). The trend line (*figure 2*) of rainfall of March and April shows a steady increase that is also most likely to persist for some time to come. The implication of this on a tropical location like the Jos Plateau may manifest in obscure and uncertain growing seasons as well as an upsurge in pests and disease incidences (Hijmans, *et al,* 2000 ,Sparks *et al,* 2014).

5. Conclusion

With newly introduced, untested and disease susceptible planting materials in circulation as well as the emergence of more sole cropping production practices coupled with unpredictable weather conditions, occasioned by climate change, it is pertinent to consider the tested and time honored traditional practices like intercropping which provides the assurance against total crop failure and ensures more efficient utilization of resources.

References

[1]   Allard, R.W. (1961) Relationship between genetic diversity and consistency of performance in different environments. *Crop Science* 1: 127-133

[2]   Bindi, M.(2007) How climate change affects potato crops. Forth assessment report of climate Change report of the Intergovernmental Panel on Climate Change (IPCC), 2007

[3]   Bouws, H and Finckh, M.R. (2008) The effect of strip intercropping of potato with non-hosts On late blight severity and tuber yield in organic agriculture. *Plant Pathology* Vol 57 (5) 916-927

[4]   Chuwang, P. Z. (2006) Productivity of Potato-Maize intercrop as Influenced by N levels and Planting Arrangement. Ph.D dissertation submitted to the Post graduate school of the Ahmadu Bello University (A.B.U.) Zaria, March 2006.

[5]   Chuwang, P.Z., Odion, E.C. and Aliyu, L, (2007)The response of potato and maize yields to N fertilizer and planting pattern in Jos, Plateau state of Nigeria. *Journal of League of Researchers in Nigeria (JOLORN)* Vol 8 (2) 22-30

[6]   Chuwang, P.Z. (2010) Maize/Potato intercrop as influenced byN levels and planting patterns On the Plateau savanna of Kuru, Jos, Nigeria *ACTA Agronomica Nigeriana* Vol. 19 (2) 22-30

[7]   Chuwang, P.Z. and Odion, E.C.(2010) Nitrogen Use Efficiency of potato/ maize intercrop as Affected by N levels and arrangement in a mid- altitude location of Jos Nigeria. *Agriculture Business and Technology Journal* Vol 8 (2) 144-155. ISSN 2007-0807

[8] Climate change (2007) Synthesis Report of the Intergovernmental Panel on Climate change (IPCC) xxvii (Valencia Spain 12th -17th Nov., 2007

[9] Daily Trust (2014) Newspaper report with caption Nigeria: Disease ravages over 500 hectares of Irish potato farms in Plateau State. 29th May 2014

[10] FAOSTAT (2012) The Food and Agriculture Organization production statistics for 2012 Forbes,G.A. and Simon, R. (2007) Implication for a warmer, wetter world on the late pathogen: How CIP efforts can reduce risk foe low input potato farmers. International Potato Cente, Apartodo 1558 Lima, Peru *Icrisat ejournal*, 2007.

[11] Gregor, P. and Author, RM. (1994) Monocropping, intercropping or crop rotation? An economic case study of west African Guinea Savanna with special reference to risk. *Agricultural Systems* 45(1994): 123-143

[12] Harris, PM. (1990) Potato crop radiation use: A justification for intercropping. *Field Crops Research* 25: 25-39

[13] Hijmans, RF., Forbes, GA. and Walker,TS (2000) Estimating the global severity of potato late blight with GIS- linked disease forecast models. *Plant Pathology* Vol 42 (6) 697-705

[14] NRCRI (2012) The annual Potato Programme Report of the National Root Crops Research Institute, Umidike (2012).

[15] Okwonko, JC., Ene, LSO. and Okoli, OO(1995) Potato Production in Nigeria. NCRCI, Umudike, Nigeria.

[16] Shankar, KS (2014) Understanding the threat of potato late blight under climate change from Ecuador to Nepal. International Potato Center (IPC). *Agricultural Research for development Newsletter*.

[17] Sparks, A.H.,Forbes, G.A.,Hijmans, R.J. and Garrett, K. A, (2014) Climate change effects on the global risk of potato late blight gisweb.ciat.cgiar.org/RTB Maps/Docs potato late blight also at onlinelibrary.wiley.com/doi/10.1111/gcb.12587.

[18] WPA (2006) World Potato Atlas: Africa-Archives, country chapter, Nigeria. Edited by Kelly Theisen of CIP Lima Peru.

17

Induction of callus and somatic embryogenesis from cotyledon and leaf explants of Yeheb (*Cordeauxia edulis* Hemsl)

Yohannes Seyoum[1, *], Firew Mekbib[2], Adefris Teklewold[3], Belayneh Admassu[4], Dawit Beyene[4], Zelalem Fisseha[1]

[1]Dryland Crop Research Department, Somali Region Pastoral and Agro-pastoral Research Institute (SoRPARI), Jijiga, Ethiopia
[2]School of Plant sciences, Haramaya University (HU), Dire-Dewa, Ethiopia
[3]Crop Research Directorate, Ethiopian Institute of Agriculture Research (EIAR), Addis Ababa, Ethioipia
[4]Holetta biotech Laboratory, Ethiopian Institute of Agriculture Research (EIAR), Holetta, Ethiopia

Email address:
yohanes1195@gmail.com (Y. Seyoum)

Abstract: 'Yeheb' (*Cordeauxia edulis* Hemsl) is a multipurpose and evergreen shrub and endemic to southeastern corner of Ethiopia and Somalia. It is adapted to low and irregular rainfall and survives a very long dry season. It has enormous economic and food security roles to the pastoralist of Somali Region State in Ethiopia. However, the plant is threatened with extinction due to over exploitation and its' poor natural regeneration capacity. The aim of this was to explore the potential for in vitro rapid regeneration of 'yeheb' from cotyledon and leaf explants on Murashige and Skoog (MS) media supplemented with 1.0 – 8.0 mg l^{-1}concentrations of 2, 4-D for callus induction and 2.0 and 3.0 mg l^{-1}concentration of N6-benzylaminopurine (BAP), thidiazuron (TDZ) and kinetin (Kin) with combination of 4.0 mg l^{-1} of 2, 4-Dichlorophenyl acetic acid (2, 4-D) for embryo induction. The result of these studies revealed that the highest percentage of callus induction (89%) were obtained from both leaf and cotyledon explants on MS media supplemented with 4.00 and 8.00 mg l^{-1} 2, 4-D, respectively. The highest percentage of embryo regeneration responses (88.89 and 77.78%) were obtained from leaf and cotyledon explants on same media: MS media supplemented by 3.00 mg l^{-1} TDZ +4.00 mg l^{-1} 2, 4-D. As a conclusion; this is the first attempt for callus and embryo *in vitro* regeneration of *C. edulis* and permissible result for mass propagation and cryopreservation.

Keywords: Callus, Cotyledon, Embryo, Explants, In Vitro

1. Introduction

Cordeauxia edulis Hemsl, belongs to the family *Leguminosae* and the subfamily *Caesalpinioideae* and is locally known as 'Yeheb' [1]. It is among the priority edible wild food plants in Ethiopia [2]. The species is a much branched ever green shrub or small tree up to 2.5m height, and is endemic to restricted localities in eastern Ethiopia and parts of central Somalia [1].

'Yeheb' is a multi-purpose plant where most parts of the plant are used. The seeds are edible and eaten fresh, roasted, boiled or dried. The seed of the species is potentially a valuable protein source with high sugar and fat contents. It has high energy value (0.39 - 1.87 MJ/Kg). The leaves are also

rich in energy (5.59 – 5.86 MJ/Kg dry matter) [3]. In addition, leaves have been used to dye cloths, calico and wool since the cordeauxiaquinone forms vividly colored and insoluble combinations with many metals [4]. Another major use of the species is its contribution of up to half of the biomass in the area that make it important dry season browse to camel and goat. The estimated average forage production is 325-450 kg/ha (1.4-2 kg/plant) [5]. In semi-arid and arid areas the species represent an economical interest. As it is adapted to low and irregular rainfall and survives a very long dry season, it could indeed represent an enormous advantage in the fight against hunger. The development of cultivation of such plants for the Sahelian zone could also constitute an interesting food supplement in an area poor in protein supply [6]. The species constitutes the staple food of the pastoralist of Somali region in Ethiopia. Moreover, the nut is sold on the market and even exported to the coastal cities of Somalia.

Even if the species has such and many other uses and has a potential to play a role in ensuring food security in the region, the plant is threatened with extinction due to overexploitation of the shrub by long term heavy grazing pressure, harvesting of seeds, cutting and fire. In addition, erosion, drought and war in the region has led to poor or none natural regeneration [4, 7 and 8].

Some reference [9 and 10] reported the decline and progressive destruction of the stands of C. edulis due to over grazing, and had recommended protection and use of the plant. Likewise [11 and 12] reported that C. edulis plant is in great danger of extinction and speedily narrowing distribution area because of the increase in population and their herds. Unlike many other plants, 'yeheb' shrubs flowers just before the onset of rains and the seeds mature when the plant moisture content is at its peak [5]. 'Yeheb' seeds have been reported not to retain viability for more than a few months, even if they are stored under ideal conditions and the recommendation has therefore been to sow them immediately. Some pilot studies made regarding vegetative propagation but finalized without greater success [12]

Hence, tissue culture technique is a power full tool for mass multiplication and germplasm conservation for this valuable species which is on the verge of extinction. In vitro regeneration via somatic embryogenesis has been success in many species, such as dahlia [13] canola [14], roselle [15], papaya [16] and etc. Nevertheless, limited work has been done so far vis-à-vis in vitro regeneration from callus in C. edulis.

Therefore, the aim of this study was to develop the basic protocols for the establishment of callus culture and induction of somatic embryogenesis from cotyledon and leaf explants of C. edulis.

2. Materials and Methods

2.1. Description of the Experimental Area

The experiment was conducted at the Plant Biotechnology Laboratory of Holetta Agricultural Research Center (HARC). The center is located 29 km west of Addis Ababa at an altitude of 2400 meter above sea level, 9^0 00'N latitude, 38^0 30'E longitude.

2.2. Experimental Material

2.2.1. Explant Selection

As the shrubs had not produced seed during the experimental period due to lack of rain in the region, the seeds that were full size and dried were collected from local market of 'Boh', Warder Zone of Ethiopia Somali Regional State (ESRS) in June, 2011. Healthy seeds were selected carefully and used for this study.

2.2.2. Explant Preparation

The seeds used for this experiment were washed under running tap water for 30 min. This was followed by immersing 70% (v/v) ethanol for 3 min, and later rinsed three times (3 min each) with sterile distilled water. After sterilization, seeds were soaked in sterile distilled water for 12 hr. The seed coats were then removed and subjected to surface disinfection with 5.00 % sodium hypochlorite for 5 min and then rinsed three times (3 min each) with sterile distilled water. Sterile seeds were inoculated in Murashige and Skoog (MS) media for germination.

Both cotyledon and leaf were used as explants for this experiment. Cotyledon explants were obtained from 21 days old seedlings grown in aseptically. Leaf explants were also obtained from aseptically grown seedlings.

2.3. Callus Induction

Callus induction studies were carried out by culturing the sterile cotyledon and leaf explants of C. edulis on MS media supplemented by different concentration of 2, 4-D (0.0, 1.0, 2.0, 3.0, 4.0, 5.0, 6.0, and 8.0 mg l^{-1}). Media contained 500 mg l^{-1} of casein hydrolysate (C7290, Sigma), 3% sucrose, and 0.7% agar. The media were adjusted to pH 5.7 after addition of the plant growth hormone, but prior to adding agar. The media was later poured into magenta (40ml) before autoclaving at 120 ^{0}C for 15 minutes. One explant was placed per magenta, and each treatment had nine explants. All the cultures were incubated in a growth room adjusted at room temperature of 25 ± 2^0C in the dark. The morphology, degree of callogenesis and days of callus formation were recorded after 8 weeks of culture.

2.4. Embryo Induction

Those calli produced on cotyledon and leaf explants were transferred to MS media supplemented with 2.0 and 3.0 mg l^{-1} concentrations of N6-benzylaminopurine (BAP), thidiazuron (TDZ) and kinetin (Kin) with combination of 4.0 mg l^{-1} of 2, 4-D and free plant growth regulator (as control), each type and level of concentration of cytokinin were combined with the Auxin to use as treatment. Media contained 500 mg l^{-1} of casein hydrolysate (C7290, Sigma), 3% sucrose, and 0.7% agar. The media were adjusted to pH 5.7 after addition of the plant growth hormone, but prior to adding agar. The media were later poured into magenta (40 ml) before autoclaving at 120^0C for 15min. One explant was placed per magenta, and each treatment had nine explants. All the cultures were incubated in a growth room adjusted at room temperature of 25 ± 2^0C in the dark. Data on embryo regeneration percentage was recorded from both explants after 8 weeks after transferred to embryo induction media.

2.5. Experimental Design and Data Analysis

Treatments in all the experiments were arranged in a completely randomized design (CRD) with three replications. The data was subject for analysis of variance (ANOVA) using SAS [16] and significant differences among mean values were compared using Duncan's Multiple Range Test (DMRT) at p<0.05. Logarithmic transformation was used for percentages data to fulfill the normality test before doing analysis of variance.

3. Results and Discussion

3.1. Callus Induction

The analysis of variance result revealed that MS media supplemented by different concentration of 2, 4-D had a highly significant effect (p<0.01) on percentage of callus induction on both explants. It was possible to induce callus successfully from both explants used on media supplemented by 4.00, 5.00, 6.00 and 8.00 mg l^{-1} of 2, 4-D. Low concentration of 2, 4-D (1.00, 2.00 mg l^{-1}) and control (hormone free) did not induce callus. Leaf explants were better than cotyledon explants in terms callus induction (Table 1).

Table 1. Effect of 2, 4-D on callus induction on cotyledon and leaf explants of 'yeheb'

Explants	2, 4-D (mg l^{-1})	Days to induce callus	Percentage of callus induction (%)	Callus size	Callus color
Cotyledon	0.00	60	0 ± 0.00^c	-	-
	1.00	60	0 ± 0.00^c	-	-
	2.00	60	0 ± 0.00^c	-	-
	3.00	60	11 ± 0.19^c	+	-
	4.00	44-46	44 ± 0.20^b	++	White
	5.00	45-43	56 ± 0.20^b	++	White
	6.00	44-45	67 ± 0.19^{ab}	+++	White
	8.00	40-42	89 ± 0.19^a	++++	Yellow
Mean			38.10		
CV (%)			2.06		
Leaf	0.00	60	0 ± 0.00^d	-	-
	1.00	60	0 ± 0.00^d	-	-
	2.00	58-60	44 ± 0.20^{bc}	+	White
	3.00	46-48	67 ± 0.19^b	+	White
	4.00	37-38	89 ± 0.19^a	++++	White
	5.00	37-38	44 ± 0.19^{bc}	+++	White
	6.00	37-38	33 ± 0.00^c	+	White
	8.00	35-36	33 ± 0.00^c	+	White
Mean			44.44		
CV (%)			1.61		

Means with same letter (s) in the same column are not significantly different at 1% according to Duncan's Multiple Range Tests (DMRT). CV= coefficient of variation (%), Callus color data was recorded after 2 month of inoculation, +=0.01mm size callus.

3.1.1. The Effect of 2, 4-D on Cotyledon Explants

The result reviled in Table 1 indicated that the highest percentage of callus induction (89%) was observed on MS media supplemented by 8.00 mg l^{-1} 2, 4-D. While poor (11%) or no response was observed on media supplemented with 3.00 mg l^{-1} and less concentration of 2, 4-D. Similar result was reported by [17] on *Vigna radiata*. Bigger callus size with shorter period of time was observed, when 2, 4-D concentration increase from 4.00 to 8.00 mg l^{-1}. The highest concentration (8.00 mg l^{-1}) of 2, 4-D produced the biggest callus size within 40-42 days after inoculation. Lower concentration (4.00 mg l^{-1}) of 2, 4-D resulted in the least callus induction. Callus production from cotyledon explants of woody plants was reported on *Parkia biglobosa* [18] and *Fraxinus pennsylvanica* [19].

3.1.2. The Effect of 2, 4-D on Leaf Explants

The two highest callus induction percentage (89% and 67%) was recorded on MS media supplemented with 4.00 and 3.00 mg l^{-1} 2, 4-D, respectively. While poor (33%) or no response was observed on media supplemented with 8.00 and 6.00 mg l^{-1} or 1.00 mg l^{-1} and free 2, 4-D. Similar biggest callus size was obtained on 4.00 mg l^{-1} of 2, 4-D within 37-38 days of after inoculation, while lower callus size was observed on the concentration 2.00, 3.00, 6.00, and 8.00mg of 2, 4-D (Table 1).

Generally callus production was observed only at the cut edges of both explants and on the abaxial surface, even when placed upside down. This result showed that callus formation is affected among other factors is orientation of the explants on the culture medium [20]. Reference [21] also reported similar observation with the leaf explants of *Cuphea ericoides*.

3.2. Embryo induction

The analysis of variance revealed that the effect of plant growth hormone had highly significant effect on embryo regeneration from both types of explant. The mean callus derived from leaf explants (47.61±0.16) resulted in highest embryo regeneration percentage than callus derivative from cotyledon explants (34.92±0.16) (Table 2 and Fig. 1a and b).

As indicated in Table 2, the highest percentage of embryo regeneration response was obtained from media supplemented by 3 mg l^{-1} TDZ +4 mg l^{-1} 2, 4-D on both cotyledon (77.78 ±0.19) and leaves (88.89 ±0.19) explants after eight week of cultured (Fig 2a and b) followed by 3 mg l^{-1} BAP +4 mg l^{-1} 2, 4-D and 2 mg l^{-1} TDZ + 4 mg l^{-1} 2, 4-D; while none embryogenic response was observed from hormone free (control) medium (Table 2). Reference [15] reported similar observation on cotyledon explants of Hibiscus sabdariffa on MS media supplemented with TDZ + 2, 4-D had better embryo regeneration potential than BAP + 2, 4-D or Kin + 2, 4-D

Table 2. *Effect of cytokinins and auxin concentration on embryo development*

Plant growth hormone	Embryo regeneration response from callus (%)	
	Cotyledon	Leaf
0.00	0±0[d]	0± 0[e]
2 mg l⁻¹ BAP + 4 mg l⁻¹ 2, 4-D	33.33 ±0[bcd]	44.44 ±0.19[cd]
2 mg l⁻¹ Kin + 4 mg l⁻¹ 2, 4-D	22.22 ±0.38[bcd]	22.22 ±0.19[de]
2 mg l⁻¹ TDZ + 4 mg l⁻¹ 2, 4-D	44.44 ±0.19[abc]	55.56 ±0.19[bc]
3 mg l⁻¹ BAP + 4 mg l⁻¹ 2, 4-D	55.56 ±0.19[ab]	77.78 ±0.19[ab]
3 mg l⁻¹ Kin + 4 mg l⁻¹ 2, 4-D	11.11 ±0.19[cd]	44.44 ±0.19[cd]
3 mg l⁻¹ TDZ + 4 mg l⁻¹ 2, 4-D	77.78 ±0.19[a]	88.89 ±0.19[a]
Mean	34.92±0.16	47.61±0.16
CV (%)	8.10	5.50

Means with same letter (s) in the same column are not significantly different at 1% according to Duncan's Multiple Range Tests (DMRT). CV= coefficient of variation (%).

(a)   (b)

Figure 1. *Embryo induction media supplemented with 3 mg l⁻¹ TDZ + 4 mg l⁻¹ 2, 4-D before 4 weeks of culture a) on cotyledon explant; b) on leaf explant*

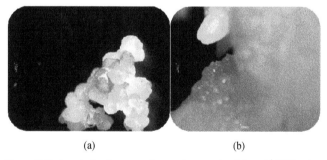

(a)   (b)

Figure 2. *Embryo induction on media supplemented with 3 mg l⁻¹ TDZ + 4 mg l⁻¹ 2, 4-D after 8 week of culture a) from leaf explant (10x magnified); b) from cotyledon (10x magnified).*

4. Conclusion

'Yeheb' (*Cordeauxia edulis* Hemsl.) is a multi-purpose plant where most parts of the plant are useable. Even if the species has multitude uses and has a potential to play a role in ensuring food security in the region, the plant is threatened with extinction due to overexploitation, this it has led to poor or none natural regeneration. This study is the foremost protocols for indirect *in vitro* regeneration of *C. edulis* from cotyledon and leaf explants. It has also indicated callogenic capacity of cotyledon and leaf explants of *C. edulis* and the possibility of inducing somatic embryogenesis from the induced calli. Further research will be needed in order to advance the embryoids to plantlets and their establishment in the field. In a nut shell, this is the first attempt and found a

permissible result to rescue this rare, endemic, and endangered species by mass and continuous plantlet production within short period of time through in vitro propagation. In addition it is used as a baseline for *ex situ* conservation through cryopreservation.

References

[1] Ali, H.M., 1988. *Cordeauxia edulis*: Production and Forage Quality in Central Somalia. Thesis for the degree of Master of Science in Rangeland Resources, National University of Somalia, Somalia.

[2] Teketay, D. and Eshete, A., 2004. Status of indigenous fruits in Ethiopia. In: Chikamai B, Eyog-Matig O, Mbogga M (eds.) Review and Appraisal on the Status of Indigenous Fruits in Eastern Africa: A Report Prepared for IPGRI-SAFORGEN in the Framework of AFRENA/FORENESSA, Kenya Forestry Research Institute, Nairobi, Kenya, pp 3-35.

[3] Miège, J. and Miège, M.N., 1978 *Cordeauxia edulis* a Caesalpinaceae of Arid Zones of East Africa, Caryologic, blastogenic and biochemical features. Potential aspects for nutrition. *Economic Botany,* 32: 336-345.

[4] Booth, F.E.M. and Wickens, G.E., 1988. Non-timber uses of selected arid zone trees and shrubs in Africa. FAO Conservation Guide 19, 52-58.

[5] Brink, M., 2006. *Cordeauxia edulis* Hemsl Record from Protobase. PROTA (Plant resources of tropical Africa / Ressourcesvégétales de l'Afriquetropicale), Wageningen, Netherlands http://database.prota.org/search.htm (Accessed on September 14, 2011)

[6] N.A.S (National Academy of Science), 1979. Tropical legumes: Resource for the Future, Nat Acad. Sci. Washington DC, 261 pp.

[7] FAO, 1988. Traditional food plants, Food and nutrition paper 42:224-27.

[8] Assefa, F., Bollini, R. and Kleiner, D., 1997. Agricultural potential of little used tropical legumes with special emphasis on *Cordeauxia edulis* (Ye-eb nut) and *Sphenostylis stenocarpa* (African yam bean). *Giessener Beiträgezur Entwicklungsforschung*, 24:237–242.

[9] Bally, P.R.O., 1966. Miscellaneous notes on the flora of Tropical East Africa, 29. Enquiry into the occurrence of the Yeheb nut (*Cordeauxia edulis* Hemsl.) in the Horn of Africa. *Candollea* 21 (1), 3-11.

[10] Hemming, C.F., 1972. The vegetation of the northern region of Somalia Republic. *Proceeding of Linnaeus Social London*, 177:173-250.

[11] Drechsel, P. and Zech, W., 1988. Site conditions and nutrient status of *Cordeauxia edulis* (*Caesalpiniaceae*) in its natural habitat in central Somalia.*Economic Botany,* 42: 242–249.

[12] Mussa, M., 2010.*Cordeauxia edulis* (Yeheb): resource status, utilization and management in Ethiopia. Thesis for the degree of Philosophiae Doctor, University of Wales.

[13] Fatima, B.,Usman M., Ashraf, T., Waseem, R. and Ali, M.A., 2007. *In vitro* shoot regeneration from cotyledon and hypocotyl explants of dahlia cultivars. *Pak. J. Agri. Sci.*,44(2):312-316

[14] Al-Naggar, A.M.M., Shabana, R., Rady,M.R.,Ghanem, S.A., Saker, M.M., Reda, A.A., MatterM.A., and Eid, S.A.M., 2010. *In vitro* callus initiation and regeneration in some canola varieties. *International Journal of Academic Research* 2(6): 357-362

[15] Sié, R.S., Charles, G.,,Sakhanokho, H.F., Toueix, Y., Djè ,Y., Sangaré, A. and Branchard, M., 2010. Protocols for callus and somatic embryo initiation for *Hibiscus sabdariffa* L. (*Malvaceae*): Influence of explant type, sugar, and plant growth regulators. *Australian Journal of Crop Science* 4(2):98-106.

[16] SAS Institute Inc., 2002. Statistical Analysis Software, Version 9.0. Cary, North Carolina, USA.

[17] Rao, S., Patil, P. and Kaviraj, C.P., 2005. Callus induction and organogenesis from various explants in *Vigna radiata* (L.) Wilczek. *Indian Journal of Biotechnology*, 4:556-560.

[18] Amoo, S.O. and Ayisire, B.E., 2005. Induction of callus and somatic embryogenesis from cotyledon explants of *Parkiabiglobosa* (Jacq.) Benth.*African Journal of Biotechnology*, 4(1):68-71.

[19] Du, N., and Pijut, P.M., 2008. Regeneration of plants from *Fraxinuspennsylvani*ca hypocotyls and cotyledons. *Scientia Horticulturae*, 118:74–79.

[20] Warren, G., 1991. The regeneration of plants from cultured cells and tissues. In: Plant Cell and Tissue Culture. A. Stafford and G. Warren (Eds.) p. 85.

[21] Rita, I. and Floh, E.I.S., 1995. Tissue culture and micro propagation of *Cupheaericoides*, a potential source of medium-chain fatty acids. *Plant Cell, Tissue and Organ Culture,* 40:187-189.

18

Assessment of decomposition rate and soil nutrient status under different woody species combination in a tree plantation

I. O. Faboya[1], S. I. Adebola[2, *], O. O. Awotoye[2]

[1]Department of Forestry, Ministry of Environment Ekiti State, Ekiti State, Nigeria
[2]Institute of Ecology and Environmental Studies, Obafemi Awolowo University, Ile-Ife, Nigeria

Email address:

fisrealoludare@yahoo.com (I. O. Faboya), segunawotoye@yahoo.co.uk (O. O. Awotoye), adebolasamuel@gmail.com (S. I. Adebola)

Abstract: Forest Litter is the major input determining the nutrient accumulation within the forest soil ecosystem which goes a long way in determining forest stand productivity. To better understand this, the study investigated the litter decomposition rate and soil nutritional status under different woody species combinations in tree plantation established in 1998. Four different pocket of tree combinations *Terminalia sp* and *Tectona grandis* (1); *Gmelina arborea* and *Tectona grandis* (2); *Khaya sp* and *Tectona grandis* (3); *Theobroma cacao* and *Cola sp.* (4) were used, while undisturbed natural forest served as the control. Three plots (25 m x 25 m) were randomly mapped out of each site in which fresh litter were collected with litter trap (1 m x 1 m) and 45 litter bags were placed and 90 composite soil samples to the depths of 0-15 cm and 15-30 cm collected using a stainless steel auger. These collections followed the principle of co-location in each of the plots. Litter bag technique was used for Litter decomposition rate. The results of the litter accumulation in the forest plantations were in the magnitude of *Tectona grandis* and *Gmelina arborea* (1249.2 kgha^{-1}) > Teak and *Khaya sp.* (899.42 kgha^{-1}) > Teak and *Terminalia sp.*, (867.58 kgha^{-1}) > natural forest (489.96 kgha^{-1}) Cocoa and Cola (199.87 kgha^{-1}). The decomposition rates under *Tectona grandis* and *Khaya sp.*, *Tectona grandis* and *Gmelina arborea* mixtures were higher than other tree species mixtures. The rate of decomposition under *Tectona grandis* and *Gmelina arborea* mixtures was 5.3 times higher than that of *Tectona grandis* and *Terminalia sp.*, Cocoa and Cola combinations and natural forest at 6 weeks. At 15-30 cm soil depth, the C/N ratio was in the magnitude of *Tectona grandis* and *Gmelina arborea* (8.6:1) < Cocoa *and* Cola (9.3:1) < *Tectona grandis and Khaya sp. (9.8:1) < Tectona grandis* and *Terminalia sp.* Natural forest (11.7:1). The organic carbon and available nitrogen at 0-15cm soil depth under *Tectona grandis* and *Khaya sp.* combinations were significantly lower compared with other trees species combinations. However, the available phosphorus was significantly higher under *Tectona grandis* and *Terminalia sp.* compared with other tree species combinations. The dendogram indicated that the soil characteristics in the various tree species combinations plot were similar up to 50% with four clusters. The observed relative nutrient availability within the structurally different forested ecosystem in the study area might not be unconnected to the litter mixtures emerging from different tree combinations.

Keywords: Tree Species Combinations, Litter Decomposition, Plantation Forestry, Natural Forest, Soil Nutrient Status

1. Introduction

The replacement of native forests by exotic tree plantations can cause important changes in diversity and community composition at local and regional scales (Brockerhoff *et al.*, 2001). Forest leaves through photosynthesis absorb and retain nutrients in foliage and through abscission follow by decomposition return the nutrient into the ecosystem. Nutrient cycling is clearly related to decomposition and the availability of nutrients in any given soil is due in large part to the decay dynamics of the organic matter in that soil (Berg and McClaugherty, 2008). Forest litter is the dominant input of organic matter in the forest ecosystem and its' decomposition represent major pathway through which nutrient is being released into the environment.

The amount of organic matter present within the system is controlled by the relative rates of litter accumulation and loss (Baldock *et* al., 2004). Decomposition of leaf and needle litter is critical to forest nutrient cycling (Cadisch and Giller 1997). The success of tree species mixture in a plantation has been based primarily on those species with complimentary characteristics such as growth rate variation, crown structure, foliar phenology and root depth (Forrester *et al.,* 2004). Variation in this characteristic among these species may result in efficient biomass production and at the same time maintain a nutrient balance in the ecosystem. Rothe and Binkley (2001) noted that quantity and quality of litter fall in forest ecosystem are primarily based on stand species composition. Ola Adams, (1978) and Okeke and Omaliko, (1992) also reported that litter decomposition depends on the species in the forest or the composition of the plantation. Exotic species could be used as catalyst for secondary growth in mixed plantations (Parrota, 1992). There are strong concerns that the conversion of tropical forests into land for agriculture or plantation has negative effects on the carbon budget (Hertel *et al.,* 2009). This implies that the conversion of natural forest to mono specific or mixed plantation could not but have its impact on the ecosystem. Singh *et al.* (1985) in their studies on changes in soil properties under different plantations reported that there are alterations in the number of soil chemical properties such as lower pH and increased nutrient availability.

Various studies have shown a reciprocal influence of different tree species or plant communities on the soil ecosystem in different ecological zones (Oyeniyi and Aweto, 1986; Adejuwon and Ekanade, 1988; Awotoye *et al.,* 2009). The accumulation of organic matter in soil can greatly increase the cation exchange capacity and have positive impacts on the nutrient holding capacity of that soil (Berg and McClaugherty, 2008). The study of Michel *et al.,* (2010), reported significant reduction in soil organic carbon content and pH in multispecies tree plantation, cocoa plantation and mixed-crop fields compare to natural forest. According to Parrotta (1999), mixed plantation has the tendency to modify degraded soil and enhanced nutrient recycling regardless of the tree species origin.

In order to avoid the consequence of bad silviculture management; there has been an increased interest among foresters on the value of a proper mixture as a factor in the successful establishment and management of plantations (Ojo, 2005). Mixed forest types are currently recommended by foresters in order to improve the stability and biodiversity value of forest ecosystems (Hooper *et al.,* 2005). Therefore, this study was set out to assess the litter decomposition rate and determine the nutrient levels of soils under different tree species combinations.

2. Materials and Methods

The study was carried out in Ogbese Forest Reserve, Ekiti State of South-Western Nigeria. The reserve covers an area of 72.52 km^2 and lies approximately between Lat. 7^0 31^1 and 7^0 49^1N and Long. 5^0 7^1 and 5^0 27^1E. The forest reserve composed of plantation of various mixtures of tree species such as (*Terminalia sp* and *Tectona grandis; Gmelina arborea* and *Tectona grandis; Khaya sp* and *Tectona grandis; Theobroma cacao* and *Cola sp.*) established in 1998 by

manual clearing of the former vegetation. The existing vegetation of the undisturbed natural forest is characterized with abundant forest trees species of different families and general. The vegetation is dominated by mature tree species with moderate numbers of climbers, shrubs and herbs.

2.1. Soil Nutrient Measurements

A total of 90 soil samples were collected from all the five tree species combinations using a stainless steel auger. In each of the site, the three plots (25 m x 25 m) were randomly mapped out from which two composite soil samples to the depths of 0-15 cm and 15-30 cm were collected by simple random technique method. The composite samples were dried and sieved using 2 mm sieve for laboratory analysis.

The soil bulk density was determined by core method and particle size was determined by hydrometer method. The organic carbon was determined by Walkley-Black wet oxidation method (Nelson and Sommers, 1982), while the total nitrogen was determined by the micro-Kjeldal digestion method (Bremner and Mulvaney, 1982). Also, the soil pH was measured electrometrically in water at 1:2 soil /water ratio using pH meter, available phosphorus was determined by Bray P1 method while exchangeable cations (K$^+$, Na$^+$, Ca^{2+}, and Mg^{2+}) were determined using 1M NH$_4$OAc buffered at pH 7.0 as extractant (Thomas, 1982). The K$^+$ and Na$^+$ concentrations were read by flame photometry while Ca^{2+} and Mg^{2+} concentrations were read using atomic absorption spectrophotometer. The exchangeable acidity (Al^{3+}and H$^+$) was determined by standard NaOH titration method.

2.2. Litter Collection and Decomposition Rate Determination

Freshly senesced leaf material was collected with twelve 5 m × 5 m litter trap mounted in each of the five sampling sites. The litters were weighed and dried (70 ^{0}C for 48 hours). 10 g of the oven dried leaf material per site were placed in litter bag. Litter decomposition was determined by litter bag technique. We used a total of 45 nylon litter bags, three replicate litter bags (20 cm × 20 cm, 1 mm nylon mesh) for each of the three plot (25 m x 25 m) in each sites. These were randomly placed in the corresponding tree species combination plots and their rate of decomposition were monitored at 2, 4, 6 and 8 weeks and the kinetic of litter decay were monitored by determining weight loss during decomposition. Litter decomposition rate was determined with the formula

$$Mt = Mo \exp(-kt)$$

expressed as

$$In (Mo/Mt) = kt$$

Where
Mo = Mass of litter at time o,
Mt = Mass of litter at time t,
t = time of incubation in weeks, and

k = decomposition rate constants (Wood, 1974).

The results of the soil tests carried out were subjected to inferential statistics using one way ANOVA and Duncan multiple range test were used to find the differences in the mean scores. An index of deterioration was determined for the vegetation and soil variables according to the procedure outlined by (Ekanade *et al.*, 1991).

3. Results

The litter accumulation in the forest plantation was in the magnitude of *Gmelina arborea* and *Tectona grandis* (1249.20 kg ha^{-1}) > *Khaya sp* and *Tectona grandis* (899.42 kg ha^{-1}) > *Terminalia sp and Tectona grandis* (867.58 kg ha^{-1}) > natural forest (489.96 kg ha^{-1}) > *Theobroma cacao* and *Cola sp* (199.87 kg ha^{-1}). In the natural forest the litter accumulation was reduced by 77.00 % compared with agro-forestry plantation.

The rate of litter decomposition at two weeks varied among the different species combinations (Figure 1). However, rates of decomposition were similar at four and six weeks under *Tectona grandis* and *Gmelina, Theobroma cacao* and Cola combinations and natural forest. The *Tectona grandis* and *Khaya*, Teak and *Gmelina*

combinations had higher rate of decomposition than other tree species combinations. The decomposition rate of *Tectona grandis* and *Gmelina arborea* was 5.3 times higher than that of *Tectona grandis* and *Terminalia sp, Theobroma cacao* and cola combinations and natural forest at 6 weeks. However, decomposition rate of *Tectona grandis* and *Gmelina arborea* decline suddenly at 8 weeks by 74.25 %.

Table 1 shows the influence of the different tree species combinations on the soil physical properties at 0-15 cm and 15-30 cm depths. *Tectona grandis* and *Khaya sp* plot had the highest sand content at 0-15 cm depth while *Tectona grandis* and *Terminalia sp* plot had the least value. There were significant differences in the sand contents between the plots, even though the tree combinations were both fast growing exotic and indigenous species. The clay content in *Tectona grandis* and *Khaya, Tectona grandis* and *Gmelina* combinations and natural forest did not show any significant difference. However, high clay content were observed in both *Tectona grandis* mixed with *Terminalia sp;* and *Theobroma cacao* mixed with cola plots. The silt content under *Tectona grandis* mixed with *Gmelina arborea* plot was significantly higher than the other tree species combinations. There was no significant difference in the bulk density of the soil across the various plots.

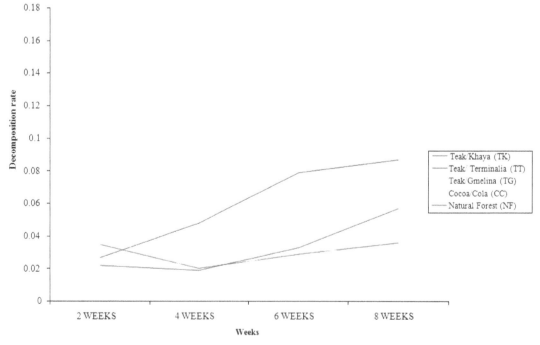

Figure 1. *Litter decomposition rate under different species combinations at 2, 4, 6 and 8 weeks*

The sand content in *Tectona grandis* and *Khaya sp* plot at 15-30 cm depth showed a similar relationship with that of the top soil, while *Theobroma cacao* and cola had lowest sand content. The sand content at 15-30 cm soil depth in the tree crop plantation was low compared to the top soil. The pattern of significant differences due to different tree species combination as observed at 0-15 cm depth was repeated at 15-30 cm. Generally, the sand content in the various plots decrease with increase in depth, but for textural classes *Tectona grandis* and *Khaya sp* plot was

observed to be sandy soil. The textural class changed from loamy sand to sandy loam under *Tectona grandis* and *Terminalia sp* mixture and *Theobroma cacao* and cola mixture with increase in soil depth. Similar change was observed under natural forest except that the sandy soil at 0-15 cm was modified to sandy loam at 15-30 cm. The soil bulk densities under *Tectona grandis* and *Khaya sp* and natural forest were significantly lower than other tree species combinations.

Table 1. Particle size analysis and bulk density of soil at 0-15 cm and 15-30 cm under different tree species combinations

Soil Depth	Tree species combinations	% Sand	% Clay	% Silt	Textural Class	Bulk Density
0-15cm	Teak/Khaya (TK)	89.65ᵃ	4.37ᵇ	5.96ᵈ	Sandy	1.20ᵃ
	Teak/ Terminalia (TT)	76.73ᶜ	7.87ᵃ	15.40ᵇ	Loamy sand	1.21ᵃ
	Teak/Gmelina (TG)	78.23ᶜ	4.54ᵇ	17.23ᵃ	Loamy sand	1.25ᵃ
	Cocoa/Cola (CC)	77.07ᶜ	7.48ᵃ	15.68ᵇ	Loamy sand	1.13ᵃ
	Natural Forest (NF)	86.23ᵇ	4.64ᵇ	9.12ᶜ	Sandy	1.10ᵃ
15-30cm	Teak/Khaya (TK)	87.50ᵃ	4.90ᵇ	7.60ᵉ	Sandy	1.20ᶜ
	Teak/ Terminalia (TT)	73.80ᶜ	10.00ᵃ	16.20ᶜ	Sandyloam	1.45ᵃ
	Teak/Gmelina (TG)	77.40ᵇ	4.52ᵇ	18.10ᵇ	Loamysand	1.40ᵃ
	Cocoa/Cola (CC)	70.00ᵈ	9.60ᵃ	20.40ᵃ	Sandyloam	1.45ᵃ
	Natural Forest (NF)	85.30ᵃ	4.80ᵇ	9.90ᵈ	Loamysand	1.19ᵇ

Means with the same letter(s) are not significantly different by Duncan's Multiple Range Test at P <0.05

Table 2 shows the soil chemical characteristics at 0-15 and 15-30 cm respectively under different tree species combinations. Soil pH ranged between 6.76 in Teak and *Gmelina*, cocoa and cola, and 6.90 in Teak *and Khaya*. The soil pH within the different plots was significantly different at (P < 0.05) although they were all near neutral pH. The organic carbon and available Nitrogen at 0-15 cm soil depth under Teak and *Khaya* combinations were significantly lower compared with other trees species combinations. The C/N ratio showed no significant difference between the plots at 0-15 cm.

Table 2. Chemical characteristics at 0-15 cm and 15-30 cm soil depths under different tree species combinations

Soil Depth	Tree Species Combinations	Soil pH	Organic Carbon g kg⁻¹	Avail Nitrogen g kg⁻¹	C:N	Avail P mg kg⁻¹	Ca	Mg	Na	K	H+Al	ECEC
							cmol kg⁻¹					
0-15cm	Teak/Khaya (TK)	6.90ᵃ	1.78ᵉ	0.15ᵉ	11.8:1ᵃ	2.70ᵃ	10.44ᵇ	1.83ᵃ	0.32b	0.11ᶜ	0.04ᵇ	12.74ᶜ
	Teak/ Terminalia (TT)	6.84ᵃᵇ	2.06ᵈ	0.18ᵇ	11.4:1ᶜ	0.95ᵈ	10.52ᵈ	1.96ᵃ	0.40ᵃ	0.12ᵇᶜ	0.04ᵇ	13.04ᶜ
	Teak/Gmelina (TG)	6.76ᶜ	2.36ᵇ	0.20ᵇ	11.8:1ᵃ	2.43ᵇ	12.24ᵇ	1.37ᵇ	0.34ᵇ	0.14ᵇᶜ	0.04b	14.12b
	Cocoa/Cola (CC)	6.76ᶜ	2.20ᶜ	0.19ᵇ	11.6:1ᵇ	2.10ᶜ	11.43ᵇ	1.00ᶜ	0.42ᵃ	0.13ᵇᶜ	0.04ᵇ	13.02ᶜ
	Natural Forest (NF)	6.80ᵇᶜ	3.18ᵃ	0.27ᵃ	11.8:1ᵃ	0.95ᵈ	24.46ᵃ	1.39ᵇ	0.33ᵇ	0.19ᵈ	0.06ᵃ	26.43ᵈ
15-30cm	Teak/Khaya (TK)	6.93ᵃ	0.88ᶜᵇ	0.09ᵇ	9.8:1ᵃᵇ	2.02ᶜ	8.88ᵇ	0.88ᵇ	0.28	0.07ᶜ	0.05ᵇ	7.78ᵇ
	Teak/ Terminalia (TT)	6.87ᵃ	1.17ᵇ	0.10ᵇ	11.7:1ᵃ	6.55ᵃ	7.41ᵇ	0.91ᵇ	0.36ᵇ	0.14ᵃ	0.04ᵇ	8.86ᵇ
	Teak/Gmelina (TG)	6.49ᵇ	0.77ᶜ	0.09ᵇ	8.6:1ᶜ	3.33ᵇ	9.07ᵇ	0.91ᵇ	0.29ᶜᵈ	0.09ᶜ	0.63ᵃ	10.08b
	Cocoa/Cola (CC)	6.87ᵃ	1.11ᶜᵇ	0.12ᵇ	9.3:1ᵇ	2.66ᶜᵇ	8.88ᵇ	1.64ᵃ	0.33ᶜᵇ	0.12ᵇ	0.04ᵇ	11.00ᵇ
	Natural Forest (NF)	6.92ᵃ	1.76ᵃ	0.15ᵃ	11.7:1ᵃ	2.63ᶜᵇ	13.83ᵃ	1.72ᵃ	0.40ᵃ	0.14ᵃ	0.04ᵇ	29.13ᵃ

Table 2. Continued

Soil Depth	Tree Species Combinations	Base Salt %	Cu mg kg⁻¹	Zn	Fe	Mn
0-15cm	Teak/Khaya (TK)	99.67ᵇ	0.63ᶜ	8.22c	2.58ᵇ	57.65a
	Teak/ Terminalia (TT)	99.67ᵇ	2.03ᵃ	9.07ᵇ	1.89c	45.75b
	Teak/Gmelina (TG)	99.71ᵃᵇ	0.70ᶜ	3.99ᶜ	0.66ᶜ	27.59ᵈ
	Cocoa/Cola (CC)	99.72ᵃᵇ	1.29ᵇ	734ᵈ	1.45d	39.22ᶜ
	Natural Forest (NF)	99.78ᵈ	0.75ᵈ	12.05ᵃ	3.55ᵃ	13.93ᶜ
15-30cm	Teak/Khaya (TK)	99.41ᵇ	0.40ᵇᶜ	8.73ᵇ	2.81ᵃ	56.93ᵃ
	Teak/ Terminalia (TT)	99.57ᵃ	0.78ᵃ	6.70ᶜ	1.56ᵇ	57.63ᵃ
	Teak/Gmelina (TG)	99.35ᵇ	0.31ᶜ	7.26ᶜ	0.67ᶜ	35.96ᵇ
	Cocoa/Cola (CC)	99.67ᵃ	0.69ᵇᵃ	2.27ᵈ	1.94ᵇ	60.15ᵃ
	Natural Forest (NF)	99.71ᵃ	0.25ᶜ	17.29ᵃ	2.91ᵃ	37.77ᵇ

Available phosphorus was significantly higher under Teak and *Khaya* compared with other tree species combinations. The Ca^{2+} and K^+ contents under the different tree combinations were significantly lower than the natural forest; however, the other exchangeable bases such as Mg^{2+} and Na^+ did not follow a definite pattern. The ECEC under natural forest was between 80-110 % higher than the other different tree species mixtures. Similarly, Zn^{2+} and Fe^{2+} content under the natural forest were significantly higher than other tree species combinations. In contrast, Cu^{2+} and Mn were significantly lower in natural forest than in other tree species combinations.

At 15-30 cm soil depth the mixture of Teak and *Gmelina* showed the least values of soil pH, organic carbon and available Nitrogen. The C/N ratio was in the magnitude of Teak and *Gmelina* (8.6:1) < cocoa *and* cola (9.3:1) < Teak *and Khaya (9.8:1)* < Teak and *Terminalia* (11.7:1) Natural forest. The subsoil level under Teak and *Terminalia* was significantly rich in available P (6.55 mg kg^{-1}). However, the exchangeable bases did not follow this pattern. The natural forest had significantly high value in respect of Ca^{2+}, while cocoa and cola plantation with natural forest were higher in Mg than other tree species plantations. The ECEC under natural forest was also significantly (29.13 mg kg^{-1}) higher than other plantations. Other elements such as Cu, Zn and Fe did not follow a definite pattern with different species combinations.

Table 3. Deterioration index of soil chemical properties at 0-15 cm and 15-30 cm soil depths under different tree species combinations

Soil Depth	Tree Species Combinations	Soil pH	Carbon g kg^{-1}	Nitrogen g kg^{-1}	Avail P mg kg^{-1}	Exchangeable Bases					
						Ca	Mg	Na	K	H+Al	ECEC
						cmol kg^{-1}					
0-15cm	Teak/Khaya (TK)	-0.01	0.44	0.45	-1.84	0.57	-0.32	0.03	0.42	0.33	0.52
	Teak/ Terminalia (TT)	-0.01	0.35	0.33	0	0.57	-0.41	-0.21	0.37	0.33	0.51
	Teak/Gmelina (TG)	0.01	0.26	0.26	-1.56	0.5	0.01	-0.03	0.26	0.33	0.47
	Cocoa/Cola (cc)	0.01	0.31	0.3	-1.21	0.53	0.28	-0.27	0.32	0.33	0.51
15-30cm	Teak/Khaya (TK)	-0.01	0.5	0.4	0.23	0.36	0.49	0.3	0.5	-0.25	0.73
	Teak/ Terminalia (TT)	0.01	0.34	0.33	-1.49	0.46	0.47	0.1	0	0	0.7
	Teak/Gmelina (TG)	0.06	0.56	0.4	-0.27	0.34	0.47	0.28	0.36	-14.75	0.65
	Cocoa/Cola (cc)	0.01	0.37	0.2	-0.01	0.36	0.05	0.18	0.14	0	0.62

Table 3. Continued

Soil Depth	Tree Species Combinations	Base Salt %	Cu	Zn	Fe	Mn
			mg kg^{-1}			
0-15cm	Teak/Khaya (TK)	0.001	-1.52	0.32	0.27	-3.14
	Teak/ Terminalia (TT)	0.001	-7.12	0.25	0.47	-2.28
	Teak/Gmelina (TG)	0.001	-1.8	0.67	0.81	-0.98
	Cocoa/Cola (cc)	0.001	-4.16	0.39	0.59	-1.82
15-30cm	Teak/Khaya (TK)	0.003	-0.6	0.5	0.03	
	Teak/ Terminalia (TT)	0.001	-2.12	0.61	0.46	
	Teak/Gmelina (TG)	0.003	-0.24	0.58	0.77	
	Cocoa/Cola (cc)	0.001	-1.76	0.87	0.33	

Table 4. Correlation coefficient of soil chemical properties with rate of decomposition at 4 ,6 and 8 weeks.

Elements	Rate of Decomp. at 4 weeks	Rate of Decomp. at 6 weeks	Rate of Decomp. at 8 weeks
pH	-0.23	-0.57**	-0.48**
Ca	-0.13	0.22	-0.14
Mg	-0.15	-0.33*	-0.35*
Na	0.21	0.37*	0.58**
K	-0.15	0.27	0.03
Pc	0.02	-0.09	0.09
Avail P	-0.40**	-0.99	0.04**

* Significant at p <0.05
** Significant at p < 0.01

Table 3 shows deterioration indices of some soil chemical properties at 0-15 cm and 15-30 cm soil depths respectively. At 0-15 cm soil depth, the mixture of the tree species of Teak with *Khaya* plot and Teak with *Terminalia* plot did not deteriorate in terms of soil pH. Similarly, the mixture of Teak with *Gmelina,* and cocoa with cola did not have pronounced negative effect on the soil pH at 0-15 cm depth. The deterioration index of organic carbon and available nitrogen were lower under Teak mixture (0.026, 0.26) compared with other tree species combinations. There was no negative effect of different mixture of tree species on the available soil phosphorus. In contrast, the exchangeable bases (Ca^{2+}, Mg^{2+}, Na^+ and K^+), the exchangeable acidity (Al^{3+} and H^+) and ECEC showed different levels of deterioration under different tree species mixtures. Similarly, deterioration indices of Zn and Fe varied among the tree species mixtures. However, the different mixtures of tree species had no deterioration effect on the soil Cu and Mn at 0-15 cm depth. At 15-30 cm depth, the trend in the deterioration indices were similar to that of the top soil except for available P under Teak and *Khaya* mixture where pronounced effect of the mixture was observed on the soil P.

The result of the dendograme (Figure 2) showed that the soil characteristics in the various tree species combination plot were similar up to 50 % when clustered into four groups as: (1) Sand, ECEC, Zn and Fe, (2) pH, Mg, Na and Cu (3) Ca, K, BS, PC and PN and (4) Clay, Silt, H-Al, Available P and Mn. Furthermore, the first two groups were related up to 40 % in their soil characteristics while the first three groups showed low relationship up to 30%. However, the whole variables were related only up to 10%. Table 5 gave positive relationship between the soil chemical properties and rate of decomposition of litter falls of the species combinations.

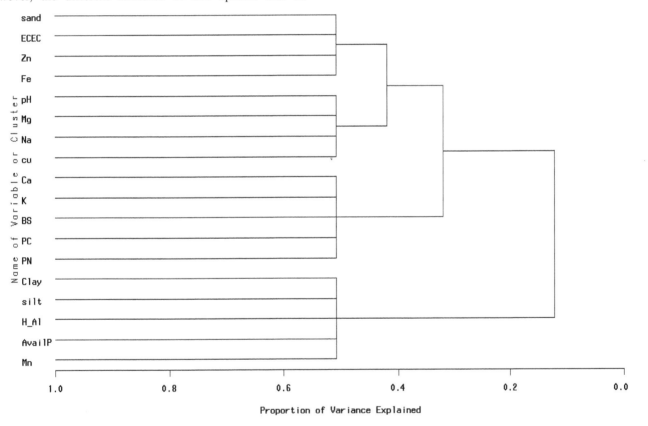

Figure 2. Inter-Relationship between the Soil Properties under the Various Species Combination

4. Discussion

The litter accumulation pattern observed in this study showed that Teak in combination with other tree species had higher litter accumulation. This agrees with Ojo (2005) who observed a massive litter accumulation in Teak plantation in Akure forest reserve. However, variation in litter accumulation among plantations could be attributed to some factors such as the number of trees in the ecosystem, type of species and size. Weaner *et al.,* (1987) observed that trees of greater timber size tend to have more litters.

Litters act as input-output system of nutrients and their decomposition is the primary mechanism by which organic matter and nutrients are returned to the soil for reabsorption by the growing plants. The fast rate of decomposition observed in the litters of tree species at the early stage decomposition in this study is in line with the decomposition characteristics of tropical tree species which had been reported by many workers (Kumar and Deepu, 1992 and Awotoye *et al.,* 2009). The contributing factors had been attributed to favourable moisture and temperature regimes (Sreekala *et al.,* 2001). However, the pronounced variation in the pattern of decomposition rate observed in the plantations as expressed in Teak and *Khaya sp*, Teak and *Gmelina arborea* compared to Teak and *Terminalia sp*, cocoa and cola mixtures may not be unconnected with different foliar chemical compositions of different tree species within the plantations. Decomposition of organic material is affected by its biochemical properties (Parrotta, 1999). Some species have been reported to have lignin content which can render some of the cellulose and the other constituents inaccessible to the microflora.

Soil physical and chemical properties have been proposed by some workers as indicators for assessing the effect of land use management (Janzen *et al.*, 1992; Bremer *et al.*, 1994; Alvarez and Alvarez, 2000). Top soil morphological characteristics between the various tree species combination plots showed no significant difference in bulk density. This could be explained by the low activities such as hunting, fuel wood gathering, and picking of snail within the forest plantations. In contrast to natural forest, most artificial forest plantations in the tropics are free from human and animal interferences because of reduced undergrowth. Trampling of pasture by cattle has been shown to increase bulk density of active pasture over that of undisturbed forest. The low sand content at 15-30 cm soil depth in different tree species combination plots as against the top soil conformed to the findings of Onyekwelu *et al.*, (2006) at Oluwa and Omo forest reserves. They reported a decrease in sand content in *Gmelina arborea* plantation sites as the soil depth increases and, an increase in the clay and silt contents as you go down the soil profile.

The soil chemical characteristics under the various species combinations showed significant differences in properties. Significant difference in pH may be due to leaching of base elements. Onyekwelu *et al.*, (2006) observed that nutrients are swiftly leached by heavy precipitation in the tropical rainforest. The organic matter and available nitrogen contents under various tree species combination was significantly lower to that of natural forest, this may be due to lack of understorey vegetation that often associate with teak plantation even in a mixture with other species. Juo and Manu (1996) found that growing vegetation tended to decrease soil pH, with low nutrient stocks. This often exposes the soil to soil erosion.

Tree plantations are often used in forest rehabilitation and restoration to foster regenerating tree species, protect soil and increase the concentration of soil nutrients such as N, P, K, Ca and Mg (Parrotta, 1992 and Leopold *et al*, 2001). The result of this study has revealed that planting *Tectona grandis* in combination with other tree species especially indigenous fast growing species has potentials for maintaining soil nutrient level. The C/N ratio of the soil under different tree species combination did not show any significant difference to that of natural forest. This implies that a mixed plantation could naturally sustain its nutrient level under efficient silvicultural practice. In a study of 5 year old pure and mixed species plantations, Montagnini (2000) observed that mixed plantations will take longer time to diminish soil nutrients.

Also, Parrotta (1999) found that soil K, Mg, Na, and Fe concentrations were greater in mixed plantations than each species in plantation separately.

The low Ca and K contents under the different tree combinations as against that of natural forest contradict the findings of (Nwoboshi, 1970). Nwoboshi (1970) reported high Ca concentration in soil especially at peak leaf fall. Similarly, Singh *et al.*, (1985) reported high concentration of exchangeable Ca, Mg and K under teak plantation. This variation may likely due to the mixture of teak with other tree species. Tucker and Murphy, 1997 and Hartey, 2002 were of the opinion that mixed species plantations could further accelerate the restoration process of degraded soil. However, the ability of tree species combination to improve the nutrient status of the soil would depend on a good and an appropriate choice of tree species combinations. When indigenous species with lesser growing characteristics is mixed with fast growing exotic species in a plantation, this may mitigate the demand of soil nutrient by fast growing species. Narong *et al.*, (2007) opined that teak should be mixed with other tree species during the commencement of plantation establishment.

5. Conclusion

The differences in the litter accumulation and rate of decomposition in the tree mixture induced significant soil nutritional differences among the combinations. Teak and *Gmelina arborea* had the highest quantity of litter and faster rate of decomposition, which is an index to high nutrient release to the soil under this combination. The observed relative nutrient availability within the structurally different forested ecosystem in the study area might not be unconnected to the litter mixtures emerging from different tree combinations.

Acknowledgments

The authors are grateful to the Management of the Ekiti State Ministry Environment, Department of Forestry for allowing us to conduct the research in their forest reserve.

References

[1]   Adejuwon, J.O. and Ekanade, O. (1988). Soil Changes Consequent upon the Replacement of Tropical Rainforest by Plantations of *Gmelina arborea*, *Tectona grandis* and *Terminalia superba*. *Journal of World Forest Resources Management* 3: 47-59.

[2]   Alvarez, R. and Alvarez, C. R. (2000). Soil organic matter pools and their association with carbon mineralization kinetics. *Journal of America society Soil Science* 64(1): 184-189.

[3]   Awotoye, O. O., Ekanade O. and Airouhudion O. (2009). Degradation of the physicochemical properties resulting from continuous logging of *Gmelina arborea* and *Tectona grandis* plantations. *African Journal of Agricultural Research.* 4(11): 1317-1324.

[4] Baldock J.A., Masiello C.A., Gelinas Y., Hedges J.I.,(2004). Cycling and composition of organic matter in terrestrial and marine ecosystems, *Marine Chemistry* 92 (2004) 39– 64

[5] Berg, B., and McClaugherty, C. (2008). Plant Litter; *Decomposition, Humus Formation, Carbon Sequestration* second edition, Springer-Verlag Berlin Heidelberg1-340. E-book available on www.springer.com

[6] Binkley, D., and Resh, S. C., (1999). Rapid changes in soils following eucalyptus afforestation in Hawaii. *Journal of America Soil Science Society* 63:222–225.

[7] Bremner, J. M. and Mulvaney C. S. (1982). Nitrogen- Total. In: Methods of soil analysis. Page, A. L. et al. (eds). Methods of Soil Analysis. Part 2. Agron. Monogr. 9. Second Edition. pp 595-624. ASA and SSSA. Madison, Wisconsin, U.S.A.

[8] Brockerhoff, E. G., Ecroyd, C. E. and Langer, E. R., (2001). Biodiversity in New Zealand plantation forests: policy trends, incentives, and the state of our knowledge. *New Zealand Journal of Forestry* 46: 31–37.

[9] Ekanade, O.; Adesina, F.A. and Egbe, N.E. (1991). Sustaining Tree Crop Production under Intensive Land Use: An Investigation into Soil Quality Differentiation under Varying Cropping Patterns in Western Nigeria. *Journal of Environmental Management*. 32: 105-113.

[10] Forrester, D.I.; Bauhus, J. and Khanna, P.K. (2004). Growth Dynamics in a Mixed Species Plantation of *Eucalyptus globules* and *Accacia mearnsii*. *Forest, Ecology and Management*. 193: 81-95.

[11] Hartley, M.J. (2002). Rationale and Methods for Conserving Biodiversity in Plantation Forests. *Forest, Ecology and Management*. 155: 81-95.

[12] Hertel, D., Harteveid, M. A. and Leuschner, C. (2009). Conversion of a Tropical Forest into Agroforest Alters the Fine root-related Carbon Flux to Soil. *Soil Biology and Biochemistry*. 41: 481-490

[13] Hooper, D.U., Chapin, F.S., Ewel, J.J., Hector, A., Inchausti, P., Lavorel, S. et al. (2005). Effects of biodiversity on ecosystem functioning: a consensus of current knowledge. Ecol. Monogr., 75, 3–35.

[14] Janzen, H. H., Campbell, C. A., Brant, S. A., Lanfond, G. P. and Townley, S. (1992). Light-fraction organic matter in soils from long-trm crop rotations. *Journal of Soil Biology and Biochemistry*. 8: 200-213.

[15] Juo, A. S. R. and Manu, A. (1996). Nutrient effect on modification of shifting cultivation in West Africa. *Journal of Agriculture, Ecosystem and Environment*. 58: 49-60.

[16] Kumar, and Deepu, J.K. (1992). Litter Production and Decomposition Dynamics in Moist, Deciduous Forests of the Western Ghats in Peninsular India. *Forest, Ecology and Management*. 50:181-201.

[17] Leopold, C.; Andrus, R.; Firkeldey, A. and Knowles, D. (2001). Attempting Restoration of Wet Tropical Forests in Costa Rica. *Forest, Ecology and Management* 142: 243-249.

[18] Michel, K. Y., Pascal, K. T. A., Souleymane, K., Jerome, E. T., Yao, T., Luc, A. and Danielle, B. (2010). Effects of land use types on soil organic carbon and nitrogen dynamics in mid-west Cote d'Ivoire. *European Journal of Scientific Research* 2: 211-222

[19] Montagnini, F. (2000). Accumulation in above Ground Biomass and Soil Storage of Mineral Nutrients in Pure and Mixed Plantations in Humid Tropical Lowland. *Forest, Ecology and Management*. 134:257-270.

[20] Narong, K.; Katsutoshi, S. and Sota, T. (2007). Composition and Diversity of Woody Regeneration in a 37-year Old Teak (Tectonal Grandis L.) Plantation in Northern Thailand. *Forest, Ecology and Management*. 247: 246-254.

[21] Nelson, D. W. and Sommers L. E. (1982). Total carbon, organic carbon and organic matter. In: Page, A. L. et al. (eds). Methods of Soil Analysis. Part 2. Agron. Monogr. 9. Second Edition. pp 539-579. ASA and SSSA. Madison, Wisconsin, U.S.A.

[22] Nwoboshi, L.C. (1970). Studies on Nutrient Cycle in Forest Plantations: Preliminary Observations on Litter Fall and Macro-Nutrient Return in a Teak (*Tectona Grandis L.F.*) Plantation. *Nigerian Journal of Science* 4(2): 231-237.

[23] Ojo, A.F. (2005). Organic Matter and Nutrient Dynamics of the Natural Rainforest and Teak Plantation in Akure Forest Reserve, Nigeria. PhD Thesis in the Department of Forestry and Wood Technology, Federal University of Technology, Akure, Nigeria. 130 pp.

[24] Okeke, A.L. and Omaliko, C.P.E. (1992). Leaf Litter Decomposition and Carbon dioxide Evolution in some Agroforestry Fallow Species in Southern Nigeria. *Forest, Ecology and Management*. 50: 103-116.

[25] Ola Adams, B.A. (1978). Litter fall and Disappearance in a Tropical Moist Semi-deciduous Forest. *Nigerian Journal of Forestry* 8:31-36.

[26] Onyekwelu, J.C.; Mosandi R., Stimm B. (2006). Productivity, Site Evaluation and State of Nutrition of *Gmelina arborea* Plantations in Oluwa and Omo Forest Reserve. *Nigerian Forest Ecology and Management*. 229: 214-227.

[27] Oyeniyi, O.C. and Aweto A.O. (1986). Effects of Teak Planting on Alfisol Topsoil in Southwestern Nigeria. *Singapore Journal of Tropical Geography*. 7:145-149.

[28] Parrota,J.A. (1992). The Role of Plantation Forests in Rehabilitation Degraded. *Tropical Ecosystem, Agricultural Ecosystems and Environment*. 41: 115-133.

[29] Parrotta, J .A. (1999). Productivity, Nutrient Cycling, and Succession in Single and Mixed-Species Plantations of *Casuarina equietifolia, Eucalyptus robusta* and *Leucaena leucocephala* in Puerto Rico. *Forest, Ecology and Management*. 124: 47-77.

[30] Rothe A., and Binkley D. (2001). Nutritional Interactions in Mixed Species Forests: A Synthesis, *Can. J. Res*. 31:1855-1870

[31] Singh, S.B.; Nath, S.; Pal, D.K. and Banarjee (1985). Changes in Soil Properties under Different Plantations of the Darjeeling Forest Division. *India Forester* 111(2): 90-98.

[32] Sreekala, N.V.; Mercy George, V.K.G.; Unnithan, P.S. and John, R. (2001). Decomposition Dynamics of Cocoa Litter under Humid Tropical Conditions. *Journal of Tropical Agriculture*. 39: 190-192.

[33] Thomas G. W. (1982). Exchangeeable cations. In: Page A. L. *et al*. (eds). Methods of soil analysis. Part 2. Agron. Monogr. 9. Second Edition. pp 159- 165. ASA and SSSA. Madison, Wisconsin, U.S.A.

[34] Tucker, N.I.J. and Murphy, T.M. (1977). The Effects of Ecological Rehabilitation on Vegetation Recruitment: Some Observations from the Wet Tropics of North Queensland. *Forest, Ecology and Management*. 99: 133-152.

[35] Walkley, A., and Black, I. A., (1934). An Examination of the Degtgareff method for determining soil organic matter and proposed of modification of the chromic acid titration method. *Journal of Soil Sciences*. 37: 29-33.

[36] Weaner, P.L.; Bindsay, R.A. and Lugo, A.E. (1987). Soil Organic Matter in Secondary Forest in Puerto Rico. *Biotropica* 19(1): 17-23.

[37] Wood, T.G., (1974). Field investigations on the decomposition of leaves of Eucalyptus delegatensis in relation to environmental factors. Pedobiologia 14, 343–371.

19

Nematode fauna of Rajaji National Park, with First record of *Granonchulus subdecurrens* Coetzee, 1966 (Mononchida: Mylonchulidae) from India

Vinita Sharma, Alka Dubey[*]

Zoological Survey of India, Northern Regional Centre, Dehradun, Uttarakhand, India

Email address:

alkabioinfo964@gmail.com (A. Dubey) vinitascb@gmail.com (V. Sharma)

Abstract: A total 26 species of terrestrial nematode (15 from order Dorylaimida and 11 from order Mononchida) has been recorded from Rajaji National Park (RNP), Uttarakhand, India. All are being reported first time from RNP. Of these, *Granonchulus subdecurrens* Coetzee, 1966 is being recorded from first time from India.

Keywords: Dorylaimida, Mononchida, *Granonchulus*, New Records, Rajaji National Park, India

1. Introduction

India is a rich faunal diversity country. In invertebrate fauna, nematode constitute second largest group after arthropods. Significant work on terrestrial nematode already being done in India by reporting 2100 species of nematodes from different ecosystem [9]. A number of Indian Nematologist [1-3,5-7] made their contribution to work out nematode fauna of Protected areas viz., Keoladeo National Park, Kaziranga National Park, Rajaji National Park and Ranthambore National Park and Silent Valley National Park.

In Uttarakhand, Rajaji National Park (RNP) is one of the important conservation area, because it has rich bio-diversity. In nematode fauna only 5 species have been already recorded from by the present authors [5] from this protected area. In the present paper an account of identified 26 species of nematode from this protected area is presented, of which *Granonchulus subdecurrens* is first reported from country.

2. Materials and Methods

During survey for nematode fauna of RNP in 2013, soil samples were collected soil around the of forest tree species from Chilla Range, Jhabar and Hazra Beats, RNP, Haridwar, were processed by sieving and decantation and modified Baerman's funnel techniques. Extracted nematodes were heat-killed and fixed in hot 4% formalin. The nematodes will be transferred from fixative to a solution of 5 parts of glycerin and 95 parts of 30% alcohol in a cavity block and dehydrated by slow evaporation method [8] and mounted in anhydrous glycerin. The mounted nematodes were later measured by ocular micrometer and drawn using drawing tube attached to Olympus BX-51 DIC Microscope. LM photographs were taken using Olympus digital camera.

3. Results

During nematode survey 26 species has been recorded from RNP, belonging to 11 genera, 7 families and 2 orders, of Dorylaimida and Mononchida. Four species of *Xiphinema* are plant parasitic in nature.

All 26 species are recorded for first time from this protected area, of which *Granonchulus subdecurrens* Coetzee, 1966 is being reported from India. All the specimens are registered in National Zoological Collection, NRC, ZSI, Dehra Dun, India.

Order Dorylaimida
Family Aporcelaimidae Heyns, 1965
1. *Aporcelaimellus capitatus* (Thorne & Swanger, 1936) Heyns, 1965
2. *Aporcelaimellus clamus* Throne, 1974
3. *Aporcelaimellus invisus* Tjepkema, Ferris and Ferris, 1971
4. *Aporcelaimellus obscurus* (Thorne & Swanger, 1936) Heyns, 1965
Family Qudsianematidae Jairajpuri, 1965
5. *Eudorylaimus chauhani* (Baqri & Khera, 1975) Andrassy, 1986

6. *Allodorylaimus irritans* (Cobb in Thorne & Swanger, 1936) Andrassy, 1986
7. *Discolaimus major* Thorne, 1939
8. *Discolaimus tenax* Siddiqi, 1964
9. *Discolaimus texanus* Cobb, 1913
10. *Discolaimus similis* Thorne, 1939
Family Xiphinematidae Dalmasso, 1969
11. *Xiphinema americanum* Cobb, 1913
12. *Xiphinema inaequle* Khan & Ahmad, 1974
13. *Xiphinema insigne* Loos, 1949
14. *Xiphinema opisthosternum* Siddiqi, 1961
Family Mydonomidae Thorne, 1964
15. *Dorylaimoides micoletzkyi* (de Man, 1921) Thorne & Swanger, 1936
Order Mononchida Jairajpuri, 1969
Family Mylonchulidae Jairajpuri, 1969
16. *Mylonchulus armus* Khan & Jairajpuri, 1979
17. *Mylonchulus brachyurus* (Butschli, 1873) Andrassy, 1958
18. *Mylonchulus hawaiensis* (Cassidy, 1931) Andrassy, 1958
19. *Paramylonchulus mashhoodi* (Khan & Jairajpuri, 1979) Jarajpuri & Khan, 1981
20. *Paramylonchulus mulveyi* (Khan & Jairajpuri, 1979) Jarajpuri & Khan, 1981
21. *Granonchulus subdecurrens* Coetzee,1966
Family Iotonchidae Jairajpuri, 1969
22. *Iotonchus indicus* Jairajpuri, 1969
23. *Iotonchus parabasiodontus* Mulvey & Jensen, 1967
24. *Iotonchus trichurus*(Cobb, 1917) Andrassy, 1958
25. *Iotonchulus longicaudatus* (Baqri, Baqri & Jairajpuri, 1978) Andrassy, 1993
Family Bathyodontidae Clark, 1961
26. *Bathyodontus mirus*(Andrassy, 1956) Andrassy in Hooper & Cairns, 1959

Abbreviations

L = Total body length; a = Body length/greatest body width; b = Body length/neck length; c = Body length/tail length; c' = Tail length/body width at anus; V = Distance of vulva from ant. end x 100/body length

Description and Diagnosis

1. *Aporcelaimellus capitatus (*Thorne & Swanger, 1936) Heyns, 1965
Morphological taxonomic calculation: Female: L=2.0mm; a=27; b=3.7; c=53; c'=1.1; V=54; Odontostyle=20μm; Odontophore= 40μm
Description: Female: Body ventrally arcuated upon fixtation. Lip region discoid. Lips with prominent papillae. Vulva a transverse slit. Reproductive system amphidelphic. Tail dorsally convex.
Male: Not found.
Locality and habitat: Soil around the roots of *Shorea robusta*.
2. Aporcelaimellus clamus Throne, 1974

Morphological taxonomic calculation: Female: L=2.3mm; a=29; b=4.0; c=59; c'=0.9; V= 53; Odontostyle=18μm; Odontophore= 38μm
Description: Female: Body slightly arcuate upon fixation. Lips region set off by slightly narrowing of neck. Vulva a transverse slit. Reproductive system amphidelphic. Tail bluntly conoid.
Male: Not found.
Locality and habitat: Soil around the roots of *Shorea robusta* from Hazra Beat.
3. *Aporcelaimellus invisus* Tjepkema, Ferris and Ferris, 1971
Morphological taxonomic calculation: Female: L=2.1mm; a=26; b=3.7; c=55; c'=0.9, V=54; Odontostyle=20μm; Odontophore= 40μm.
Description: Female: Body slightly arcuate upon fixation. Lips region set off by slightly narrowing of neck. Odontostyle aperture 63% of odontostyle length. Vulva a transverse slit. Reproductive system amphidelphic. Tail long, bluntly conoid.
Male: Not found.
Locality and habitat: Soil around the roots of unidentified trees from Hazra Beat.
Remarks: The measurements of present specimens fit well with the specimen described by Tjepkema, Ferris and Ferris, 1971 except odontophore (length obscure in type).
4. *Aporcelaimellus obscurus(*Thorne & Swanger, 1936) Heyns, 1965
Morphological taxonomic calculation: Female: L=2.2mm; a=28; b=3.6; c=68; c'=0.76, V=58; Odontostyle=21μm; Odontophore= 40μm
Description: Female: Body ventrally arcuated upon fixtation. Lips distinct, well separated. Vulva a transverse slit. Reproductive system amphidelphic. Tail convex-conoid, tip blunt.
Male: Not found.
Locality and habitat: Soil around the roots of *Ehretia laevis* from Jhabar Beat.
5. *Eudorylaimus chauhani* (Baqri & Khera, 1975) Andrassy, 1986
Morphological taxonomic calculation: Female: L=1.5mm; a=29; b=4.0; c=24; c'=2.3; V=50; Odontostyle=18μm; Odontophore= 30 μm
Description: Female: Body curved ventrally upon fixation. Lip region well offset by constriction. Vulva a transverse slit. Reproductive system amphidelphic. Tail dorsally convex–conoid with sub-acute tip.
Male: Not found.
Locality and habitat: Soil around the roots of *Tectona grandis* from Hazra Beat.
1. *Allodorylaimus irritans* (Cobb in Thorne & Swanger, 1936) Andrassy, 1986
Morphological taxonomic calculation: Female: L=1.2mm; a=23; b=3.7; c=27; c'=1.8; V=51; Odontostyle=20μm; Odontophore= 45μm
Description: Female: Body curved ventrally upon fixation. Lip region offset. Lips well separated. Odontostyle aperture

is 50% of odontostyle length. The expanded part of the oesophagus occupies 1/3 of total oesophageal length. Vulva a transverse slit. Reproductive system amphidelphic. Tail dorsally convex and tail tip acute.

Male: Not found.

Locality and habitat: Soil around the roots of unidentified plants from Hazra Beat.

2. Discolaimus major Thorne, 1939

Morphological taxonomic calculation: Female: L=2.2mm; a=37; b=4.4; c=55; c'=1.2, V=50; Odontostyle=21µm; Odontophore= 43µm

Description: Female: Body slightly curved ventrally upon fixation. Lip region discoid, set off from body. Vulva a transverse slit. Reproductive system amphidelphic. Tail dorsally convex-conoid, with rounded tip.

Male: Not found.

Locality and habitat: Soil around the roots of Shorea robusta from Chilla Range.

3. Discolaimus similis Thorne, 1939

Morphological taxonomic calculation: Female: L=1.3mm; a=45; b=3.7; c=50; c'=1.3; V=56; Odontostyle=14µm; Odontophore= 20µm

Description: Female: Body slightly curved ventrally upon fixation. Lip region discoid, set off from body. Odontostyle aperture 54% of odontostyle length. Vulva a transverse slit. Reproductive system amphidelphic. Tail dorsally convex-conoid, with rounded tip.

Male: Not found.

Locality and habitat: Soil around the roots of Ficus benghalensis from Hazra Beat.

4. Discolaimus tenax Siddiqi, 1964

Morphological taxonomic calculation: Female: L= 1.5 mm; a=39; b=4.5; c=53; c'=1.5; V= 54; Odontostyle=16µm; Odontophore= 26µm

Description: Female: Body ventrally arcuate upon fixation. Lip region set off by constriction. Vulva a transverse slit. Reproductive system amphidelphic. Tail convex conoid, rounded.

Locality and habitat: Soil around the roots of Dalbergia sissoo from Chilla Range.

5. Discolaimus texanus Cobb, 1913

Morphological taxonomic calculation: Female: L=1.2mm; a=41; b=3.9; c=46; c'=1.2; V=37; Odontostyle=14µm; Odontophore =28µm

Description: Female: Body ventrally arcuate upon fixation. Lip region set off by constriction. Vulva a transverse slit. Reproductive system amphidelphic. Tail convex conoid, rounded.

Male: Not found.

Locality and habitat: Soil around the roots of Terminalia bellirica Roxb. from Hazara Beat.

6. Xiphinema americanum Cobb, 1913

Morphological taxonomic calculation: Females: L=1.6-1.9mm; a=46-50; b=5.2-5.7; c=42-63; c'=1.4-1.9; V=50-55; Guiding ring=70-75µm; Odontostyle=75-88µm; Odontophore= 48-53µm; Prerectum= 143µm

Description: Females: Body C shaped upon fixation. Lip region rounded and set off from body. Amphid stirrup shaped with slit like apertures and located at the base of lip region. Basal bulb of oesophagus 23-25% of the length. Reproductive system amphidelphic. Tail short, convex-conoid.

Male: Not found.

Locality and habitat: Soil around the roots of Dalbergia sissoo from Chilla Range.

7. Xiphinema inaequle Khan & Ahmad, 1974

Morphological taxonomic calculation: Females: L=2.2-2.4mm; a=65-66; b=5.9-6.0; c=26-29; c'=3.8; V=28-3; Guiding ring=100µm; Odontostyle=108µm; Odontophore= 58µm; Prerectum= 250 µm

Description: Females: Body C shaped upon fixation. Lip region rounded almost continuous with body. Amphid stirrup shaped with slit like apertures. Basal bulb of oesophagus 23-25% of the length. Reproductive system amphidelphic. Tail short conoid,

Male: Not found.

Locality and habitat: Soil around the roots of Shorea robusta and Dalbergia sissoo from Chilla Range.

8. Xiphinema insigne Loos, 1949

Morphological taxonomic calculation: Females: L=2.1-2.4mm; a=64-66; b=5.9-6.0; c=23-29; c'=3.7-3.9; V= 28-29; Guiding ring=88-100µm; Odontostyle=108-118µm; Odontophore= 58- 63µm; Prerectum= 386-540µm

Description: Females: Body almost slightly curved upon fixation Lip region almost flat or rounded and slightly set off from body. Amphid stirrup shaped with slit like apertures and located at the base of lip region. Basal bulb of oesophagus 22-25% of the length. Reproductive system amphidelphic. Tail narrow, elongate conoid,

Male: Not found.

Locality and habitat: Soil around the roots of Trewia nudiflora Linn from Chilla Range.

9. Xiphinema opisthosternum Siddiqi, 1961

Morphological taxonomic calculation: Female: L=1.4mm; a=61; b=5.4; c=43; c'=2.3; V=67; Guiding ring=56µm; Odontostyle=69µm; Odontophore= 38µm

Description: Female: Body C shaped upon fixation. Lip region rounded and set off from body. Amphid stirrup shaped. Basal bulb of oesophagus 22% of the length Reproductive system amphidelphic. Tail elongate-conoid.

Male: Not found.

Locality and habitat: Soil around the roots of Dalbergia sissoo from Chilla Range.

10. Dorylaimoides micoletzkyi (de Man, 1921) Thorne & Swanger, 1936

Morphological taxonomic calculation: Female: L=1.3mm; a=36; b=6.3; c=20; V=40; Odontostyle=8µm

Description: Female: Body cylindroids arcuate upon fixation. Head slightly constricted, lips smooth. Amphids cup shaped. Odontostyle dorylaimoid, arcuate. Odontophore arcuate, Reproductive system amphidelphic. Tail elongate conoid, terminus dorsally bent.

Male: Not found.

Locality and habitat: Soil around the roots of *Bahunia varigata* from Chilla Range.

11. *Mylonchulus armus* Khan & Jairajpuri, 1979

Morphological taxonomic calculation: Female: L= 81mm; a=29; b=3.1; c=25; c'= 1.8; V= 58

Description: Female: Lip region 18μm x 8μm. Buccal cavity 20μm x 13μm. Dorsal tooth of median size, its apex at 15μm. from base of buccal cavity. Subventral walls with 5 transverse rows of denticles. Submedian teeth absent. Reproductive system amphidelphic. Tail conoid with clavate terminus. Caudal glands grouped. Spinneret terminal.

Male: Not found.

Locality and habitat: Soil around the roots of *Ehretia laevis* Roxb.from Jhabar Beat.

12. *M. brachyurus* (Butschli, 1873) Andrassy, 1958

Morphological taxonomic calculation: Female: L=1.0 mm; a=22; b=3.0; c=20; c'= 2.2; V= 60%

Description: Female: Lip region 25μm x 10μm. Buccal cavity 22μm x 14μm. Dorsal tooth of massive, its apex at 15μm from base of buccal cavity. Subventral walls with 6 transverse rows of denticles. Submedian teeth present. Reproductive system amphidelphic. Tail conoid with blunt terminus. Caudal glands grouped. Spinneret subterminal.

Male: Not found.

Locality and habitat: Soil around the roots of *Tectona grandis* from Chilla Range.

13. *M. hawaiensis* (Cassidy, 1931) Andrassy, 1958

Morphological taxonomic calculation: Females: L=0.82-0.86 mm; a=23-27; b=3.2-3.3; c=19-29; c'= 1.4-1.9; V= 58-59

Description: Females: Lip region 22-29μm x 5-10μm. Buccal cavity 16-23μm x 10-14μm. Dorsal tooth of massive, its apex at 13-15μm from base of buccal cavity. Subventral walls with 7 transverse rows of denticles. Submedian teeth present. Reproductive system amphidelphic. Tail tip slightly clavate. Caudal glands tandem. Spinneret terminal.

Male: Not found.

Locality and habitat: Soil around the roots of *Shorea robusta* from Chilla Range.

14. *Paramylonchulus mashhoodi* (Khan & Jairajpuri, 1979) Jarajpuri & Khan, 1981

Morphological taxonomic calculation: Female: L=1.1mm; a=37; b=3.3; c=23; c'= 2.1; V= 76

Description: Female: Lip region 20μm x 5μm. Buccal cavity 18μm x 11μm. Dorsal tooth of median size, its apex at 15μm from base of buccal cavity. Subventral walls with 3 or 4 transverse rows of denticles. Submedian teeth absent. Reproductive system mono-prodelphic. Posterior uterine sac absent. Tail conoid, tapering sharply in posterior half, with a slightly rounded tip. Caudal glands grouped. Spinneret terminal.

Male: Not found.

Locality and habitat: Soil around the roots of *Cassia fistula* Linn. from Hazra Beat.

15. *Paramylonchulus mulveyi* (Khan & Jairajpuri, 1979) Jarajpuri & Khan, 1981

Morphological taxonomic calculation: Female: L=1.0mm; a=34; b=3.2; c=21; c'= 2.1; V= 78

Description: Female: Lip region 20μm x 8μm. Buccal cavity 15μm x 9μm. Dorsal tooth of massive, its apex at 14μm from base of buccal cavity. Subventral walls with 4 transverse rows of denticles. Submedian teeth absent. Reproductive system mono-prodelphic. Posterior uterine sac absent. Tail elongate conoid. Caudal glands grouped. Spinneret terminal.

Male: Not found.

Locality and habitat: Soil around the roots of *Cordia oblicua* from Hazra Beat.

16. Granonchulus subdecurrens Coetzee(1966)

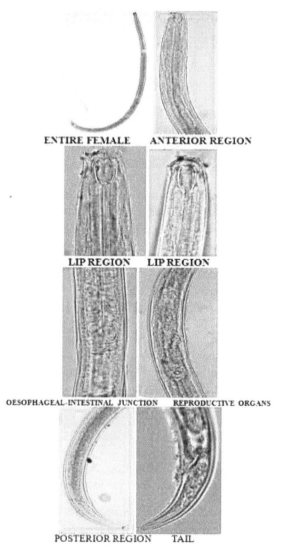

Fig 1. Female Granonchulus subdecurrens

Morphological taxonomic calculation: Females: L=1.3-1.4mm; a=33-37; b=4.3-4.5; c=19-21; c'=2.7, V= 59

Description: Females: Body ventrally arcuated upon fixtation. Lip region slightly offset, bearing papillae, 23-24μm x 9-13μm. Buccal cavity barrel shaped, 21-22μm x 11-13μm. Amphid cup shaped, situated slightly anterior to dorsal tooth apex; aperture slit like, about 4μm wide. Dorsal tooth moderately developed, anteriorly developed, its apex at

63-67% of the buccal cavity length from its base, sub ventral small denticle, opposed to dorsal tooth, arranged in two groups, an anterior transverse row in line with the dorsal tooth apex and posterior irregular arranged group extending to base of stoma, oesophagus–intestinal junction non-tuberculated. Reproductive system amphidelphic. Valve transverse slit, vagina sclerotized. Pre-rectum absent. Tail conoid, ventrally arcuated with rounded tip. Caudal glands three, arranged in tandem, opening terminal.

Male: Not found.

Locality and habitat: Soil around the roots of *Shorea robusta* from Chilla Range.

Distribution: South Africa.

Remarks: This species was originally described from South Africa (Coetzee, 1966) and has been recorded thereafter. The description given agrees well with original description.

This is the first report from India and after original its description.

17. *Iotonchus indicus* Jairajpuri, 1969

Morphological taxonomic calculation: Female: L=2.1mm; a=29; b=4.6; c=6; c'= 9; V= 58

Description: Female: Lip region 45µm x 15µm. Buccal cavity 45-55µm x 30µm. Dorsal tooth small and basal, its apex at 9µm from base of buccal cavity. Subventral walls with 4 transverse rows of denticles. Submedian teeth absent. Reproductive system amphidelphic. Tail elongate conoid. Caudal glands poorly developed. Opening subterminal dorsally.

Male: Not found.

Locality and habitat: Soil around the roots of *Tectona grandis* from Chilla Range.

18. Iotonchus parabasiodontus Mulvey & Jensen, 1967

qMorphological taxonomic calculation: Female: L=1.6mm; a=40; b=4.6; c=3.6; c'= 18; V= 59

Description: Female: Lip region 26µm x 5µm. Buccal cavity 26µm x 17µm. Dorsal tooth minute and basal, its apex at 7µm from base of buccal cavity. Submedian teeth absent. Reproductive system prodelphic. Tail conoid than cylindroids. Caudal glands present. Opening terminal.

Male: Not found.

Locality and habitat: Soil around the roots of *Bahunia varigata* from Chilla Range.

19. *Iotonchus trichurus*(Cobb, 1917) Andrassy, 1958

Morphological taxonomic calculation: Female: L=1.6mm; a=39; b=5.3; c=4; c'= 17; V= 58

Description: Female: Lip region 28µm x 10µm. Buccal cavity 30µm x 18µm. Dorsal tooth small and basal, its apex at 8µm from base of buccal cavity. Subventral walls with 4 transverse rows of denticles. Submedian teeth absent. Reproductive system mono-prodelphic. Tail long. Caudal glands present.Terminal opening present.

Male: Not found.

Locality and habitat: Soil around the roots of *Ficus religiosa* Linn. from Jhabar Beat.

20. *Iotonchulus longicaudatus* (Baqri, Baqri & Jairajpuri, 1978) Andrassy, 1993

Morphological taxonomic calculation: Female: L=1.3mm; a=37; b=4.5; c=3.8; c'= 15; V= 59

Description: Female: Lip region 25µm x 15µm. Buccal cavity 18 µm x 10µm. Dorsal tooth medium size, in anterior half of buccal cavity, its apex at 19µm from base of buccal cavity. Reproductive system mono-prodelphic. Tail long, filiform. Caudal glands present. Terminal opening present.

Male: Not found.

Locality and habitat: Soil around the roots of *Shorea robusta* from Chilla Range.

21. *Bathyodontus mirus*(Andrassy, 1956) Andrassy in Hooper & Cairns, 1959

Morphological taxonomic calculation: Female: L=1.1mm; a=34; b=3.5; c=45; c'= 0.8; V= 57

Description: Female: Lip region 23µm x 5µm, set off by a deep constriction, lips rounded. Buccal cavity 20µm x 6µm. Reproductive system amphidelphic. Tail rounded. Caudal glands present. Spinneret terminal.

Male: Not found.

Locality and habitat: Soil around the roots of unidentified tree from Chilla Range.

4. Discussion

The present authors have already reported five species of *Mylonchulus* from RNP [5]. Taking into account 5 already reported species, total terrestrial nematode species is now 31.

Granonchulus helicus was recorded from Bareilly, Uttar Pradesh, India [10]. *Granonchulus subdecurrens* Coetzee, 1966 is second species recorded from India.

Acknowledgements

The authors wish to thanks Director, Zoological Survey of India, Kolkata and the Officer–in-Charge, NRC, ZSI, Dehradun for proving necessary facilities and authors also grateful for Uttarakhand State Council for Science and Technology, Dehradun, Uttarakhand, India for funding this scientific research project.

References

[1] W. Ahmad, Md. Banyamuddin and U. Tauheed, *Rhinodorylaimus kazirangus* gen. n., sp. n. (Dorylaimida: Dorylaimidae) from Kaziranga National Park, Assam, India, *Nematology*, 12, 2010, pp. 149-155.

[2] Md. Baniyamuddin and W. Ahmad, Two new and a known species of dorylaim nematodes (Dorylaimida: Nematoda) from Kaziranga National Park, Assam, India 45, 2011, pp. 2965-2980.

[3] P. Bohra and Q.H. Baqri, Plant and soil Nematodes from Ranthambore National Park, Rajasthan, India, *Zoos Print Journal*, 22, 2005, p. 2126.

[4] V. Coetzee, "Species of the genera *Granonchulus* and *Cobbonchus* (Mononchidae), occurring in Southern Africa," *Nematologica* 12, 1966, pp. 302-312.

[5]   A. Dubey and V. Sharma. Morphological Studies of Five Known Nematode Species Via Taxa-Informatics Approaches. *Science Innovation.* Vol. 2, No. 1, 2014, pp. 7-10. doi: 10.11648/j.si.20140201.12.

[6]   R. Khan, A. Husain, R. Sultana and Q. Tahseen, Description of two new Monohystrid species, (Nematoda) from Keoladeo National Park, Rajasthan, India, *Nematode medit.*, 33, 2005, pp. 67-73.

[7]   T. Nusrat, A. Anjum and W. Ahmad, Mononchida (Nematoda) from Silent Valley National Park, India, *Zootaxa* 6535, 2013, pp. 224-236.

[8]   J.W. Seinhorst, A rapid method for transfer of nematodes from fixative to anhydrous glycrine. *Nematologica,* 1959, 117-128.

[9]   V. Sharma and Q.H. Baqri, *Plant and Soil Nematodes of India: A Checkist.* Bishen Singh Mahinder Pal Singh, Dehradun,2014, pp 266.

[10] R. K.Sharma and V. Saxena, "*Granonchulus helicus* sp.n. (Nematoda : Mononchida) from north India, *Nematol. medit.* 9, 1981, pp. 159-162.

20

Multitemporal land use changes in a region of Pindus mountain, central Greece

Apostolos Ainalis[1], Ioannis Meliadis[2], Konstantinos Tsiouvaras[3], Katerina Ainali[4], Dimitrios Platis[5], Panagiotis Platis[6]

[1]Directorate of Coordination and Inspection of Forests, Decentralised Administration Macedonia-Thrace, 46th Agriculture School St, Thessaloniki

[2] Lab. of Remote sensing and GIS, Forest Research Institute, N.AG.RE.F., Vassilika, Thessaloniki

[3]Laboratory of Range Management (236), Dept. of Forest and Natural Environment, Aristotle University of Thessaloniki, Thessaloniki

[4]Lab. of Geoinformatics, Rural and Surveying Engineering, National Technical University of Athens, Zografou Campus, Iroon Polytechniou 9, Athens

[5]Lab. of Ecology, Dept. of Agriculture, Aristotle University of Thessaloniki, Thessaloniki

[6]Lab. of Range science, Forest Research Institute, N.AG.RE.F., Vassilika, Thessaloniki

Email address:

aainalis@hotmail.com (A. Ainalis), meliadis@fri.gr (I. Meliadis), tsiouvar@for.auth.gr (K. Tsiouvaras), kainali30@gmail.com (K. Ainali), dimitris_pl@hotmail.com (D. Platis), pplatis@fri.gr (P. Platis)

Abstract: Natural ecosystems are renewable resources with special environmental, social and economical attributes and characteristics. The increasing need of the human beings for a better environment leads to the use of new technologies that offer many advantages in detecting changes in the ecosystems. In this study the integration of remote sensing tools and technology and the spatial orientation analysis of Geographical Information Systems (G.I.S.) combined with *in situ* observations were used in determining any changes in land cover categories along an 18 year period. The study area of 9,287 ha extends to Pindus mountain, in the municipality of Plastira, central Greece. The results have shown that the current technologies can be used for the modelling of environmental parameters improving our knowledge on its attributes, characteristics, situation, trends and changes of natural ecosystems. The multitemporal changes that were observed are mostly due to vegetation evolution and less to socioeconomic reasons. The basic management strategy for the specific area should combine forest, pasture and livestock in such a way that each component produces usable products, while in the same time preserves sustainability.

Keywords: G.I.S., Image Processing, Multitemporal Analysis, Range Management, Land Use, Livestock Fluctuation

1. Introduction

The internal observation of vegetation changes requires a series of parameters which can be met not only with terrestrial methods of the landscape observation, but also with the use of remote sensing and digital image processing. All these, offer the possibility of developing and applying methods and techniques for studying environmental problems and phenomena of monitoring changes in vegetation cover classes, specifically in areas with significant importance to nature conservation. The changes in these characteristics can be recorded over time in order to understand the multitemporal dynamics. To perceive this objective, we must first identify the vegetation classes and the land use at the reference time. Moreover, an important element in the temporal classification of vegetation is the livestock fluctuation. The reduction or increase of grazing animal number differentiates succession of vegetation, bringing about positive or negative results (Noitsakis et al. 1992, Milchunas 2006).

A wide range of remote sensing and G.I.S. applications in geotechnical sciences refers to multitemporal studies (Richards 1993, Lillesand and Kiefer 1994). Examples of applications in environmental studies refer to land cover changes (Chavez and MacKinnon 1994, Mas 1999, Houvardas et al. 2001, Ainalis et al. 2006, Ainalis et al. 2007, Platis et al. 2009), forest fires and deforestation (Kuntz and

Karteris 1993), coastal changes in the prefecture of Magnesia, Greece (Perakis et al. 1997), habitat changes of protected areas (Meliadis et al. 2004, Platis et al. 2004) and rangelands inventory (Platis et al. 2001). Various evaluation methods have been proposed by researchers on the assessment of land cover changes (Sunar 1998, Mas 1999, Wrbka et al. 1999).

The purpose of this study was the recording of multitemporal land use changes, especially vegetation changes, in the area of the municipality of Plastira and specifically in Pindus mountain, central Greece, using satellite images and *in situ* observations.

2. Materials and Methods

The study was conducted in an area of Pindus mountain, located in the municipality of Plastira, central Greece (Fig. 1). It extends from 200 m to 1,200 m above sea level and its largest part is covered with grasslands, shrublands and broadleaf forests.

The survey has covered an area of 9,287 ha in the west – southwestern part of Karditsa prefecture, located at the north side of Lake Plastiras. Part of the study area is designated as a Special Protection Area (SPA).

The prevailing parent rocks in the study area are colluvium shale located around the lake. These are deposits of the Tertiary period in the lowland areas and flysch in the other regions (Barntovas and Rantogianni - Tsiampaou 1983). The soil is sandy-clay and clay with moderate organic matter (Nakos 1977). The climate is continental with cold, rainy and

several times snowy winters and hot, dry summers (Mavrommatis 1980).

Information concerning the annual rainfall for the period 1989 – 2007 were taken from the nearest meteorological station of Kalambaka region situated at an altitude of 450 meters (60 km NW of the municipality of Plastira) (H.N.M.S. 2013). The region belongs to the Mediterranean mountain ecological zone (*Quercetalia pubescentis*) (Dafis 1973).

Figure 1. General map of the study area in Pindus mountain, central Greece.

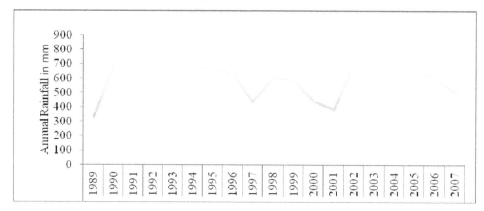

Figure 2. Annual rainfall in the period 1989 – 2007 from the meteorological station of Kalambaka in central Greece.

The data sources used are from satellite images that were taken at three different times, during an 18 year period: in 1989, 2000 and 2007. The images of 1989 were obtained from the satellite LANDSAT 5 TM (Thematic Mapper) with code 184 – 033 and date July16, 1989, while the images of 2000 were obtained from the newest satellite LANDSAT 7 ETM+ (Enhanced Thematic Mapper Plus) with code 184 - 032 and date May 02, 2000. Finally, images of 2007 were obtained from the satellite LANDSAT 7 ETM with code 184 - 033 and date June 24, 2007 (Fig. 3).

The following vector and point data such as: roads, hydrographical network, contour lines, boundaries settlement, and local names were used as auxiliary elements. The

recording was at an average scale (1:50,000) detailing the corresponding pixel size of the satellite images used (30x30m).

The methodology was distinguished in four separate stages: a) the preliminary processing of satellite data (corrections, georeference, etc), b) the classification of satellite data, c) the detection of temporal changes and d) the use of a Geographic Information System (G.I.S.), namely ArcGis 9.3 for the processing of the maps. In the final stage of the investigation, the thematic maps generated by the second stage combined with the results of temporal changes in a common geographic database.

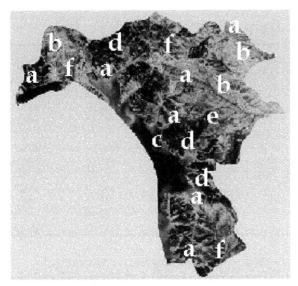

Figure 3. Satellite image processing for the year 2007 showing vegetation changes in the study area of Pindus mountain, central Greece (a: rangelands, b: agricultural lands, c: lake surface, d: forests, e:urban areas , f: flooded lands).

The images were geometrically rectified and registered to the same projection namely GGRS87 (Greek Grid Reference System) to lay them over each other.

The initial (1989) the middle (2000) and final (2007)

Landsat imageries were subjected to a classification of zones. Supervised classification was utilized to classify the images to different land use categories.

The classification of the 1989 image was collated with the images of 2000 and 2007 and then the classification - application of the majority filter was followed (Meliadis et al. 2009). Finally, the maps of the three years (1989, 2000 and 2007) were compared, using as reference year the 2007. Ground-based observations were taken *in situ* during 2007. Livestock numbers, animal species, grazing period, grazing system, shed position, site quality and range condition were recorded and then marked on maps. Range condition was classified following the criteria on table (1) (Papanastasis 1989).

The sheep and goats were the main livestock species grazing in the study area. A few cattle were also grazing in this area, but cattle farming were mostly limited. Grazing was performed for seven months per year, following a traditional continuous grazing system. The average stocking rate in the study area was estimated as the number of small ruminants per ha. One cattle was assumed equal to five small ruminants (Holechek et al. 1989). Grazing capacity was evaluated taking into account the dry matter production of the rangelands, the total rangeland area and the forage needs of domestic animals (Athanasiou et al. 2007).

Table 1. Criteria for range condition classification (After Papanastasis 1989).

Class	Desirable plants	Vegetation cover	Shrub characteristics		Erosion
			High	Cover	
Good	≥ 70%	≥ 2/3	≤ 1m	≤ 40%	No evidence of erosion
Fair	40% - 70%	1/3 – 2/3	≤ 1m	≤ 70%	No evidence of accelerated erosion
Poor	< 40%	< 1/3	≥ 1m	≥ 70%	Signs of accelerated erosion

3. Results and Discussion

Livestock capital in the area was averaged to 8,200 small animal units, during the period of 1994 to 2008, with a mean grazing pressure of 0.18 a.u.m./ha on land available for grazing, while the grazing capacity did not exceed 0.14 a.u.m./ha for seven months grazing (Athanasiou et al. 2007).

The majority of rangelands in the study area were classified in fair condition (Table 2). Especially, in the grasslands, some desirable plant species for livestock (*Festuca valesiaca* Schleich.; *Poa bulbosa*, L.; *Trisetum flavescens*, (L) Beauv.; *Trifolium hirtum*, All.; *T. campestre*, Schreb.; *T. nigrescens*, Viv.; *T. glomeratum*. L.; *T. angustifolium*, L.; *Medicago minima*, Crufb.; *M. lupulina*. L.; *Vicia sativa*, L.) participated in relatively low percentages in the vegetation composition (Athanasiou et al. 2007). A significant part of the total rangeland area was in poor condition (28%) with signs of accelerated erosion, especially in high altitudes and steep slopes (Table 2). Only 9% of the total rangelands were classified in good condition and were present mainly in medium elevated sites with slopes not exceeding 15%.

The prevailing site quality in the rangelands was site class I and II (Table 3). The 62% of the rangelands in the study area was classified as site class I having a deep soil (>30cm) (Table

3), 33% was classified as site class II (with soil depth between 15-30cm), while the rest area was classified as site class III (Table 3).

Table 2. Range condition classes in the study area of Pindus mountain.

Range class	Area	
	ha	%
Good	131.36	9.00
Fair	919.47	63.00
Poor	408.65	28.00
Total	1,459.48	100.00

Table 3. Site quality classes in the rangelands in the study area of Pindus mountain.

Site class	Area	
	ha	%
I*	904.88	62.00
II	481.62	33.00
III	72.98	5.00
Total	1,459.48	100.00

* I: soil depth >30cm, usual slope <15%, II: soil depth 15-30cm, usual slope 15-30%, III: soil depth <15cm, usual slope >30%

The monitoring implementation has shown that the dominant vegetation for livestock grazing was classified in the type of evergreen broadleaf shrublands, followed by the abandoned fields, the grasslands and the grazed open forests. The area changes in hectares per land use category over the years are shown in table (4). An increase in the coverage of land uses was recorded at the end of the study period for the categories of grasslands (158.3%), forests (14.9%), agricultural land (14.4%), lake water surface (10.2%) and the urban areas (17.7%), corresponding to a decrease recorded for the categories of abandoned fields (67.2%), evergreen broadleaf shrublands (37.5%), grazed open forests (68.3%) and barren areas (61%) (Table 4). The forests covered 46% of the total area and were mainly coppice forests of deciduous broadleaf species, such as *Quercus conferta,* Kit. and *Q. petraea,* L. (*Q. sessiliflora,* S.) (Table 4).

Table 4. *Land uses (in ha) in the study area of Pindus mountain, central Greece, during the period 1989 - 2007.*

Land use categories	Years		
	1989	2000	2007
Rangelands	1933.63	1502.04	1459.48
Abandoned fields	485.12	292.28	158.89
Grasslands	231.09	321.03	596.93
Evergreen Shrublands	1031.10	775.88	644.63
Grazed Open Forests	186.32	112.85	59.03
Forests	3723.12	4089.27	4277.45
Other Areas	3630.26	3695.70	3550.08
Agricultural lands	2159.90	2323.25	2470.00
Flooded lands	761.12	605.18	296.71
Lake surface	690.12	744.97	760.87
Urban areas	19.12	22.30	22.50
Total	9287.01	9287.01	9287.01

Table 5. *Multitemporal land use differences (positives and negatives) (in ha) between the years 1989- 2000, 2000-2007 and 1989-2007.*

Land cover categories	Differences between 1989 - 2000		Differences between 2000 - 2007		Differences between 1989 - 2007	
	Positive	Negative	Positive	Negative	Positive	Negative
Rangelands	89.94	-521.53	275.90	-318.46	365.84	- 839.99
Abandoned fields		-192.84		-133.39		326.23
Grasslands	89.94		275.90		365.84	
Evergreen Shrublands		-255.22		-131.25		386.47
Grazed Open Forests		-73.47		-53.82		127.29
Forests	366.15		188.18		554.33	
Other Areas	221.38	-155.94	162.85	-308.47	384.23	-464.41
Agricultural lands	163.35		146.75		310.10	
Flooded lands		-155.94		-308.47		464.41
Lake surface	54.85		15.90		70.75	
Urban areas	3.18		0.20		3.38	
Total	677.47	- 677.47	626.93	- 626.93	1304.40	- 1304.40

Throughout the period of 18 years (1989 – 2007) the land uses were diversified (Table 5). Among the increased land uses, forests contributed by 54% to the increase for the period 1989 – 2000, by 30% for the period 2000 – 2007 and by 42.5% for the whole period of 1989 - 2007 (Table 5). These changes were probably due to the gradual decline of transhumance grazing in mountain areas (Fig. 2).

More specifically, the gradual decrease of sheep number (19.8%), especially during the period between 2000 and 2008 (Fig. 4) as well as the gradual decrease of goats (26.7%) for the whole period (Fig. 2), contributed to the decrease of grazing pressure especially in the shrublands. As a result the evergreen shrublands were grown in height and thicken taking the form of forests contributing to the increase of the forest area (Papanastasis 2003, Athanasiou et al. 2007). The

decrease of evergreen shrublands contributed by 37.7% and 20.9%, respectively, to the total negative change of land use area in the two periods (1989 – 2000) and (2000 – 2007) and by 29.6% as a total for the whole period (Table 5).

The increase of agricultural land use contributed by 24.1% and 23.4%, respectively, to the total positive change of land use area in the two periods (1989 – 2000 and 2000 – 2007) or 23.8% for the whole period (Table 5). The increase of agricultural land was mainly due to the reduction of the abandoned fields, which was in a large degree a result of the return of land owners back to farming.

The reduction of the abandoned fields contributed by 28.5% and 21.3%, respectively, to the total negative change of land use area in the two periods (1989 – 2000 and 2000 – 2007), and by 25% as a total for the whole period (Table 5).

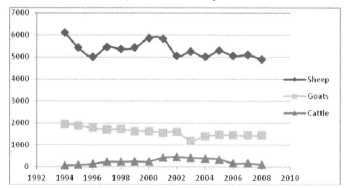

Figure 4. *Multitemporal annual changes of animal number, grazing in the study area of Pindus mountain, central Greece, during the period of 1994–2008.*

The increase of grasslands contributed by 13.3% and 44%, respectively, to the total positive change of land use area in the two periods (1989 – 2000 and 2000 – 2007) and by 28% as a total for the eighteen years period. This probably happened due to the reduction of annual rainfall in the region within the period of 2000 and 2007 (Fig. 2) and the withdrawal of flooded soils which led to the increase of grasslands (Fig. 5). The reduction of grazed open forests by 10.8% and 8.6% during the two periods and by 9.8% as a total for the whole period (Table 5) was due to the restoration of density and growth in height of deciduous broadleaf species. This forest restoration was favored by the gradual decrease of grazing animal number, especially, in mountainous areas (Table 4 and 5). Furthermore, an increase of 8.1% and 2.5% in lake water surface was recorded, which is probably attributed to the date (early May 2000) of receiving the satellite imagery (moving along the shoreline of the lake during the year).

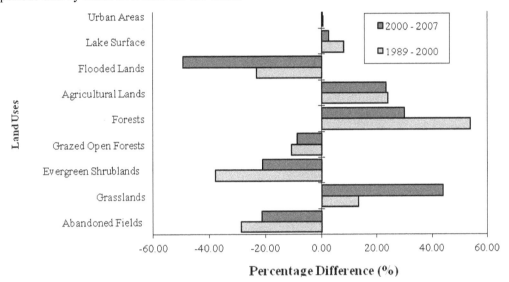

Figure 5. *Land use changes (%) in the study area of Pindus mountain, central Greece in the periods 1989 – 2000 and 2000 - 2007.*

Overall, the reduction of total grazing areas as a part of an overall reduction of land uses accounted to 63.7% in the first period (1989 – 2000) and to 6.8% in the second period (2000 – 2007) with a corresponding increase in forest and agricultural lands (Table 5 and Fig. 5).

4. Conclusions

The land use changes observed over time in a region of Pindus mountain, central Greece, are likely due to both, the evolution of vegetation and socioeconomic factors. However, some changes to a limited extent are probably due to errors in the classification algorithm. This might be a result of the spectral similarity of land use classes. The gradual dominance of forests in mountainous areas is the result of the reduction of transhumance grazing, a fact that along with the good site quality, has favored the forest cover increase. The study and the continuous monitoring of land use cover in the Plastira municipality, region of Pindus mountain, in central Greece, is imperative for the coming years in order to obtain the scientific information necessary for proper land management in the area.

Acknowledgements

This work was funded by the Ministry of Education, under the Progamme "Environment - Archimedes II" and was carried out by the Department of Forestry in Karditsa, Technological Education Institute of Larissa, Greece with collaboration of Dept. of Forestry and Natural Environment, Laboratory of Range Management, Aristotle University of Thessaloniki and the Forest Research Institute, NAGREF.

References

[1] Ainalis A., I. Meliadis, P. Platis and K. Tsiouvaras (2006).Classification and multitemporal monitoring of rangelands in the watershed of torrent Bogdana, prefecture of Thessaloniki: V.P. Papanastasis and Z.M. Parisi (eds), Rangelands in dry and warm areas. Proceedings of 5th Panhellenic Rangeland Congress. Heraklion, 1-3 November 2006. Greek Range and Pasture Society, Rep. No. 13, p. 221-226.

[2] Ainalis A., P. Platis, I. Meliadis, S.X. Papadopoulou, P. Sklavou and K. Tsiouvaras (2007). Monitoring of vegetation changes of rangelands in the watershed of torrent N. Apollonia, prefecture of Thessaloniki and the insect species which occur in the region. Proceedings of 13th Panhellenic Forestry Conference. Kastoria 7-10 October 2007.Volume II, p. 402-410.

[3] Athanassiou Z., B. Tantos, K. Tsiouvaras, P. Platis, A. Ainalis I. Meliadis, H. Karmiris and P. Sklavou (2007). Rangeland organization and development in semimountainous and mountainous ranges of Karditsa, Larissa, Magnesia Greece - Create a system for monitoring implementation of Program Setting - Archimedes II 'Strengthening research groups in the Higher Technological Institute of Larissa, Final report p. 52.

[4] Barntovas E. and I. Rantogianni - Tsiampaou (1983). Geological map of Greece (2nd edition). Institute of Geological and Mineral Research, Athens.

[5] Chavez P.S. and D. MacKinnon (1994). Automatic detection of vegetation changes in the southwestern United States using remotely sensed images. Photogrammetric Engineering and Remote Sensing, 60, pp. 571-583.

[6] Chouvardas D., I. Ispikoudis and V. Papanastasis (2006). Analysis of temporal changes in the landscape of the basin of Lake Koronia using GIS.Grasslands of lowland and semimountain areas: rural development lever (P. Platis, A. Sfougaris, T. Papachristou and A. Tsiontsis, eds). Proceedings of 4th Panhellenic Rangeland Congress, Volos, 10-12 November 2004. Greek Range and Pasture Society. Rep. No. 12, p. 253-261.

[7] Collins J. and C. Woodcock (1996). Explicit Consideration of Multiple Landscape Scales While Selecting Spatial Resolutions. Spatial Accuracy Assessment in Natural Resources and Environmental Sciences: Second International Symposium, USDA-Forest Service, Ft. Collins, CO.

[8] Dafis S. (1973). Classification of forest vegetation of Greece. Scientific Yearbook of Agriculture and Forestry School, Vol. IE, part B, Thessaloniki.

[9] H.N.M.S. (2013). Hellenic National Meteorological Service. Elements from Meteorological station of Kalambaka in central Greece.

[10] Holechek J.L., R.D, Pieper,. and C.H., Herbler (1989). Range management principles and practices. Prentice Hall Inc., USA.

[11] Kuntz S. and M. Karteris (1993). Fire risk modeling based on satellite remote sensing and GIS. In: "Satellite technology and GIS for Mediterranean forest mapping and fire management", International Workshop. Thessaloniki Greece, 4-6 November 1993 European Commission, pp. 165-177.

[12] Lillesand T.M. and R.W. Kiefer (1994). Remote sensing and image interpretation. Third Edition. John Willey and Sons Inc. 750 p.

[13] Mas J-F (1999). Monitoring land-cover changes: a comparison of change detection techniques. International Journal of Remote Sensing, 20 (1) :139 -152.

[14] Mavrommatis G. (1980). The bioclima of Greece. Relations of climate and natural vegetation. Bioclimatic Map of Forest Research, Volume I (Appendix).

[15] Meliadis I., K. Radoglou and S. Kazantzidis (2004). Monitoring changes in habitat area of Special Protection Mountains Antichasia - Meteora using temporal digital satellite images. NAGREF - F.R.I., Thessaloniki, Independent issue, p. 61.

[16] Meliadis I., P. Platis, A. Ainalis and M. Meliadis (2009). Monitoring and analysis of natural vegetation in a Special Protected Area of Mountain Antichasia - Meteora, central Greece. Environmental Monitoring and Assessment, 153 (1-4): 1-11.

[17] Milchunas D.G. (2006). Responces of plant communities to grazing in the southwestern United States. Gen. Tech. Rep. RMRS-GTR-169. Fort Collins, CO: U.S. Department of Agriculture, Forest Service, Rocky Mountain Research Station, p.p. 35-37.

[18] Nakos G. (1977). General Soil Map of Greece. Forest Research Institute of Athens.

[19] Noitsakis B., I. Ispikoudis, Z. Koukoura and V. Papanastasis (1992). Relation between successional stages and productivity in a mediterranean grassland. Proc. of Intern. Symp. in 'Natural Resource Development and Utilization', (CEC-DG12), Wageningen.

[20] Papanastasis V. (1989). Rangeland survey in Greece. Herba, 2:17-20.

[21] Papanastasis V.P. (1994). Grassland and Environment. Proceedings of the Workshop "Meadows and Rural Development". Greek Range and Pasture Society, No. 1, p. 71-81.

[22] Papanastasis V.P. (2003). Ecology and Management of Subalpine Rangelands. In: P. Platis and T. Papachristou (eds), Rangeland and Development of Mountainous Areas. Proceedings of the 3rd Panhellenic Rangeland Congress. Karpenissi, 4-6 September 2002. Greek Range and Pasture Society, No. 10, p. 437-445.

[23] Perakis K., H. Beriatos and I. Gkeskou (1997). "Study of temporal changes characteristic of coastal areas of Magnesia in the last two decades based on maps and satellite images." 4th National Cartographic Conference, Cartography and Maps on the Promotion and Protection of the Environment, p. 103-111.

[24] Platis P., T. Papachristou and V. Papanastasis (2001). Possibilities of using the program inventory pasture management Prairie Region of Epirus. The Rangeland the threshold of the 21st century (T. Papachristou and O. Dini-Papanastasi, eds). Proceedings of the 2nd Panhellenic Rangeland Congress. Ioannina, October 4-6 2000.Greek Range and Pasture Society. Rep. No. 9, p. 43-49.

[25] Platis P., I. Meliadis, T. Papachristou, D. Trakolis, S. Kazantzidis, K. Mantzanas, A. Makras, A. Dimalexis and S. Bourdakis (2004). Longitudinal monitoring of changes in habitat in the mountains Akarnanian using satellite images for sustainable management and protection. Final Report (Issue A) Program "Environmental Protection and Sustainable Development". E.T.ER.P.S. -, NAGREF - Forest Research Institute. Thessaloniki. Separate edition, p. 58.

[26] Platis P., A. Ainalis, I. Meliadis and K. Tsiouvaras (2009). Multitemporal monitoring of rangelands in the watershed of Vambakia, prefecture of Thessaloniki. Scientific Yearbook of the Department of Forestry and Environmental Management and Natural Resources, Democritus University of Thrace, Volume II, p. 167-176.

[27] Richards J.A. (1993). Remote Sensing Digital Image Analysis:An Introduction. Second Edition. Springer-Verlag, Berlin Heidelberg. p. 340.

[28] Sunar F. (1998). An analysis of changes in a multi-date data set: a case study in the Ikitelli area, Istanbul, Turkey. International Journal of Remote Sensing, 19 (2): 225-235.

[29] Wrbka T., K. Reiter and E. Szerencists (1999). Landscape structure derived from satellite images as indicator for sustainable land use. In: Operational Remote Sensing for Sustainable Development (Nieeuwenhuis, GJA, RAVaugham and M. Molenaar, eds), 18th EARseL symposium, p.119-127.

21

Response of Maizeto FYM, Gypsum and Pore Volume of Leaching Water in Saline Sodic Soil of Bisidimo, Babile District, Eastern Lowlands of Ethiopia

Assefa Adane[1], Heluf Gebrekidan[2], Kibebew Kibret[2]

[1]Chemistry Department of Hawassa College of Teacher Education, Hawassa, Ethiopia
[2]School of Natural Resources Management and Environmental Sciences of Haramaya University, Dire Dawa, Ethiopia

Email address:
assefatsion@gmail.com (A. Adane)

Abstract: A green house experiment was conducted during Oct-Dec2012 to evaluate the efficiency of selected treatment combinations of FYM, gypsum and pore volume (PV) of leaching water on growth parameters (number of leaves, plant height, leaf area, fresh and dry biomass) of maize (Zea mays L.) crop. Treatments included the combinations of the two rates (0 and 20 t ha^{-1}) of FYM, four rates of gypsum (0, 50, 75 and 100% gypsum requirement, GR) and three (1.0, 2.0 and 3.0) PV of leaching water arranged in complete randomized design with three replications. The results indicated that growth parameters of maize showed significant ($p < 0.005$) response to combined application of treatments. Similarly, the responses of growth parameters to combined application of gypsum and PV of water were also significant. Maximum growth parameters were observed in the plots that received 20 t FYM ha^{-1} + 100% GR + 3.0 PV of water compared to other combinations. Results also indicated that increasing the GR by 25% showed consistent improvement in crop growth parameters across each PV of leaching water. Analysis of the post harvest soils showed that soils received combined applications of treatments decreased pH, ECe and SAR of saline sodic soils. However, significantly ($p < 0.01$) higher decrease in pH, ECe and SAR were recorded in the combined application of 20 t FYM ha^{-1} + 100% GR + 3.0 PV of water. Combination of 20 t FYM ha^{-1} + 50% GR + 3.0 PV of leaching water reduced pH, ECe and SAR by 7.5, 23.5 and 10.0% over the control, respectively. This combination is deemed suitable for improving soil properties to agriculturally permissible limits and for optimal maize crop production. Hence, this combination can be recommended for the production of economically optimal maize crop production in saline sodic soil of Baile low lands.

Keywords: Gypsum Requirement, Growth Parameters, Biomass

1. Introduction

Maize (Zea mays L.), which is a sensitive crop for soil salinity and sodicity problems, is an important crop usually grown in the low lands of Ethiopia. In Babile (Bisidimo area) cultivable land is mainly occupied by maize and groundnut crops which give direct benefit to farmers in terms of food and cash. However, the productivity of the cultivable land in the area is declining due to the development of salinity/sodicity problems (Gizaw 2008). As a result, marginal lands have now been turned into crop fields due to the introduction of irrigation facilities, however, the irrigation water contains soluble salts in amounts that are harmful to plants or have adverse effects to convert soils into saline/sodic which require improvements in existing soil management systems.

Studies in different areas of semi arid regions of the world have compared the effectiveness of various amendments in improving physic-chemical properties of saline sodic soils (Hanay et al., 2004; Amezketa et al., 2005). The relative effectiveness of gypsum has received most attention because it is widely used as a reclamation amendment. It is however, blamed for its slow reaction but being still much popular due to its low cost and availability (Heluf, 1995).On the other hand, FYM and compost have been investigated for their effectiveness on improving the physical conditions of soils for crop growth besides their role as fertilizers (Wahid et al., 1998; Sardina et al., 2003; Liang et al., 2005; Tajada et al., 2006). Hence, combined application of organic and gypsum treatments could improve the saline sodic soils for sustainable maize production.

Research information with regard to the role of combining organic and chemical treatments in improving saline sodic properties and their residual effect on maize production is inadequate particularly in the Babile District, eastern Ethiopia. The present study was, thus, conducted to evaluate the response of maize to application of FYM and gypsum in combination with different pore volume of leaching water as well as to determine the residual effects of treatments on selected chemical properties of a saline sodic soil.

2. Materials and Methods

2.1. General Description of the Study Area

The study was conducted at Bisidimo station, in Babile District located in the Oromia Regional State, Eastern lowlands of Ethiopia lying between $8^0 21$'-$9^0 11$'Nlatitude and $42^0 15$'-$42^0 55$'E longitude with an altitude of 900-2000 meters above sea level. The study site is situated at about 30 and 90 km from Harar and Jijiga towns respectively. According to the Ethiopian agro-climatic zonation (MOA, 1998), the study area falls in the lowland and mid altitude region. The ten years (2001-2011) climatic data of Bisidimo area indicated an average annual rainfall of 650 mm which is characteristic of bimodal rainfall pattern. The annual mean maximum and minimum temperatures were 30.9 and 23.5 ^{0}C, respectively.

According to FAO (1998) classification, the soil class of the study area is Regosol and RegoArenosol association, Lithosol, Luvisol/Nitosol and their association, and Fluvisoland its association are the major soil types found in the district. The soil is dominantly sandy loam with pocket areas of clay and clay loam. Major crops commonly grown in the study area are sorghum, groundnuts and haricot beans. Sorghum, maize and haricot bean are cultivated for food consumption whereas groundnuts and chat are grown as cash crops. Crop production is based on rain fed agriculture and harvested usually once in a year but farmers around Bisidimo area practice irrigation agriculture using Errerriver and ground water sources. Agriculture production in the district is constrained by small land holdings, high price of inputs and inadequate credit service. The livestock raised includes cattle, camel, goats, chickens and donkeys. The major vegetation groups found in the study area includes: woodland, acacia, bush and shrub.

2.2. Experimental Materials and Procedures

The soil used for this study was saline sodic having alkaline pH of 8.5; EC_e 4.7dS m^{-1}, ESP 22.5, SAR 16.7 and clay loam texture (Table 1.1). Composite soil sample (0-30 cm) was collected from the experimental site and treated with combination of two rates of FYM (0 and 20 t ha^{-1}), four rates (0, 50, 75 and 100% GR) of gypsum and three PV (1, 2 and 3) of leaching water. The experiment was combined in 2 x 4 x 3 factorial arrangements. In addition, a control without gypsum, FYM and leaching water was included as a treatment. The factorial arrangements of the various treatments of the three factors yielded a total of 25 treatment combinations arranged in CRD with three replications. Subsequently, each plastic pot

was filled up with 3 kg air-dried soil after mixing with required doses of FYM and gypsum.

Table 1.1. Physical and chemical properties of the surface soil (near profile 3)

Parameter	value
Texture	Cay loam
Clay (%)	39
Silt (%)	36
Sand (%)	27
Bd (g cm^{-3})	1.2
GR(t ha^{-1})	20.7
pH	8.5
EC (d Sm^{-1})	4.7
Ex. Na (Me L^{-1})	8.7
ESP (%)	22.5
CEC (cmol$_{(+)}$ kg^{-1})	39.8
SAR (cmol L^{-1})$^{1/2}$	16.7

Bd = bulk density; CEC= cation exchange capacity; ECe = electrical conductivity of pest extract; Ex,Na= exchangeable sodium; GR= gypsum requirement; Me L^{-1}= mille equivalent per litter; SAR= sodium adsorption ratio; t ha^{-1}= ton per hectare

The pots were rewetted regularly with water every other day to maintain field capacity. All pots were incubated in a greenhouse for a week to aid dissolution of the treatments and facilitate the reaction of the soil and treatments before applying the predetermined volume of leaching water was applied to the soil in each pot. Following this, the soils in each pot were left to drain and partially dry. The partially dried soils were loosened and six healthy maize seeds ('Alemaya' variety) were sown into each pot. The pots were watered every third day until harvesting and weeds were removed by hand; no additional fertilizer was applied. Numbers of days from sowing to emergence of 25, 50 and 100% seed were recorded. The number of germinating seeds were counted and recorded to determine the effect of the treatments on germination. Following full seedling development stages, the plants in each pot were thinned to two plants.

The experiment was conducted from Oct.23 to Dec.23, 2012. Harvesting time of maize is different for grain production and biomass determination. In the case of grain production harvesting time is 90-120 days (Haque, 2003), but for biomass determination harvesting time is 55-60 days (Motalib, 2003).Above and below ground part of the crop in each pot was harvested to determine the growth parameters.

2.3. Growth Parameters

The growth parameters (plant height, number of leaves, fresh and dry biomass per pot) were recorded at the harvesting time. The plant height was measured from the base of the plant to the growing tip at harvesting time and mean values were expressed in centimeter. Before cutting the maize plant, the average number of leaves per plant was counted from each pot. The leaf area (cm^2) was measured using average length and width of a leaf (base, middle and tip of the leaves).

The above and below ground parts of the maize plant were measured by uprooting them carefully at 60 days after sowing when 50% of plants were in the flowering stage. Uprooted plants were washed thoroughly to remove adhering soil/dirt.

Immediately after harvesting, the fresh biomass (g pot^{-1}) for each treatment was recorded after which the plants were chopped into small pieces and kept in open paper bag to enable air drying before drying in the oven at 70°C for 24 hrs to achieve constant weight. Fresh and dry weights of the plant were expressed in g pot^{-1}.

2.4. Post Harvest Soil Sampling

After harvesting the maize plants, soils treated with the same treatments rates but leached with different PV of water were mixed and 8 composite soil samples were prepared. These samples were brought to the laboratory, dried, before grinding to pass through a 2 mm sieve, labeled and were stored for analysis. Soil reaction (pH) and EC_e were measured in 1:1(w/v) soils: water suspension using pH meter and conductivity meter, respectively, according to the method given in Ryan *et al.* (2001). Soluble Na^+ and K^+ were determined by flame photometer, while Ca^{2+} and Mg^{2+} concentrations were determined by atomic absorption spectrometry. The SAR in soil extract was calculated using the following expression.

$$SAR = \frac{Na}{\sqrt{\frac{(Ca+Mg)}{2}}}$$

Where the concentration of Na, Ca and Mg are in mile equivalent per liter (Me L^{-1})

3. Results and Discussion

3.1. Response of Maize to FYM, Gypsum and PV of Water

Treatment combinations significantly ($P < 0.05$) increased the maize crop height, number of leaves, fresh and dry biomass per pot. Significantly higher values of these parameters were recorded when 20 t FYM ha^{-1} and gypsum 100% GR rate applied together than their applications alone across the 3 PV of water.

Leaching the soil by 3.0 PV of water gave significantly higher growth parameters than leaching by 1.0 and 2.0 PV of water across all applied combined FYM and gypsum. The statistically significant growth parameters response of maize to combined application of FYM + gypsum + 3 PV of water is due to the replacement of exchangeable Na by Ca^{2+} and leaching of the released Na^+ below the root zone. Relatively better growth of maize by the sole application of gypsum treatment over the sole use of FYM on saline sodic soils could be due to the ability of gypsum to help decreased soil sodicity in the root zone. This result was supported by Mohammad et al, (2010) who noted that gypsum was effective in lowering the chemical parameters that might be due to substitution of exchangeable Na by Ca that produced more soluble salts (NaCl, or Na2SO4) and was leached by the irrigation water.

3.1.1. Maize Crop Seed Germination Rate
Irrespective of the difference in applied treatments, 25 and 50% seed emergence did not differ among all the treated soils i.e. 25 and 50% of the seedlings emerged from almost all pots

within the same range of days (Table 1.2).

The effects of some treatments on the rate of seed germination were almost similar (Table 1.2). For example, sole applications of gypsum at 50 and 75% GR rates, respectively, on the soils and each leached with 1.0 pore volume of water showed similar germination rate (83.3%). Similarly, when 20 t FYM ha^{-1} mixed with gypsum at 50% GR rate applied in the soil and leached with the same volume (1.0 PV) of water, the rate of seed germination was also the same (83.3%). When the soils leached with 2.0 PV water, similar seed germination rates were observed in the soils treated with combined treatments such as:20 tFYM ha^{-1} and gypsum at 75% GR rate, 20 t FYM ha^{-1} and gypsum at 100% GR rate and sole application of gypsum at 100% GR rate. However, sole application of each rate of gypsum and their combinations with FYM were superior to the application of FYM and the control (without FYM, gypsum and PV of water).

The rate of seed germination showed consistent increment as the PV of leaching water increased (Table 1.2). For example, sole application of gypsum at 50 and 75% GR rates when leached with 1.0 PV of water showed similar (66.7%) seed germination but when the amount of leaching water increased to 2.0 and 3.0 PV, seed germination rates increased to 83.3 and 100.0%, respectively. Similar trend of increment in seed germination rates was observed for soils treated with combined use of gypsum+ FYM and PV water (Table 1.2).

Table 1.2. Interaction effects of FYM, gypsum and PV water on seed germination rate (%).

FYM tha^{-1}	gypsum % GR	PV of water				
		0.0	1.0	2.0	3.0	mean
0	0	66.7	50.0	33.3	33.3	45.8d
	50		72.2	72.2	100.0	81.5b
	75		72.2	83.3	100.0	85.2b
	100		83.3	100.0	100.0	94.4a
20	0		50.0	61.3	83.3	64.8c
	50		72.2	83.3	100.0	85.2b
	75		83.3	100.0	100.0	94.4a
	100		83.3	100.0	100.0	94.4a
	Mean	66.7cd	70.8c	79.2b	89.6a	

Means across column and row followed by the same letter are not significantly different at P < 0. 05
Interaction effect of FYM*gypsum* PV of water are significant at p < 0.05

The mean seed germination rate of all soils treated with sole and mixed treatments increased from 68.8 to 89.6% when the quantity of leaching water increased from 1.0 to 3.0 PV (Table 1.2). However, the seed germination rate showed a decreasing tendency when the soils in the control leached with increasing PV of leaching water (from 1 to 2 PV). This finding is in line with the work of Heluf (1995) who suggested that the decrease in seed germination rate as a result of leaching the soil without treatments might be due to the aggravated effect of exchangeable Na on soil properties with decreasing electrolyte concentration of soils.

On the other hand, the relative increment in the seed germination rate with increasing treatment rates might be due to the reduction of the toxic concentration of Na^+ at the soil exchange site. This result was also supported by Alawi *et al,*

(1980) and Kwaer *et al,* (2006) who suggested that applied mixed organic and chemical treatments on saline sodic soils and then leached with increasing volume of water can significantly flush down the toxic concentrations of Na^+ and thereby creating favorable conditions for the seeds to germinate.

3.1.2. Maize crop Height

The height of the maize plant grown under the influence of different rates of treatments were recorded at 60 days and presented in Table 1.3. The data then indicated that sole and combined application of FYM and gypsum showed significant ($p < 0.05$) influence on this parameter. The magnitudes of the difference due to combined treatments were higher than sole application of treatments (Table 1.3). These could be highly likely related to the increase in organic matter content due to FYM and reduction of the toxic concentration of Na^+ in the soil exchange site due to gypsum and PV of leaching water (Shwetha and Babalad 2008).

Table 1.3. Interaction effects of FYM, Gypsum and pore volume of water on the root length and plant height (cm)

Treatments		A. Root length					B. Plant height				
		Applied pore volume of water					Applied pore volume of water				
FYM	Gypsum	0.0	1.0	2.0	3.0	mean	0.0	1.0	2.0	3.0	mean
	0	11.2	13.4	15.4	16.3	14.0f	33.1	39.6	46.3	51.7	45.7h
0	50		25.2	27.4	23.1	25.3d		51.8	57.5	61.3	56.9f
	75		27.4	34.3	38.6	33.4c		55.3	67.8	72.9	65.3d
	100		38.2	41.2	47.7	42.4b		61.3	69.0	78.2	69.5c
	0		18.5	20.3	22.5	20.4e		47.7	52.4	58.5	52.9g
20	50		28.3	32.5	37.4	32.8c		57.8	61.6	68.3	62.5e
	75		39.1	43.3	47.5	43.3b		64.2	71.3	82.1	72.5b
	100		44.1	47.7	53.9	48.6		82.8	93.9	96.4	90.9a
	mean	11.2d	29.3c	32.8b	36.9a		33.1d	57.5c	64.9b	71.2a	

Three factor Interaction means across all columns and rows followed by the same letter are not significantly different at P <0. 05
Interactions among FYM*gypsum* PV of water are significant at p < 0.05

In general, when PV of leaching water increased, soils treated with FYM + gypsum at various rates increased the plant height than the soils treated with either FYM or gypsum alone. This finding was also supported by various researchers (Chonkar, 2003; Ghuman and Sur, 2006 and Shwetha and Babalad 2008) in different areas of the world who concluded that application of FYM + gypsum in saline sodic soil and leaching improved soil chemical properties.

3.1.3. Number of Leaves and Leaf Area

The overall mean in Table 1.4 indicated that increasing the rates of gypsum by 25% GR increased the number of leaves plant^{-1} and the increment was consistent across overall PV of water. When PV of leached water averaged, maximum (12.77) number of leaves plant^{-1} was recorded in the 20 t FYM ha^{-1} + gypsum at 100% GR rates, whereas; the minimum (5.11) number of leaves plant^{-1} was recorded in the control and was significantly affected by most of the treatments.

Soils treated with 20 tons FYM ha^{-1} increased the mean number of leaves plant^{-1} by 8.45% and the leaf area plant^{-1} by 17.24% over the control, while 17.07, 39.31 and 63.79% increment in mean number of leaves plant^{-1} counted when gypsum at 50, 75 and 100% GR rates used, respectively, compared to the control (Table 1.4). Combined 20 tons ha^{-1} and gypsum at 50% GR rate showed 42.23% increment in number of leaves plant^{-1} and 25.02% of leaf area plant^{-1} than the soils treated with gypsum at 50% GR rate and leached with 3 PV of water. Similarly, combined application of 20 t FYM ha^{-1} and gypsum at 100% GR rate showed 113.90% more number of leaves plant^{-1} and 92.72% leaf area plant^{-1} than sole application of gypsum at100% GR rate and leached with 3 pore volume of water (Table 1.4).

The larger number of leaves and leaf area in the combined FYM and gypsum than sole applications at the same PV of water could be due to better root growth of plants in the former can extract nutrients due to improved soil properties (Minhas *et al.,* 1994 and Balyan *et al.,* 2006).

Table 1.4. Interaction effects of FYM, Gypsum and PV of water on the number of leaf and leaf area (cm^2)

Treatments		a. Number of leaf per pot					b. leaf area per pot				
		Applied pore volume of water					Applied pore volume of water				
FYM	Gypsum	0.0	1.0	2.0	3.0	mean	0.0	1.0	2.0	3.0	mean
	0	5.2	5.7	5.8	6.0	5.8g	47.4	49.2	71.5	70.8	63.8f
0	50		6.5	6.7	7.2	6.8e		71.4	72.2	80.8	74.8e
	75		7.5	8.5	8.4	8.1d		86.4	86.7	90.1	87.8d
	100		9.4	9.5	9.8	9.5b		96.5	99.4	101.1	99.1c
	0		6.1	6.3	6.5	6.3f		72.4	58.5	72.2	67.7f
20	50		7.8	8.1	8.7	8.2c		85.1	87.0	88.5	86.8d
	75		9.2	9.5	10.0	9.6b		93.0	108.4	129.2	110.2b
	100		10.5	12.5	12.8	11.9a		124.9	128.7	136.4	130.0a
	mean	5.2d	7.8c	8.3b	8.7a		47.4d	84.9c	89.1b	96.1a	

Three factor Interaction means across all columns and rows followed by the same letter are not significantly different at P <0. 05
Interactions among FYM*gypsum* PV of water are significant at p < 0.05

This could be due to the fact that the pots treated with combined treatments had avoided the osmotic influence as well as specific ion toxicities and supplied the crop with ample nutrition thereby increasing reproductive growth to a greater extent as compared to control. This finding was supported by Minhas et al, (1994) who suggested that integrated organic and chemical amendments can improve vegetative nourishment thus produced higher number of leaves and leaf are.

In general, the data in Tables 1.4 and 1.5 indicate that combined use of FYM and gypsum was relatively better than either one alone in increasing plant height, number of leaves and leaf area. These increments could be due to the reduction of Na^+ toxicity and or increased soil fertility due to FYM. The findings regarding the growth parameters are in agreement with the work of Rezende et al., (1994) and Prapagal et al., (2012) who concluded that number of leaves increased with increasing nitrogen rate due to the application of FYM as a soil amendment for improving the soil health and plant growth.

Table 1.5. Interaction effects of FYM, Gypsum and pore volume of water on the root length and plant height (cm)

Treatments		a. Fresh biomass per pot					b. dry biomass per pot				
		Applied pore volume of water					Applied pore volume (PV) of water				
FYMt ha^{-1}	Gypsum(% GR)	0.0	1.0	2.0	3.0	mean	0.0	1.0	2.0	3.0	mean
0	0	8.5	33.33	37.53	45.43	36.20e	3.72	3.79	3.96	4.06	3.88e
	50		46.77	48.06	52.50	49.11d		4.96	5.21	5.21	5.03d
	75		54.30	60.70	62.97	59.32c		6.34	6.32	5.33	5.68c
	100		60.20	78.33	97.87	78.80b		6.52	6.84	8.18	6.57b
20	0		44.80	45.90	54.20	48.30d		4.85	4.87	5.35	4.95d
	50		65.70	60.90	54.30	60.30c		5.32	6.23	6.57	5.71c
	75		76.80	74.00	92.93	81.24b		6.78	6.74	7.98	6.56b
	100		81.30	87.20	101.60	90.03a		7.52	7.7	6.11	7.07a
	mean	8.5e	28.5d	7.9c	61.58b		3.72e	4.72cd	5.89c	6.11b	

Three factor Interaction means across all columns and rows followed by the same letter are not significantly different at P <0. 05
Interactions among FYM*gypsum* PV of water are significant at p < 0.05

3.2. Fresh and Dry Biomass Accumulation

The fresh and dry biomasses of the crop produced in saline sodic soils treated with sole and combined treatments are given in Table 1.6. The highest fresh weight (101.6 g pot^{-1}) obtained from maize crop grown in the soil treated with gypsum at 100 % GR plus 20 t FYM ha^{-1}, while the lowest (8.3 g pot^{-1}) obtained from the control.

The analysis of variance showed that combined use of FYM and gypsum increased the fresh weight of the maize plant from 60.3 to 90.03 g pot^{-1}when the dose of gypsum combined with 20 t FYM ha^{-1} increased from 50 to 100% GR. The positive response of the fresh weigh of the crop with increasing levels of combined treatments could be attributed to the effect of improved soil fertility due to FYM on vegetative growth (Rezende et al., 1994).

The mean dry matter weight of maize plant produced from the soils treated with all treatments were also significantly (P < 0.05) different from each other. Similarly, the effect of varying PV of leaching water on fresh and dry matter weight was also relatively higher when combined application were used as reclaiming material than either gypsum or FYM alone (Table 1.6). For example, sole applications of 20 ton FYM ha^{-1} as well as gypsum at 50% GR rate, significant (P < 0.05) increased the mean fresh weight by 33.4 and 35.66%, and dry matter by 2.75 and 5.09%, respectively, over the control.

Whereas, 66.58 and 34.32% increment in mean fresh and dry matter weights, respectively, recorded when the two treatments combined and leached with the same PV of water. Similar findings were also reported by Ahmed et al, (2011),

Izhar et al, (2007), Ghuman et al.,(2006) and, Swarp et al, (2004) who observed that combining FYM with gypsum helped in increasing the fresh and dry matter weights of wheat, which may be attributed directly nutritional effect as well as indirectly through improving soil properties.

The data in Table 1.6 again indicated that increasing the amount of PV water by 0.5 showed significant increase on the mean fresh and dry matter weight of the maize crop. For example, when the soil treated with full recommended gypsum (100% GR) rate and leached with 1.0 PV of water, 60.2 g pot^{-1} fresh weight and 7.52 g pot^{-1} dry matter weight were recorded, however, 78.33 and 97.87 g pot^{-1} fresh weight and 6.84 and 8.18 g pot^{-1} dry matter weight, respectively, recorded when the amount of leaching water increased from 2.0 to 3.0 pore volume, respectively. When the volume of leaching water averaged, applied gypsum at 75% GR rate resulted 20.79 and 13.52% fresh and dry matter weights, respectively, more than the weights recorded at half recommended rate of gypsum (50% GR).

Considering the mean fresh weight of maize across all levels of treatments, leaching the soil with 3.0 PV of water gave maximum fresh weight (70.23g pot^{-1}) compared to control (28.5g pot^{-1}) produced without treatments. Among the possible reasons may be the improvement in porosity and hydraulic conductivity due to treatments and the increasing PV of water that could have enhanced the leaching of soluble salts for greater improvement in soil health the maximum vegetative growth of the maize plant. This result was also supported by Alawi et al, (1980) who suggested that combined application of FYM and gypsum on saline sodic soils and then leached with increasing volume of leaching water can significantly flush down the toxic concentrations of Na^+ and other soluble ions thereby increasing the vegetative growth of crops. The improvement in physicochemical properties of saline soil, as observed in the earlier section could be the major reason for enhancement of the fresh and dry matter weights.

3.3. Post Harvest Selected Soil Chemical Properties

The data in Table 1.6 shows that soils collected after sole application of FYM or gypsum or their combination improved some soil chemical properties (soil pH, EC_e and SAR). The minimum decrease in soil pH and EC_e were recorded in the control soil (treated with neither of the treatments) while the greater decrease in soil pH and EC_e over the control were recorded in the soil treated by combined application of 20 t FYM ha^{-1} and 100% GR.

Table 1.6. Effects of sole and combined treatments on selected soil chemical properties

Treatments	pH	ECe	SAR
Control	8.3	4.3	14.8
FYM at 20 ton ha^{-1}	8.2	4.2	14.4
Gypsum at 50% GR	7.9	3.9	13.6
Gypsum at 50% GR + FYM at 20 ton ha^{-1}	7.7	3.3	13.3
Gypsum at 75% GR	7.5	3.7	13.4
Gypsum at 75% GR + FYM at 20 ton ha^{-1}	7.3	3.2	12.8
Gypsum at 100% GR	7.5	3.3	12.7
Gypsum at 100% GR + FYM at 20 ton ha^{-1}	7.2	3.0	11.9
Mean	7.8	3.8	14.0

ECe= electrical conductivity of pest extract; SAR =sodium adsorption ratio

Therefore, relatively maximum decrease in soil pH and EC_e were observed in the soils treated with combined than sole application of treatments (Table 1.6). For example, the soil treated with 20 t FYM ha^{-1} decreased the pH by 1.45% and ECe by 2.45% while soils treated by gypsum at 75% GR decreased the pH by 9.7 and ECe by 13.5%, respectively, over the control, whereas, when the two treatments combined (rate + 20 tons FYM ha^{-1} + gypsum at 75% GR), a decrease of 11.7 and 30.6% of soil pH and ECe, respectively, were recorded. This result was supported by Muhammed et al, (210) who conclude that the decrease might be the result of improved infiltration due to FYM and gypsum addition.

These results suggested that growing maize crop after applied combined FYM with gypsum were superior to either one alone. However, soils collected after treating with sole application of gypsum at different rates also decreased soil pH and ECe over the control and sole application FYM. For example, soil treated by gypsum at 50, 75 and 100% GR reduced the soil pH by 5.2, 9.2, and 9.9%, and the ECe by 9.4, 14.1 and 22.1%, respectively, over the control.

The observed decline in soil pH suggested reduction in soil sodicity of the saline sodic soil as a result of favorable effects of FYM and gypsum This finding was supported by Wahid et al., (1998) who noted that the lowered pH increases the solubility of gypsum, thus, removing some of the Na^+ from the soil. The reduction of ECe may probably be due to leaching of soluble salts into the deeper layers of the profile. Consistent

with the results observed in this study, Niazi et al., (2001) also reported that combined application of gypsum with FYM reduce the ECe more than sole application of treatments.

On the other hand, soils collected after sole application of 20 t FYM ha^{-1} and growing maize crop decreased the SAR by 2.30%, while sole application of gypsum at 50, 75 and 100% GR decreased by 10.0, 9.2 and 14.1%, respectively, as compared to the control (Table 1.6). Anll the gypsum treatments and their combination with FYM were significant in reducing the SAR of the soil to values less than the control. The reason for comparatively less reduction of SAR due to the sole than the combined treatments may be due to slow reaction of these treatments over short term; gypsum has less solubility and takes more time for complete reclamation of sodic soil. The result also supported by Izharet al., (2007) who concluded that the reduction in SAR may be the result of increased Ca^{2+} + Mg^{2+} that help displace Na^+ from the soil exchange site.

4. Conclusion

Saline sodic soils received combined application of treatments (FYM, gypsum and PV of water) improved important soil chemical properties over the application of these treatments alone. The significant improvement in soil properties and plant growth parameters due to combined application of gypsum and FYM treatments to a saline sodic soil followed by leaching with varying PV of water could be due to flush down of the toxic concentrations of Na^+ and other soluble ions bellow the root zone. Hence application of combined FYM and gypsum followed by leaching with three PV of water is recommended to improve maize crop growth parameters on saline sodic soils. All combined application of treatments improved important soil properties in general,20 t FYM ha^{-1} + gypsum at 50% GR is preferable and economical than sole applications of gypsum at 75 and 100% GR in particular for resource poor farmers. However, further field studies are recommended to determine the optimum rates of treatments applied to reclaim saline sodic soils.

Acknowledgment

This work was supported by a grant from the SIDA Project for which the authors are grateful. They also like to acknowledge Hawassa College of Teacher Education and the SNNP Soil Laboratory Institute for providing the necessary resources to conduct this study.

References

[1]   E.A. Amezketa, R. Aragues, R. Gazol, Efficiency of sulfuric acid, mined gypsum and two gypsum by-products in soil crusting prevention and sodic soil reclamation. *Agronomy of Journal.* 97: 983-989.1988.

[2]   B.Alawi, J. Stroehlem, F. Hanlonand F. Turner. Quality of irrigation water and effect of sulphuric acid and gypsum on soil properties and Sudan grass yield. *Soil Science.* 129: 315-319.1980. 2005.

[3] I.A Ahmet, Reclamation of saline and sodic soil by using divided doses of phosphogypsum in cultivated condition. Central Research Institute of Soil, Fertilizer and Water Resources, Ankara, Turkey. 2011.

[4] R.S. Ayers, and D.W. Westcott, Water Quality for Irrigation. FAO Irrigation and Drainage Paper 29. FAO: Rome. 1985.

[5] J.K. Balyan and P. Singh, B.S. Sumpawat and L.K. Jain, Effect of integrated nutrient management on maize (*Zea mays* L.) growth and its nutrients uptake. *Current Agriculture*.30: 79-82. 2006.

[6] P.K. Chhonkar, Organic farming: Science and belief. *Journal of Indian Society of Soil Science. 5: 365-377. 2003.*

[7] FAO (Food and Agriculture Organization), Guidelines for soil profile description, 3rd edn.FAO, Rome, Italy. 1990.

[8] B.Gizaw, Characterization and Classification of the Soils and Irrigation Water Sources of the Bisidimo areas. Babile District in East Hararghe Zone of Oromia National Regional State. MSc Thesis, Haramaya University. 2006.

[9] B.S. Ghuman and H.S. Sur, Effect of manuring on soil properties and yield of rain fedwheat. *Journal of Indian Society of Soil Science.* 54: 6-11. 2006.

[10] M.M. Haque, Variety development of maize and its characteristics in Bangladesh. Production and Uses of Maize in Bangladesh. Published by BARI, CIMMYT and Integrated. 2003.

[11] A. Hanay, F. Buyuksonmez, F.M. Kizilolu and M.Y. Canbolat, Reclamation of saline sodic soils with gypsum and MSW compost. *Compost Science Utilization.* 12: 175-179. 2004.

[12] H. Izhar, B. Muhammad and F. Iqbal, Effect of gypsum and FYM on soil properties and wheat crop irrigated with brackish water. 2007.

[13] M. A. Kwaer, E.A. Sayad, M.S. Ewees, Soil and plant analysis as a guide for interpretation of the improvement efficiency of organic conditioners added to different soil in Egypt. *Comm. Soil Sci. Plant Anal.* 29: 2067–2088.

[14] Y. Liang, S. Nikolic, M. Peng, W. Chen and Y. Jiang, Organic manure stimulates biological activity and barley growth in soil subject to secondary salinization/ sodification. *Sols Biology and Biochemistry.* 37:1185-1195. 2005.

[15] MoA (Ministry of Agriculture), Agroecological Zones of Ethiopia, Natural Resources Management and Regulatory Department, Addis Ababa. 1998.P.S. Minhas, Use of saline waters for irrigation. Salinity management for sustainableAgriculture. CSSRI. Karnal, India. 205p.1994.

[16] A.B. Mohd. R. Singh and K. Anshuman, Effect of integrated use of farmyard manure and fertilizer nitrogen with and without sulphur on yield and quality of Indian mustard (*Brassica juncea* L.). Journal of Soil Society of Soil Science.55: 224-226. 2007.

[17] M. A. Motalib, Use of maize as fodder. Production and uses of maize in Bangladesh. Published by BARI, CIMMYT and Integrated Maize Development Project, Bangladesh. p 142145. 2003.

[18] M. Qadir, J. Oster, S. Schubert, A. Noble and K. Sahrawat, Remediation of sodic and saline sodic soil. *Journal advances in Agronomy.* 96: 17-247. 2007.

[19] Z.I. Raza, M.S. Rafiq and R. Abdu, Gypsum application in slots for reclamation of saline sodic soils. *International Journal of Agriculture and Biology.* 3: 281-285. 2001.

[20] G.D. Rezende and M.P. Ramalho, Competitive ability of maize and common beans(*Phaseilus vulgaris)* cultivars intercropped in different environment. *Journal of AgriculturalScience.*123: 185–190. 1994.

[21] P. R. Ryan, E. Delhaize and D. L. Jones. Function and echanism of organic anion exudation from plant roots. *Annu. Rev. Plant Physiol.* 52:527–560. 2001.

[22] M. Sardina, T. Muller, H. Schmeisky and R.G. Joerensen, Microbial performance in soils along a salinity gradient under acidic conditions. *Journal of applied soil ecology 23: 237-244.* 2003.

[23] K. Prapagar, S.P. Indraratne1 and P. Premanandharajah, Effect of Soil Amendments on Reclamation of Saline-Sodic Soil. *Tropical Agricultural Research 23: 168 –176 (2012).*

[24] Y.P. Singh, S. Ranbir and S.K. Sharma, Com binedeffect of reduced dose of gypsum and salt tolerant varieties of rice (*Oryza sativa*) and wheat (*Triticum aestivum*) on rice-wheat cropping system in sodic soils. *Indian Journal of Agronomy.* 54: 24-28. 2009.

[25] Z. Sestak, Castsky and P.G. Jarvis, Plant photosynthetic production.Manual of methods (Ed.) W. Junk N.V., Publications. The Hunghus, p.343-381.1971

[26] A.Swarp and N.P. Yaduvanshi, Effects of Integrated nutrient Management on soil properties and yield of rice in alkali soils. *Journal of Indian Society of Soil Science.* 48: 279-282. 2004.

[27] M. Tajada, C. Garcia, J.L. Gonzalea and M.T. Hernadez, Use of organic amendments as a strategy for saline soil remediation: influence on the physical, chemical and biological properties of soil. 2006.

[28] A. Wahid, S. Akhtar, I. Ali and E. Rasul, Amelioration of saline-sodic soils with OM and their use for wheat growth. Commun. *Soil Science and Plant Analalysis*.29: 2307-2318. 1998.

[29] J.J. Muhammad, T.J. Mohammad, U.K. Amin, A. Arif, Management of saline sodic soil through cultural practices and gypsum. *Pakistan Journal of Biotechnology*, 42: 4143-4155, 2010.

22

Fertilizer, lime and manure amendments for ultisols formed on coastal plain sands of southern Nigeria

Ayodele O. J., Shittu O. S.

Department of Crop, Soil and Environmental Sciences, Ekiti State University, Ado-Ekiti, Nigeria

Email address:

olubunmishittu@yahoo.com (Ayodele O. J.)

Abstract: The highly weathered and leached soils formed on Coastal Plain Sands under excessive rainfall regime in southern Nigeria are Ultisols. The appropriate management practices with which to obtain high crop yields in these soils, characterized by high acidity, nutrient deficiencies and imbalances, should be developed. Surface layer (0-15 cm) samples of soils with extreme acidity (pH 4.0-4.6) formed on Coastal Plain Sands were collected from four locations in southern Nigeria and grown to maize (SUWAN 1-SR-Y) in pots for two cycles of six weeks each to measure the direct and residual effects of applied fertilizer (90 kg N+ 36 kg P+ 60 kg K.ha^{-1}), 2.5 MT.ha^{-1} lime, 10 MT.ha^{-1} farm yard manure (FYM) compared to a control. The direct effect of FYM produced the highest dry matter yield while fertilizer and lime did not differ significantly from the control. The residual effects were significant in dry matter yield for FYM in all the soils and for lime in three soils. Lime and FYM increased soil pH and exchangeable bases, reduced iron, manganese and aluminium; fertilizer and FYM raised available P while only FYM increased soil organic matter contents. Application of lime, fertilizer and FYM in all possible combinations compared to the control in one soil showed that FYM + Fertilizer gave the highest maize dry matter yield, improved soil characteristics and would be the recommended nutrient management practice for these acid soils.

Keywords: Ultisols, Acidity, Dry Matter, Nutrient Uptake, Residual Effects

1. Introduction

One of the remaining frontiers whose contribution can be substantial for global food security is the vast humid tropical region whose soils have potentials that can be exploited for arable crop production (Sanchez and Buol, 1975; Ofori *et al.*, 1986; Buol *et al.*, 2011). Unfortunately, most of the soils are acidic, low-base status Oxisols and Ultisols characterized by poor crop growth and yields. This is because high soil-solution hydrogen ion (H^+) concentration affects the availability and uptake of several metallic ions, causes the deficiencies of calcium (Ca), magnesium (Mg), potassium (K), phosphorus (P) and molybdenum (Mo) and presence of aluminium (Al), iron (Fe) and manganese (Mn) at high to (Fageria and Baligar, 2008). Besides, clay, Al and Fe ions and oxides convert native and applied P to insoluble forms through fixation processes (Brady and Weil, 2002). Ultisols are extensive in Nigeria, being located mainly in the south where the Coastal Plain Sands parent materials and excessive rainfall regime produce highly weathered and leached soils with low organic matter and cation exchange capacity as a reflection of the kaolinitic

and Fe, Al-oxide (low activity clay) mineralogy (Juo and Wilding, 1996). The major challenge is to overcome acid soil infertility, indicated by low nutrient status (deficiencies) and nutrient imbalances, through the development of realistic management systems the farmers would adopt to obtain reasonable and sustained high crop yields.

Lime application has been recognized and used as the main practice for ameliorating strong acidity which curtails the availability of nutrients required at high amounts in soils for maximum yields (Fageria and Baligar, 2008). Liming based on the quantity needed to neutralize exchangeable Al which is the principal factor responsible for poor crop growth in acid soils, and also supply Ca and Mg (Haynes and Naidu, 1998) was beneficial to yield in soils with pH <5.5 but not in moderately acid soils or when liming targeted pH= 7.0 or more (Sumner and Yamada, 2002). Also, lime efficiency decreases at high rates and over-liming has detrimental effects on the availability of micronutrients, especially zinc (Zn) and copper (Martini and Mutters, 1989).

Fertilizer use is promoted as a vital component of the improved technology that farmers must adopt to attain high yields in crops. Unfortunately, the humid areas where acid soils are common features have a poor fertilizer use history in relation to the dominant multi-storey tree crop agroforestry systems that rely on somewhat closed nutrient cycles to maintain soil productivity (Agboola, 2002). Fertilizers have become scarce and expensive in Nigeria as a result of economic structural adjustment policies adopted by government that necessitated subsidy reduction/withdrawal which substantially reduced fertilizer use (Banful et al., 2010; Liverpool-Tasie et al., 2010) and the resultant low total nutrient content is inadequate for open arable crop production systems. Besides, declining yields have often been observed where fertilizer was used alone while continuous application at high rates increased soil acidification (Juo et al., 1996) as one of the general effects on environmental degradation.

Manure application increases crop growth and yields in acid soils (Bouwer and Powel, 1995) because of added nutrients and the ability to neutralize acidity that may arise from fertilizer use through the suppression of Al solubility. Organic manure added to the soil over one season changed pH from acidic to near neutral after 90 days (Evelyn et al., 2004). Hargrove and Thomas (1981) had observed that the benefit of increased P availability was because the metabolic products of microbial decomposition of the manure form stable complexes with Al, Fe and Mn which are responsible for P sorption or fixation through anion exchange. The phenolic and alcoholic functional groups of the organic acids (humates and fulvates) form complexes with these metal ions or compete with P on the exchange sites (Haynes and Mokolobate, 2001; Yang et al., 2013). Annual application of 5-10 metric tonnes (MT).hectare (ha^{-1}) improved soil physical properties, increased organic matter level and sustained yields in permanently cultivated sub-humid Alfisols (Mokwunye, 1980) but larger rates would probably be needed in the humid tropics with predominant Ultisols. The larger rates are compatible with applications at longer intervals which are more effective than treating soils frequently with small amounts (Müller-Sämann and Kotschi, 1994).

The potentials of these management options for ameliorating acid soil conditions to enhance the productivity of some arable crops in Nigeria have been indicated since the 1970s. Maize, seed cotton, and sorghum yields increased with organic matter application but liming only changed the soil pH and level of exchangeable Al (Heathcote, 1970). Maize growth and yield increased through liming only when nitrogen (N), P, K, Mg, sulphur and organic manure were also applied (Fore and Okigbo, 1973). The implication is that balanced nutrition through the application of lime, manure and fertilizer, to provide the nutrients whose optimum levels are needed to obtain high crop yield and responses to fertilizers in acid soils, should be considered in the development of appropriate management packages. The objective of this study is to evaluate the potentials of fertilizer, lime and manure as components of the management recommendations to be adopted for increased performance of maize in Ultisols represented by four surface soil samples taken from farmers' plots in Edo and Delta States of Nigeria.

2. Materials and Methods

The study was carried out between March and June 2011. Surface (0-15 cm) samples of soils were taken from four locations in Southern Nigeria- Oko and Ugbogiobo in Edo State; Agbarho and Effurum in Delta State- representing the upland and lowland areas of soils developed on unconsolidated sands and sandstones (Coastal Plain Sands) grouped as reddish brown soils of the Benin fasc and yellow to yellowish brown soils of Calabar fasc, respectively (Vine, 1970). The soils at Ugbogiobo and Oko were classified as Oxic Kandiudult and Arenic Kandiudult while at Oleh and Agbarho the soils belonged to Aquic Kandiudult (Ojanuga et al., 1981; FDALR, 1990).

There were two experiments involving 3 kg of air-dried and sieved (<2 mm) soil samples weighed into 5 litre plastic pots. The first experiment was an evaluation of the effects of fertilizer, lime, manure and a control on the four soils in three replicates and arranged in a split-plot design with the soils as main plots and the amendments as subplots. The amendments were 2.5 metric tonnes (MT).hectare (ha^{-1}) of lime (as 100 mesh quicklime, CaO), fertilizer (90 kg N, 30 kg P and 60 kg K.ha^{-1} supplied as urea, single superphosphate and muriate of potash, respectively) and 10 MT.ha^{-1} sawdust/cow dung compost (manure- 3.32% N, 0.58% P, 1.24% K, 0.94% Ca and 0.39% Mg). These amendments were thoroughly mixed with the soils and the mixture was watered and kept moist for seven days. Maize (SUWAN 1-SR-Y variety) seeds were sown and the seedlings thinned to two.pot^{-1} after emergence. The plants were watered daily and grown for 42 days after which the top-growth was harvested from the soil surface, oven-dried at 60°C for 48 hours and weighed. The soils were allowed to air-dry for about two weeks, sampled and cropped to maize for another cycle of 42 days, to evaluate the residual effects of the amendments.

The second experiment was a randomized complete block design of the following treatments: control, fertilizer (90 kg N, 30 kg P and 60 kg K.ha^{-1}), 10 MT.ha^{-1} manure, 2.5 MT.ha^{-1} lime, manure + fertilizer, manure + lime, lime + fertilizer, lime + fertilizer + manure applied to the soil from Oko in three replicates. Two seedlings of maize were grown in each pot for 42 days and the top-growth was harvested, oven-dried and weighed.

The soil samples were analyzed for selected physical and chemical properties using the laboratory procedures described in IITA (1979): particle size distribution (Bouyoucos' hydrometer method); pH in 1: 2 (w/v) soil-water suspension and measured by pH meter; organic carbon (Walkley-Black wet dichromate oxidation method); total N (macro-Kjeldahl digestion and titration); available P (Bray's P-1 extraction and molybdenum blue method and colour intensity read on a spectrophotometer); exchangeable cations (extraction with neutral 1N ammonium acetate, determination of Na, K and Ca with flame photometer and Mg with atomic absorption spectrophotometer); and exchangeable acidity (1N KCl extraction and titration with NaOH). The sum of total exchangeable bases and exchangeable acidity were used to calculate the effective CEC and base saturation. Available Mn and Fe were extracted with 0.1$\underline{N}$HCl and determined with atomic absorption spectrophotometer. The soil samples taken after the first and second cropping were analyzed for pH, organic carbon, exchangeable cations and available P, Mn and Fe.

Dried plant samples were ground in a Wiley mill and digested in a mixture of concentrated nitric, perchloric and sulphuric acids (25-5-5 v/v). Total Ca, Mg, Mn and Fe in the digests were determined with atomic absorption spectrophotometer while total P was determined by the

vanado-molybdate yellow method.

Data of maize top-growth dry matter yield from the direct and residual effects of the amendments were subjected to analysis of variance (ANOVA) and treatment means separated using the least significant difference (LSD) (P= 0.05) as described in Steel *et al.* (1997). Changes in soil chemical properties after adding amendments and each cropping were determined by subtraction relative to the initial values.

3. Results

Table 1. Physiographic features of the locations where soil samples were collected

	Oleh	Agbarho	Oko	Ugbogiobo
Location	5°25'N, 6°08'E	5°34'N, 5°52'E	6°25'N, 5°30'E	6°33'N, 5°37'E
Land Use	Citrus farm mixed with plantain and pineapple. Fertilizer use not practiced; water table was about 1 m to soil surface.	Multiplication Farm for SPDC Nig. Ltd. Agric Projects and Extension Services, used for cultivation of arable crops with regular fertilization; water table within 1 m of soil surface.	ADP On-Station Research Farm for cultivation of arable crops; the previous crop was maize with regular use of fertilizer.	Oil palm nursery site cleared from >7 year fallow; no fertilizers used in the site before.
Vegetation	Seasonal swamp forest	Seasonal swamp forest	Moist lowland forest	Moist lowland forest
Rainfall	2649 mm	2386 mm	2257 mm	2242 mm
Landform	Sombreiro-Warri deltaic plains	Sombreiro-Warri deltaic plains	Nearly level coastal plain terraces, long slopes with coarse-textured dendtritic drainage pattern	Undulating coastal plain terraces dissected by deep gullies and long slopes
Soil group	Yellowish brown soils of Calabar fasc	Yellowish brown soils of Calabar fasc	Reddish brown soils of Benin fasc	Reddish brown soils of Benin fasc
Soil unit	Fresh water-swamp soils	Fresh water-swamp soils	Nearly level plains on Coastal Plain Sands	Nearly level plains on Coastal Plain Sands
Soil type	Typic Tropaquent/ Aquic Kandiudult	Typic Tropaquent/ Aquic Kandiudult	Arenic Kandiudult	Oxic Kandiudult

Sources: FDALR (1990)

Table 1 shows some properties of the surface (0-15 cm) soils used for the studies. The soils were extremely acidic (pH 4.0-4.6) sandy loams to sands with medium organic matter content and very low exchangeable basic cations. Available P varied was 2.1-5.3 mg.kg^{-1} in three soils and 17.5mg.kg^{-1} in the soil from Ugbogiobo.

The top growth dry matter yields of maize during the first and second cropping are shown in Table 2. The main effects of locations and amendments were significant (P=0.05) in the first cropping. The soils from Oleh and Ugbogiobo gave the highest yields (16.97 and 15.46 g respectively) which did not differ while the soils from Agbarho and Oko gave significantly (P=0.05) lower yields. Manure application gave the highest yield (27.39 g) which was superior to the other amendments. The soil x amendment interaction was significant (P=0.05). Lime produced higher dry matter yields in the soils from Agbarho and Oko compared to the control but the increase was not significant. Responses to fertilizer varied among the soils with calculated yield increases low in Ugbogiobo (5.2%) and Agbarho (21.9%), high in Oko (77.4%) and negative in Oleh (-27.2%). Manure was the best amendment in all the soils. Maize dry matter yield was lower in the second cropping with the soils from Oleh and Ugbogiobo still significantly (P=0.05) higher. Fertilizer application reduced maize dry matter yield the most in soils from Ugbogiobo and Agbarho; lime had significant residual

effects in soils from Agbarho, Oleh and Ugbogiobo while manure was beneficial in all the soils.

Table 2. Properties of soils developed on Coastal Plain Sands used for the study

Properties	Oleh	Agbarho	Oko	Ugbogiobo
Sand, %	88.6	83.4	82.6	84.6
Silt, %	4.8	6.4	3.8	3.2
Clay, %	6.6	10.2	14.6	12.2
Textural class	S	LS	LS	LS
pH (water)	4.3	4.1	4.0	4.6
Organic carbon, %	0.92	1.06	0.72	1.12
Total N,%	0.09	0.10	0.07	0.12
Available P, mg.kg^{-1}	2.4	4.9	3.2	16.4
Exchangeable cations, cmol.kg^{-1}				
K	0.11	0.06	0.08	0.24
Ca	0.63	0.17	0.20	0.94
Mg	0.34	0.38	0.19	0.83
Na	0.02	1.31	0.34	0.53
Acidity	3.98	3.41	3.88	2.70
CEC	5.08	5.33	4.69	5.24
Base saturation, %	21.65	36.02	17.27	48.47
Al, mg.kg^{-1}	12.2	24.3	23.0	7.3
Fe, mg.kg^{-1}	1.4	1.7	1.1	2.5
Mn, mg.kg^{-1}	22.0	14.0	21.9	28.0

S = Sand; LS = Loamy sand
CEC = Cation exchange capacity

Table 3. Dry matter yield of maize top-growth as affected by soil amendments

Amendments					
Soils	Control	Fertilizer	Lime	Manure	Mean
1st Cropping					
Agbarho	2.24u	2.73u	4.06u	26.47q	8.88b
Oko	2.03u	3.67u	3.75u	22.21r	7.91b
Oleh	12.54s	9.13t	11.60s	34.60p	16.97a
Ugbogiobo	12.42s	12.52s	10.61st	26.30q	15.46a
Mean	7.31y	7.01y	7.51y	27.39x	
SE = 0.94					
2nd Cropping					
Agbarho	1.88w	1.62w	4.04rst	4.55qr	3.02b
Oko	1.65w	1.76w	1.53w	3.05u	2.00c
Oleh	3.74st	3.85st	4.55qr	5.09pq	4.31a
Ugbogiobo	3.61tu	2.40v	4.47qrs	5.53p	4.00a
Mean	2.72z	2.41z	3.65y	4.55x	
SE = 0.21					

Means for the soils and amendments in each cropping followed by same letters do not differ significantly (P=0.05)

The direct and residual effects of the amendments on maize nutrient content are shown in Table 3. Fertilizer and manure application significantly (P=0.05) increased %P in all soils during the first cropping compared to very low and similar values from lime and the control treatments. Manure still increased maize tissue %P in all soils during the second cropping while the increase from fertilizer was obvious only in the soils from Oko and Oleh. Lime increased %Ca significantly (P=0.05) and the effect was more pronounced during the second cropping while it caused highest reduction of Mn content in the first and second cropping. Fertilizer reduced Mn and Fe content compared to the control while manure application decreased Mn but gave inconsistent results for Fe content.

Table 4 shows the changes in soil properties caused by amendments after the first and second cropping compared to pre-cropping conditions. Lime increased soil pH after the first cropping and the effect was highest (3.1 units) in the soil from Oko and the least (1.8 units) in Agbarho. Lime also increased exchangeable Ca the most but the effects on available P, organic matter and exchangeable Mg were inconsistent while Al, Mn and Fe decreased in all soils. Fertilizer slightly increased soil pH in three soils (0.1-0.6 units) but decreased it in soil from Ugbogiobo (-0.3 units). It increased available P and exchangeable Ca and Mg in all soils and increased exchangeable Al and Mn in Oleh and Ugbogiobo. Manure increased soil pH, organic matter, available P and the exchangeable cations but reduced the levels of Mn, Fe and Al. The increase in available P compared to the initial values was very large. The calculated increase ranged from 780% in Ugbogiobo to 6,500% in the soil from Agbarho. Successive cropping sustained or magnified the changes in soil chemical properties as shown by the slight increase in pH and large increases in available P in all soils after the second cropping, increase in exchangeable Ca and Mg, reduction in organic matter in soils from Oko and Ugbogiobo, and the decrease in Al, Mn and Fe. Soil organic matter decreased after the second cropping compared to the previous levels attained.

Table 4. Effect of soil amendments on nutrient content of maize grown on Ultisols developed on Coastal Plain Sands

Locations	Treatments	P		Ca		Mn		Fe	
		%				mg.kg-1			
		I	II	I	II	I	II	I	II
Oleh	Control	0.10c	0.21c	0.44b	0.85b	194b	295a	1110b	1740a
	Fertilizer	0.23b	0.35b	0.32c	0.35d	285a	321a	580d	720c
	Lime	0.12c	0.16c	0.80a	1.09a	70c	150b	805c	1620b
	Manure	0.31a	0.54a	0.30c	0.62c	194b	183b	1660a	475d
Agbarho	Control	0.10c	0.14b	0.40b	0.57b	604a	812a	1400a	1800a
	Fertilizer	0.28a	0.19ab	0.31c	0.30c	285b	320b	590c	730c
	Lime	0.11c	0.16b	0.89a	1.11a	121d	130c	890b	1650b
	Manure	0.24b	0.27a	0.26c	0.70b	160c	350b	480d	456d
Oko	Control	0.14c	0.16bc	0.46b	0.70b	700a	1035a	1630a	1800a
	Fertilizer	0.44a	0.25b	0.35c	0.33c	350b	645b	702b	625c
	Lime	0.08c	0.14c	0.90a	1.14a	70d	165d	570c	630c
	Manure	0.27b	0.42a	0.37c	0.68b	130c	265c	230d	1330b
Ugbogiobo	Control	0.14c	0.20b	0.44b	0.72b	365a	335a	1600a	1650b
	Fertilizer	0.21b	0.10b	0.42b	0.32d	161b	145b	430c	750d
	Lime	0.12d	0.23b	0.88a	0.91a	130b	150b	960b	1870a
	Manure	0.28a	0.52a	0.31c	0.62c	325a	130b	445c	1475c
	S.E.	0.01	0.04	0.01	0.02	12.9	24.2	12.9	24.9

Values in each column followed by the same letters do not differ significantly (P=0.05)
I = First Cropping
II = Second Cropping

Table 5. Changes caused by application of amendments and successive cropping on some chemical properties of Ultisols developed on Coastal Plain Sands

Soils	Treatments	Organic pH I	II	matter I	II	Available P I	II	Ca I	II	Mg I	II	Al I	II	Mn I	II	Fe I	II
Oleh	Control	0.7	0.7	0.70	0.51	1.08	-0.77	0.85	0.80	0.10	0.20	-2.75	-2.75	-2.75	-4.76	-0.56	-0.23
	Fertilizer	0.6	0.3	-	-	4.70	4.65	1.23	0.80	1.92	1.71	5.40	10.80	8.24	-2.75	-0.56	-0.28
	Lime	2.7	2.7	0.42	0.20	1.05	1.10	3.50	3.55	-0.10	-0.05	-11.66	-11.66	-16.98	-16.98	-0.76	-0.76
	Manure	1.8	1,8	1.65	1.04	155.30	202.20	2.20	2.15	1.10	1.05	-11.66	-11.66	-16.98	-16.98	-0.76	-0.76
Agbarho	Control	0.3	0.4	0.44	0.41	1.88	1.91	0.30	0.25	-0.24	0.00	-4.10	-1.20	-3.98	-3.98	-0.44	-0.39
	Fertilizer	0.1	0.1	-	-	11.50	10.40	0.23	0.17	0.14	0.15	-14.85	-13.50	-17.60	-23.09	-0.55	-0.27
	Lime	1.8	1.8	1.43	1.40	2.09	3.80	3.60	3.35	-0.20	-0.05	-22.95	-23.76	-4.09	-10.99	-0.86	-1.17
	Manure	1.5	1.6	1.85	1.19	145.91	225.40	2.00	2.35	1.05	1.35 -	22.68	-23.76	-5.00	-0.24	-0.92	-0.50
Oko	Control	0.3	0.4	-0.32	-0.33	-0.94	-0.57	0.15	0.40	0.00	0.00	-3.24	-1.62	-8.00	-0.55	-0.42	-0.11
	Fertilizer	0.1	0.1	-	-	-1.66	-0.60	0.25	0.23	0.39	0.30	-6.75	-2.70	-10.90	-13.65	-0.56	0.00
	Lime	3.1	3.1	-0.12	-0.25	-1.63	0.57	3.95	4.20	-0.05	0.10 -	22.41	-22.41	-18.99	-16.98	-0.76	-0.56
	Manure	2.0	2.3	1.23	0.62	152.43	198.43	2.30	2.00	1.50	1.35 -	22.41	-22.41	-3.00	-16.98	-0.48	-0.20
Ugbogiobo	Control	0.2	0.3	-0.26	-0.31	-2.04	-6.00	0.10	0.10	-0.25	-0.10	-0.54	-0.54	-14.23	-2.01	-0.13	-0.44
	Fertilizer	-0.3	-0.3	-	-	3.00	3.00	0.29	0.10	0.82	0.60	-0.54	-0.54	21.97	10.98	-0.55	0.00
	Lime	2.4	2.6	-0.13	0.30	-5.88	-7.60	3.20	3.30	-0.45	-0.35	-6.75	-6.75	-20.86	-19.98	-1.22	-1.22
	Manure	1.4	1.6	0.34	0.24	136.50	182.50	1.35	1.50	1.25	1.10	-6.75	-5.67	-3.99	-19.98	-0.63	-0.86

The effects of all possible combinations of the amendments added to the soil from Oko on maize dry matter yields are shown in Table 5. Lime or fertilizer alone gave similar dry matter yield (6.40 g) which was 78% increase above the control (3.60 g) but was significantly lower than manure application (17.44 g). Manure + Fertilizer produced the highest dry matter yield (26.13 g) that was significantly different (P=0.05) from other treatments and represented 95% increase over the treatment that contained the three amendments. All treatments, except fertilizer alone, increased soil pH to slightly acid status after cropping. Lime + fertilizer and manure treatments gave the highest maize tissue %P concentration. The presence of lime in the treatments increased %Ca, manure and lime decreased Mn and Fe content while application of fertilizer alone gave the highest values.

Table 6. Effect of possible fertilizer, lime and manure combinations on maize dry matter yield and nutrient content in an Ultisol (pH=4.0)

Treatments	Final soil pH	Dry matter	P yield (gm)	Ca	Mn %	Fe mg.kg^{-1}
Fertilizer	3.8f	6.35d	0.21b	0.18d	1175a	2700a
Lime	6.1c	6.40d	0.10c	0.38a	550d	1560c
Manure	5.5d	17.44b	0.28ab	0.26c	425ef	700e
Lime+ Fertilizer	6.2bc	16.20b	0.32a	0.39a	425ef	1600c
Lime+ Manure	6.3b	14.20c	0.11c	0.36ab	450e	1200d
Fertilizer+ Manure	5.3e	26.13a	0.22b	0.21d	650c	2120b
Fertilizer+ Lime+ Manure	6.5a	13.40c	0.23b	0.33b	350f	1250d
Control	3.9f	3.60e	0.03c	0.22cd	1025b	2630a
SE	0.06	0.26	0.044	0.023	29.6	93.9

4. Discussion

The soil characteristics, especially nutrient fertility levels, varied in relation to the nature of land use in each site. Ugbogiobo site was a plot cleared from a previous fallow long enough for the soil to accumulate highest amounts of organic matter and respective nutrients whereas at Oko, the plot had been under continuous tillage and cultivation to annual crops in succession for over 30 years and contained the least values. Based on the soil test interpretation and soil fertility classes for Nigeria (FDALR, 2004; Anon, 2006), the soils contained medium organic matter and low amounts of available P and exchangeable cations. The unusually high available P in the soil from Ugbogiobo can be attributed to the high organic matter on account of the previous long fallow needed for nutrient fertility build-up and low exchangeable Al (Lal, 1999).

There was no significant response to liming in the soils from Oleh and Ugbogiobo probably because of the failure to adequately neutralize the factors responsible for low P availability (notably P fixation). Thus, peculiar P deficiency symptoms (purple leaf colouration) observed are a reflection of the disturbance of P nutrition in these soils despite the fact that they contained more available P than the soils from Agbarho and Oko which gave 81 and 85% increase in dry matter yield respectively. Manure application produced better maize dry matter yield than fertilizer and lime probably because it adds more nutrients as part of the observed general improvement in soil productivity (Opara-Nadi *et al.*, 2000). The response was higher in the soils from Oko and Agbarho (994 and 1,081.7%) with lower nutrient status compared to Ugbogiobo and Oleh (111.7 and 175.9%).

The lack of yield response to fertilizer in the second cropping may be expected given that nutrients in the fertilizer are in soluble forms which leave little residual effects (Tisdale *et al.*, 1993). However, the mineral fertilizer left residues as indicated by higher available P but the soil had become more acid with higher Al, Mn and Fe which must have affected P nutrition. This is unlike lime whose residual effect, as obtained in all the soils, is recognized and responses can be obtained in soils even after two years of application (Brady and Weil, 2002). Kang (1989) noted that low lime rates (0.5-2.0 MT.ha^{-1}) which serve primarily to meet Ca and Mg requirements and neutralize exchangeable Al can have considerable residual effects and sustain maize yields for six years. The manure was thoroughly mixed with the soils at the commencement of the study while subsequent watering during the first cropping followed by air-drying before the second cropping would have increased the rate of organic matter decomposition which reduced the contribution to nutrient supply with time.

Ultisols require proper nutrient management programmes for meaningful arable crop production to take place. Fertilizer alone would not be the appropriate management recommendation given the little effect on crop growth. First, the concentrations of Fe, Mn and Al were high in the soils and which lime and manure reduced significantly. Reeve and Sumner (1970) had suggested 1 mg.kg^{-1} (0.011 cmol.kg^{-1}) as the toxic level to maize such that exchangeable Al was already limiting maize growth in the soils from Agbarho and Oko. Second, P was limiting maize growth and treatments without manure showed typical deficiency symptoms. Unfortunately, farmers are rarely advised to apply lime and manure before or along with fertilizer, as components of the improved technologies being extended, such that resultant poor maize growth and low yields must have contributed to the poor fertilizer use adoption in the states.

Lime reduced exchangeable Al which is responsible for poor crop growth in soils with pH<5.0, but it did not increase yield suggesting that other factors affecting P availability were still active and not effectively neutralized in the soils. Thus, one of the benefits of manure application is reduction of this Al to non-toxic levels. The increase in soil pH causes reduction in the exchangeable acidity since Al, its main component, would precipitate to Al-hydroxide the form in which Al exists in soils with pH 6.0 and above (Ano and Agwu, 2005). Manure reduced Mn but was not as effective as lime. The role of manure was probably indirect because high Ca concentration, contained more in the lime, is the actual factor responsible for reducing the amounts of Mn in soils. Natschner and Schwetmann (1991) noted that manure increased soil pH and reduced exchangeable acidity due to Ca ions released into soil solution during microbial de-carboxylation of the manure. Fe decreased due to reduction in its solubility as pH increased and the ability of organic products of microbial decomposition to chelate exchangeable Fe (Hargrove and Thomas, 1981). Manure application had increased soil organic matter content and so would alter the metal status of soils by affecting their solubility through the high molecular weight ligands (humic substances) it contains (Nolan and McLaughlin, 2005). Humic substances consist of soluble organic matter (fulvic acid) which promotes dissolution of metals from the adsorption sites on clay minerals while the less soluble fraction (humic acid) enhances metal adsorption on soil minerals thereby reducing their bio-availability.

Thus, the approach for nutrient management would be to neutralize this Al, supply Ca and Mg followed by adequate fertilizer application. Igbokwe *et al.* (1981) suggested liming to 5.0-5.5 to deal with the acidity problems of Ultisols in southern Nigeria because above this range, water-soluble Zn and Mn were reduced to non-detectable levels while H^{+} from hydroxyl-Fe, Al compounds and organic matter would ionize resulting in complications which would culminate in the reduction of yield. The treatments which contained manure did not show P deficiency symptoms probably because of adequate supply of available P from the chelating action of organic acids (products of microbial decomposition of organic matter) on Al, Fe and Mn, responsible for P fixation and the expected contribution from native organic matter mineralization. Also, the manure provides P in organic form that is not easily fixed in the soils by the constituents, especially the metal oxides. The soils contained 1.20-1.86% organic matter which would contribute little to the available P, assuming 4% mineralization rate (Ayodele, 1986). However, the organic matter levels after cropping did not suggest sufficient input from decomposition and mineralization, and it would appear that inactivation of P fixation components ensured release of more available P for crop uptake.

Manure enhanced maize response to lime and fertilizer application through the improvement in soil pH, available P and exchangeable Ca and Mg, and reduction in Al. Fe and Mn. Busari *et al.* (2004) had noted that individual amendments were not as effective as the combined application of manure, fertilizer and lime. The Manure + Fertilizer gave better performance than the treatment containing the three amendments and should be the recommended practice for the management of these extremely acidic Ultisols.

5. Conclusion

Ultisols are located in the humid region of southern Nigeria where unconsolidated sands and sandstones (Coastal Plain Sands) parent materials and high rainfall regimes promote formation of acid soils deficient in basic cations. The potentials of these soils for maize production would be realized with management options that emphasize application of manure and complementary manure-inorganic fertilizer which promoted growth through increased availability of P and cations and reduction in toxic elements.

References

[1]   Agboola, A.A. 2002. Farming systems in Nigeria. In: Agboola, A.A. (ed). Essentials of Agricultural Production in Nigeria. Green Line Publishers, Ado-Ekiti, Nigeria: 29-71

[2]   Ano, A.O. and Agwu, J.A. 2005. Effect of animal manures on selected soil properties. *Nig. J. Soil Sci.* 15: 14-19

[3]   Anon (2006). Nigeria Fertilizer Strategy Report. Presented at Africa fertilizer Summit held at International Conference Centre, Abuja, 9-13th June, 2006. 47pp.

[4]   Ayodele OJ (1986). Effect of continuous maize cropping on yield, organic carbon mineralization and phosphorus supply of savannah soils in western Nigeria. *Biology and Fertility of Soils* 2: 151-155.

[5]   Banful, A.B., Nkonya, E. and Oboh, V. 2010. Constraints to Fertilizer Use in Nigeria: Insights from Agricultural Extension Service. IFPRI Discussion Paper No 01010. International Food Policy Research Institute, Washington. 36pp

[6]   Brady NC, Weil RR (2002). The Nature and Properties of Soils. 13th Edition, Pearson Educ. Pub. New Delhi, India. 881pp.

[7]   Brouwer J, Powel JM (1995). Soil aspect of nutrient cycling in a manure application experiment in Niger. In: Powel JM *et al*. (eds). Livestock and Sustainable Nutrient Cycling in Mixed Farming Systems of Sub-Saharan Africa. Volume II. Technical Papers. Proceedings of International Conference held 22-26 Nov. 1993. ILCA, Addis Ababa, Ethiopia. 211pp.

[8]   Buol, S.W., Southard, R.J., Graham, R.C. and McDaniel, P.A. 2011. Soil Genesis and Classification.

[9]   Busari MA, Salako FK, Sobulo RA, Adetunji MT, Bello NJ (2004). Variation in soil pH and maize yield as affected by application of poultry manure and lime. In: Salako FK *et al*. (eds). Managing Soil Resources for Food Security and Sustainable Environment. Proceedings of 29th Annual Conference of Soil Science Society of Nigeria held at University of Agriculture, Abeokuta, Nigeria, December 6-10, 2004: 139-142.

[10] Evelyn, S.K., Jan, O.S. and Jeffrey, A.B. 2004. Functions of Soil Organic Matter and the Effect on Soil Properties. CSIRO Land and Water Report. 129pp

[11] Fageria, N.K. and Baligar, V.C. 2008. Ameliorating Soil Acidity of Tropical Oxisols by Liming for Sustainable Crop Production. Advances in Agronomy 99: 345-399

[12] FDALR (1990). The Reconnaissance Soil Survey of Nigeria (1:650,000). Soils Report Volume Three (Bendel, Lagos, Ogun, Ondo and Oyo States). Federal Department of Agricultural Land Resources, Lagos. 149pp.

[13] FDALR (2004). Handbook on Soil Test-based Fertilizer Recommendations for Extension Workers. Federal Department of Agricultural Land Resources in collaboration with National Special Programme for Food Security, Abuja, Nigeria.39pp.

[14] Fore RE, Okigbo BN (1973). Yield response of maize to various fertilizers and lime on Nkpologu sandy loam soil. *Nigerian Agric. J.* 9: 124-127.

[15] Hargrove WL, Thomas GW (1981). Effects of organic matter on exchangeable aluminium and crop growth in acid soils. In: Chemistry in the Environment. ASA Special Publication No 40. Amer. Soc. Agron, and SSSA, Madison, Wisconsin.

[16] Haynes, R.J. and Mokolobate, M.S. 2001, Amelioration of Al toxicity and P deficiency in acid soil by addition of organic residues. *Nutrient Cycling. Agroecosyst.* 59: 47-63

[17] Haynes RJ, Naidu R (1998). Influence of lime, fertilizer and manure on soil organic matter content and soil physical conditions: A review. *Nutrient Cycling Agroecosyst.* 15: 123-127.

[18] Heathcote RG (1970). Soil fertility under continuous cultivation in Northern Nigeria I. The role of organic manure. *Exptl Agric* 6: 229-237.

[19] Igbokwe MC, Njoku BO, Odurukwe SO (1981). Liming effects on the response of maize to phosphate fertilizer on an Ultisol in Eastern Nigeria. *Nigerian J of Soil Science* 11: 120-130.

[20] IITA (1979). Selected Methods of Soil and Plant Analysis. Manual Series No 1. International Institute of Tropical Agriculture, Ibadan, Nigeria. 70pp.

[21] Juo ASR, Wilding LP (1996). Soils of the lowland forest of West and Central Africa. In: essays on the Ecology of the Guinea Congo Rainforest. Proceedings Royal Society of Edinburgh. Vol. 1043. Section B: Biological Science. Edinburgh, Scotland, U.K.: 15-26.

[22] Juo ASR, Franzluebbers K, Dabiri A, Ikhile B (1996). Soil properties and crop performance on a kaolinitic Alfisol after 15 years of fallow and continuous cultivation. *Plant and Soil* 180: 290-217.

[23] Kamprath EJ (1971). Potential detrimental effects from liming highly weathered soils to neutrality (A review). *Soil Crop Sci Soc Fla. Proc* 31: 7-9.

[24] Kang, B.T. 1989. Nutrient management for sustained crop production in the huid and subhumid tropics. In: van der Heide, J. (ed). Nutrient Management for Food Crop Production in Tropical Farming Systems. Institute for Soil Fertility, Haren, The Netherlands and Universitas Brwijaya, Indonesia: 3-28

[25] Lal R (1999). Land use and cropping system effects on restoring soil carbon pool of degraded Alfisols in Western Nigeria. In: Lal R *et al*. (eds) Global Climate Change and Tropical Ecosystems. Advances in Soil Science. CRC Press, Boca Raton: 157-165.

[26] Liverpool-Tasie, S., Olaniyan, B.,Salau, S. and Sackey, J. 2010. A Review of Fertilizer Policy Issues in Nigeria. Nigeria Strategy Support Program (NSSP) Report No 28, IFPRI-Abuja. 5pp

[27] Martini JA, Mutters RC (1989). Soyabean root growth and nutrient uptake as affected by lime rates and plant age. II. Ca, Mg, K, Fe, Cu and Zn. *Turrialba* 39: 254-259.

[28] Mokwunye U (1980). Interactions between farmyard manure and fertilizers in savannah soils. In: *FAO Soils Bulletin* No 43: 192-200

[29] Müller-Sämann KM, Kotschi J (1994). Sustaining Growth: Soil Fertility Management in Tropical Smallholdings. GTZ, Germany and CTA, Netherlands. Margraf Verlag, Germany. 486 pp

[30] Natshner I, Schwetmann U (1991). Proton buffering in organic horizons of acid forest soils. *Geoderma* 48: 93-106.

[31] Nolan AL, McLaughlin MJ (2005). Prediction of zinc, cadmium, lead and copper availability to wheat in contaminated soil using chemical speciation, diffuse gradients in thin films extraction and isotopic diffusion techniques. *J Environ Quality* 34: 496-507.

[32] Ofori, C.S., Higgins, G.M. and Purnell, M.F. 1986. Criteria for choice of land suitable for clearing for agricultural production. In: Lal, R., Sanchez, P.A. and Cummings, R.W. Jr (eds). Land Clearing and Development in the Tropics. A.A. Balkema, Rotterdam: 19-28

[33] Ojanuga AG, Lekwa G, Akamigbo FRO (1981). Survey, classification and genesis of acid soils. In: Udo, E.J. and Sobulo, R.A. (eds). Acid Soils of Southern Nigeria. Soil Science Society of Nigeria Special Publication Monograph No 1: 1-8.

[34] Opara-Nadi AO, Omenihu AA, Ifemedehe SN (2000). Effects of organic wastes, fertilizers and mulch on productivity of an Ultisol. In: Babalola O (ed). Proceeding of 26th Annual Conference of Soil Science Society of Nigeria held at Ibadan, 30th October-3rd November, 2000: 112-120.

[35] Reeve NG, Sumner ME (1970). Effects of Al toxicity and phosphorus fixation on crop growth in Oxisols in Natal. *Soil Sci Soc Amer. Proc.* 34: 262-267.

[36] Sanchez, P.A. and Buol, S.W. 1975. Soilsof the tropics and the world food crisis. In: Abelson, P.H. (ed). Food: Politics, Economics, Nutrition and Research. American Association for the Advancement of Science Misc. Publication 75-7: 115-120

[37] Steel RDG, Torrie JH, Dickey DA (1997). Principles and Procedures of Statistics: A Biometrical Approach. 3rd Edition. McGraw Hill, New York. 665pp.

[38] Tisdale SL, Nelson WL, Beaton JD, Havlin JL (1993). Soil Fertility and Fertilizers. 5th Edition, Macmillan, New York.

[39] Vine H (1970). Review of Work on Nigerian Soils. Report of the National Research Council Committee on Tropical Soils. London, England.

[40] Yang, S., Zhang, Z., Cong, L., Wang, X. and Shi, S. (2013). Effect of fulvic acid on the phosphorus availability in acid soils. *J. Sci. and Plant Nutri.* 13: 526-533

23

Effect of Some Micro-Catchment Water Harvesting Techniques on Some Soil Physical Properties

Azmi Elhag Aydrous[1], Abdel Moneim Elamin Mohamed[2], Hussein Mohammed Ahmed Abuzied[3], Salah Abdel Rahman Salih[4], Mohamed Abdel Mahmoud Elsheik[4]

[1]Department of Agricultural Engineering, Faculty of Agriculture, Omdurman Islamic University, Omdurman, Sudan
[2]Department of Agricultural Engineering, Faculty of Agriculture, University of Khartoum, Khartoum, Sudan
[3]Department of Landscape and dry land cultivation, Faculty of Agriculture, Omdurman Islamic University, Omdurman, Sudan
[4]Department of Agricultural Engineering, Faculty of Agriculture, Elneelain University, Khartoum, Sudan

Email address:

azmielhag@yahoo.com (A. E. Aydrous), azmielhag21@hotmail.com (A. E. Aydrous)

Abstract: The experimental work was conducted at Jebel Awlia locality 40 kilometers south of Omdurman city during 2010-2011 and 2011-2012 rainy seasons to investigate the effect of micro-catchment water harvesting techniques on some soil physical properties. Techniques used were, semi- circular, V-shaped, pits, deep ditches and land without water harvesting technique control. Soil properties studied were infiltration rate, saturation percentage, bulk density and the percentages of clay, silt and sand. The results showed that infiltration rates in all treatments were lower than that of the control, the mean differences between treatments were not significant in the first season but significantly lower means were obtained by the semi circular and pits in the second season. Saturation percentage in both seasons, were significantly lower after rainfall as compared to that before rainfall for all treatments. Except for the semi-circular and the V-shaped treatments in both seasons and deep ditches in the first season and pits in the second season, bulk density after rain fall was significantly lower than that before rainfall. Clay content in both seasons was not significantly affected by the water harvesting techniques, except under deep ditches in the second season and overall in both seasons. Silt content, in both seasons, was not significantly influenced by the technique for all treatments, except during the first season, in which the techniques before rainfall had a significantly higher mean as compared to that after rainfall. Effect of the water harvesting technique on sand content had insignificant effect, except the overall mean of the techniques during the second season, in which before rainfall was significantly higher as compared to that of the control treatments.

Keywords: Micro-Catchment Techniques, Infiltration Rate, Saturation Percentage, Bulk Density

1. Introduction

Dry lands cover about 5.2 billion hectares, a third of the land area of the globe (UNEP, 1992). Roughly one fifth of the world population lives in these areas.

Dry lands have been defined by FAO on the basis of the length of the growing season, as zones which fall between 1-74 and 75-199 growing days to represent the arid and semi-arid dry lands, respectively, and receiving rainfall between 0 – 600 mm annually (FAO, 1978).

The main feature of "dryness" is the negative water balance between the annual rainfall (supply) and the evaporative demand. Many of the world's dry lands are grazing rangelands and characterized by the need to manage and cope with erratic events of rain that constrain opportunities for development. In dry-lands, production is possible only when additional water is made available for cultivation. With declining investments in irrigation in developing countries, alternative methods, such as soil and water conservation, have become more important in recent decades (Devi et al., 2005). Water harvesting is one such technique and is based on the collection and concentration of surface runoff for cultivation before it reaches seasonal or perennial streams (Reij et al., 1988).

2. Materials and Methods

The experimental work was conducted at Jebel Awlia locality 40 kilometers south of Omdurman city and 25 kilometers from the west bank of the White Nile River during 2010-2011 and 2011-2012 rainy seasons. The experiments covered an area of 5 hectares as a part of the area designated for Khartoum New International Airport. The climate of the area is semi desert, which was characterized by high temperature of an average of $45^{\circ}C$ during the summer. Wind speed is very high evoking dust. Very sparse herbaceous plants and *Acacia* trees comprise the plant cover which is green during the rainy season. The soil is light to sandy in composition except at lane beds especially Mansourab dam. Soil changes gradually to clay and sand-clay according to level and topography.

Five water harvesting techniques were used as follows:

(1) Semi- circular water traps, (T_1) designed with 30 meters diameter, 90 cm height and 20 meters distance between one trap and the other. The water dikes were composed of 3 units, with a distance of 50 meters from the next unit of dikes.

(2) V-shaped water dikes (T_2): Each 30 meters side length and 30 meters bottom of the V-shaped width. The distance between a set of dikes and the other was 20 meters. The water trap was composed of 3shapes at the front and 2shapes at the rear at a distance of 50 meters between the front and rear.

(3) Pits (T_3): The pits were designed at 5 meters width, 10 meters length and 10 meters between pits. Pits were dug according to the land gradient. Water trap was composed of 3 pits at the front and 2 pits at the rear at a distance of 50 meters between the front and rear pits.

(4) Deep ditches (T_4): The deep ditches were dug by a motor grader. The length of each ditch was 30 meters and depth of 90cm, at a distance of 20 meters between ditches. The water trap in this design was composed of three ditches at the front and two ditches at the rear, the distance between the front and rear ditches was 50 meters.

(5) Land without water harvesting technique (control) treatment denoted by (T_0).

The infiltration rate was determined using a double ring infiltrometer as described by Michael (1978). The double ring infiltrometer consisted of two concentric cylinders each 0.25 cm thick, 30 cm height with diameter of 30 cm for the inner ring and 60 cm for the outer one. The infiltrometer was pressed firmly in the soil and hammered gently with the help of a wooden flank until it was driven to a depth of 10 cm. A filter paper was then placed at the bottom of the inner cylinder to prevent disturbing the surface of the soil. Water was then poured gently into the inner cylinder. The space between the inner and the outer cylinder was filled immediately with water after filling the inner one to prevent the lateral water movement. Readings of the depth of the ponded water in the inner cylinder were taken every 5 minutes then the rate of water intake over the time was measured as described by Michael (1978). Soil analysis

included saturation percentage (S.P), bulk density (B.D.) and soil mechanical analysis (Clay%, Silt% and sand %) were carried out in the laboratories of Al-Neelain University.

3. Results

Table 1. Mean infiltration rate under the different micro catchment techniques during 2011/2012 and 2012/2013 seasons.

treatments	Mean infiltration rate (mm/h)	
	2011/2012 season	2012/2013 season
T0	157.7	162.0
T1	107.9	90.5
T2	114.3	91.7
T3	134.4	102.1
T4	113.4	110.7

Table 1 shows the mean infiltration rates under the micro catchment techniques in both seasons. The results showed that infiltration rates in all treatments were lower than that of the control. The mean differences between treatments in the first season were not significant. The percentage of decrease in infiltration rate for the semi circular, pits, deep ditch and V-shaped micro-catchment as compared to the control was 46.2%, 38.0%, 17.3% and 39.1%, respectively. In the second season, all micro catchment treatments had lower mean of infiltration rate than the control, with significantly lower mean obtained by the semi circular and pits. The treatments of semi circular, pits, deep ditch and V-shaped catchments showed decreased infiltration rate as compared to the control by about 79.0%, 76.7%, 58.7% and 46.3%, respectively.

In both seasons, saturation percentage (SP) was significantly lower after rainfall as compared to the before rainfall for all treatments (Table 2). In the first season, (SP) after rainfall under T_0, T_1, T_2, T_3, T_4 and overall was reduced by about 54.4%, 48.4%, 38.9%, 50.4%, 43.3% and 47.9%, respectively, as compared to that before rainfall. On the other hand, in the second season, the reduction in SP for the same treatments after rainfall as compared to before rainfall was 0.8%, 33.3%,44.5%, 21.4%, 53.2% and 30.8%, respectively.

Table 2. Saturation percentage before and after rain fall under the different micro catchment techniques during 2011/2012 and 2012/2013 seasons.

Treatments	Saturation percentage			
	2011/2012		2012/2013	
	before rain fall	after rain fall	before rain fall	after rain fall
T0	20.40	44.83	45.15	45.53
T1	24.13	46.80	28.82	43.20
T2	28.60	46.80	25.44	45.80
T3	21.47	43.27	34.19	43.50
T4	25.73	45.40	22.00	46.97
Over all	24.07	45.42	31.12	45.00

Except T_0 and T_1 in both seasons and T_4 in the first season and T_3in the second season bulk density (Bd) after rain fall was significantly lower than that before rainfall (Table 3). In the first season T_2, T_5 and overall showed a reduction in (Bd) after rainfall as compared to those before rainfall of 21.6%, 25.3% and 20.6%, respectively, whereas T_4 and the overall means in the second season for the same condition reduced this bulk density by 20.7%, 3.6% and 6.6%, respectively.

Table 3. Bulk density (gm/cm³) before and after rain fall under the different micro catchment techniques during 2011/2012 and 2012/2013 seasons.

Treatments	Bulk density (gm/cm³)			
	2011/2012		2012/2013	
	before rain fall	after rain fall	before rain fall	after rain fall
T0	1.65	1.48	1.74	1.60
T1	1.57	1.25	1.62	1.55
T2	1.67	1.31	1.69	1.34
T3	1.78	1.33	1.65	1.65
T4	1.81	1.35	1.67	1.61
Over all	1.70	1.35	1.67	1.56

Clay content in both seasons was not significantly affected by the water harvesting technique, except under T_4 in the second season and overall in both seasons (Tables 4). In the second season, T_4 treatment after rainfall had a significantly higher mean of clay content as compared to the before rainfall and the percentage of increase was 41.6%. On the other hand, after rainfall in both seasons clay content for overall effect of micro catchment techniques was significantly high as compared to that before rainfall and the percentage of increase was 7.3% and 21.8% for the first and second seasons, respectively.

Table 4 shows that silt percentage in both seasons was not significantly influenced by the water harvesting technique, except during the first season, in which the overall techniques before rainfall had significantly higher means as compared to that of the control.

Table 4. Clay, silt and sand % before and after rain fall under different micro catchment techniques during 2011/2012 season.

Treatments	before rain fall			after rain fall		
	Clay%	Silt%	Sand%	Clay%	Silt%	Sand%
T0	34.52	31.10	34.37	33.74	30.72	35.54
T1	32.90	32.31	34.79	38.22	30.82	30.95
T2	35.03	31.27	33.37	37.22	29.39	33.39
T3	35.33	25.80	38.87	36.99	25.25	37.76
T4	34.58	29.89	35.53	48.67	28.82	32.51
Over all	34.47	30.08	35.39	36.97	29.00	34.03

Table 5. Clay, silt and sand % before and after rain fall under different micro catchment techniques during 2012/2013 season.

Treatments	before rain fall			after rain fall		
	Clay%	Silt%	Sand%	Clay%	Silt%	Sand%
T0	32.65	27.02	40.33	32.06	27.08	40.86
T1	27.46	34.81	37.72	35.57	31.77	32.66
T2	27.60	35.82	36.58	35.85	33.25	30.90
T3	28.74	37.08	34.18	30.71	36.63	32.66
T4	34.08	29.36	36.56	49.19	24.86	25.95
Over all						

Effect of water harvesting technique on sand content (Table 4) was insignificant, except the effect of the overall techniques during the second season. (Table 4).

4. Discussion

The mean of the infiltration rate were lower under the micro-catchment techniques as compared to the control, but statistical analysis showed that only means of both T_1 and T_2 in the second season were significantly lower than the control. The lower mean infiltration rate under the micro-catchment techniques as compared to the control may be attributed to the land preparation for construction of these measures as well as the effect of these measures on accumulation of clay by rain water flow from the natural upslope area. Elboshra (2011) observed that both holes and crescents positively affected infiltration rate of the soil where the present study was conducted. However, Hillel (1982) and Hensly and Bennie (2003) stated that micro-catchment measures as tied ridge/bund, contour furrow/bund and bench terraces enhanced infiltration rate and soil moisture storage, while the conventional farming systems reduce infiltration due to compaction, soil crusting, hard pan formation and hence reduce water holding capacity.

Both clay contents and saturation percentage under micro-catchments technique increased in the micro-catchments as compared to the control, mainly under T_4. Capturing of clay material which was carried in the flowing water from the upslope area and accumulation of it in these catchments may be the reason of higher clay content under the micro-catchments mainly the V-shaped. The study also indicated that clay content relative to depths was increased in the catchments after rainfall as compared to that before rainfall particularly at 30 – 60 and 60 – 90 cm depths. This finding is in line with that of Ali and Yazar (2007) who observed that clay content slightly increased after rainfall with depth, while sand decreased with depth. Jianxin et al., (2007) pointed out that micro-catchments gain higher moisture content due to the runoff collected and infiltration at each depth. On the other hand, the higher saturation percentage under the micro-catchments after rainfall as compared to that before rainfall may be attributed to the higher clay content in the catchments after rainfall.

The study also showed that silt, sand contents and bulk density were slightly reduced under the micro-catchments and depths after rainfall as compared to those before rainfall. The reduction in silt content in the catchments after rainfall as compared to that before rainfall may be attributed to the movement of these materials down the profile with the infiltrated water. Meanwhile the reduction in sand content under the micro-catchments after rainfall maybe due to the reduction of this substance at the soil surface during the micro-catchments construction as well as the reduction of soil erosion due to rain water runoff which was interrupted by these micro-catchments. The slight higher sand content under the control treatment after rainfall compared to that before rainfall may reflect the effect of rain water runoff on such substance when it carries out the clay and silt particles and leaving a rough soil surface.

5. Conclusion

The micro-catchment water harvesting techniques improve some soil physical properties.

References

[1] Ali, Akhtar and Attila Yazar, (2007). Effect of Micro-catchment Water Harvesting on Soil-water Storage and Shrub Establishment in the Arid Environment. International Center for Agricultural Research in the Dry Area (ICARDA), Aleppo, Syria Department of Irrigation and Agricultural Structures, Cukurova University Adana, Turkey. INTERNATIONAL JOURNAL OF AGRICULTURE and BIOLOGY 1560–8530/2007/09–2–302–306.

[2] Elboshra, M. A. (2011). Effect of Holes and Crescents Water Harvesting Techniques on Growth of Sidr (*Ziziphus Spina-Christi*) Around Khartoum New International Air Port. M.S.c. Thesis, Agric., University of Khartoum, Khartoum, Sudan.

[3] FAO (1978). Soil and water conservation in semi-arid areas. *FAO Soils Bulletin No. 57*, FAO, Rome.

[4] Hensley, M. and Bennie, A.T.P., (2003). Application of water conservation technologies and their impacts on sustainable dryland agriculture in sub-Saharan Africa. In: Beukes, D., de Villiers, M.; Mkhize, S.; Sally, H. and van Rensburg, L. (Eds.), Proceedings of the Symposium andWorkshop on Water Conservation Technologies for Sustainable Dry land Agriculture in Sub-Saharan Africa (WCT), Bloemfontein, SouthAfrica,8-11April,2003.pp.2–17.

[5] Hillel, D. (1982). Infiltration and surface runoff. In: Hillel, D. (Ed.), Introduction to Soil Physics. Academic Press, New York, pp. 211–234.

[6] Jianxin, Z., Zheng, D., Wang, D., Duan, Y. and Su, Y. (2007). Two water harvesting type within-field Rainwater harvesting measures and their effects on increasing soil moisture and crop production in north china. College of Resources and Environment, China Agricultural University, Beijing, China, 100094.

[7] Michael, A.M. (1978). Irrigation: Theory and Practices. Vikas publishing House, PVT Ltd. New Delhi, India.

[8] Reij, C.:, Scoones, I. and Toulmin, C. (1998). *Sustaining the Soil, Indigenous Soil and Water Conservation in Africa.* Earthscan, London, U.K.:

[9] UNEP (1992). Rain and Storm Water Harvesting in Rural Areas. Ed. United Nation Environmental Programmed, Dblin: Tycooly International.

[10] Devi, B.L.:, Maheshwari, B. and Simmons, B. (2005). Rainwater harvesting for residential irrigation: How sustainable is it in an urban context. Proc. of 12th International Conference on Rainwater Catchment Systems. New Delhi, India.

24

Tree Growth Response of *Pinus oocarpa* Along Different Altitude in Dedza Mountain Forest Plantation

Anderson Ndema, Edward Missanjo

Department of Forestry, Malawi College of Forestry and Wildlife, Dedza, Malawi

Email address:

andersoneendema@gmail.com (A. Ndema), edward.em2@gmail.com (E. Missanjo)

Abstract: Understanding of the effects of altitude on tree growth is central to forest management, especially in the establishment of seed source stands. A study was conducted to investigate the effect of altitude on the growth height, diameter at breast height (dbh) and volume of *Pinus oocarpa* in Malawi. Stands of *Pinus oocarpa* at the altitude of 1500m, 1700m and 1900m above the sea level (asl) were measured for total height, dbh and volume at the age of 18 years. Data obtained were subjected to analysis of variance. The results shows that there were significant ($P<0.001$) differences in total mean height, dbh and volume among the different altitudes. Higher mean height (19.2m), dbh (24.5cm) and volume ($0.417m^3$) was observed at 1500m asl, while total mean height, dbh and volume at 1700m asl and 1900m asl were 17.1m, 22.9cm, $0.322m^3$ and 15.4m, 20.8cm, $0.243m^3$ respectively. Total mean height, dbh and volume decreased with an increase of altitude. This was attributed to differences in supply of soil nutrients and specific leaf area. It is therefore, recommended that seed sources stands for *Pinus oocarpa* in Malawi and the surrounding countries should be established at 1500m to 1600m above the sea level for better genetic growth parameters.

Keywords: Total Height, Diameter at Breast Height, Volume, Specific Leaf Area

1. Introduction

Understanding of the effects of altitude on tree growth is essential for improvement of forest cover as well as central to forest management [1], especially in the establishment of seed source stands [2]. Various studied on the effect of altitude on tree growth in different species have been conducted by many researchers [1, 3 – 12]. However, mixed results have been reported. A reduction in tree height and diameter with an increase in altitude was observed in various studies [1, 3 – 5]. On the other hand, there were no significant differences in tree height and diameter growth due to different altitudes [6 – 8].

Pinus oocarpa Schiede ex Schltdl. is one of the major exotic plantation species in Malawi and other Southern African countries. This species of pine tree is native to Mexico and Central America [13]. It is a national tree of Honduras, where it is known as Ocote [14]. Trees of *Pinus oocarpa* can be recognized by their irregular crowns, thick, gray, platy bark, ovoid-shaped cones with a large thick peduncles, and needles in fascicles of five. The tree species occurs from 350 to 2500 m above the sea level in Mexico and Central America but

reaches its best development between 1200 and 1800 m above the sea level where the soils are deep and annual rainfall are above 1200mm. The growth rates in these regions are approximately 3 to 4 m^3 per hectare per year [15]. Tree size varies considerably over its range; heights up to 37 m; diameters 40 to 80 cm and occasionally 127 cm. Boles are cylindrical, straight and clear up to 15 m or more and it is fire resistant [16]. Nevertheless, the tree grows poorly and reach the height of 10 to 15 m in northern Mexico, where the climate is drier than in most parts of Central America. Trees are also often less than 10 to 12 m height where they grow at elevations below 800 to 900 m above the sea level or on shallow, eroded soil on ridge tops. The growth rate in these dry regions is approximately 1 m^3 per ha per year [15]. *Pinus oocarpa* has a seed potential of about 140 seeds per cone. The number of seeds per kg varies from 43000 to 78000 and in Guetamala the seed size was found to decrease with increasing elevation [15, 16]. In contrast, seed yields are poor for *Pinus oocarpa* planted as exotic near the equator but improve with increasing latitude [15]. For example, Arce and Isaza [17], as reported by [15], found only seven filled seeds per cone in stands established between 1360 m and 1800 m above the sea level of 12 to 21 years of age in Columbia (2° N latitude), while about 25 filled seeds per cone were found in Venezuela (10° to 11°N latitude) in stands 10 to 12 years of age when established above 800 m above the sea level. Seeds of *Pinus oocarpa* usually will begin to germinate in 7 to 10 days using standard nursery or laboratory techniques. Seedlings reach a planting height of 20 to 25 cm in 5 to 7 months [15]. *Pinus oocarpa* is an important timber tree. The wood is easy to work with hand and machine tools. In Mexico and Central America, the wood is used for plywood, construction lumber, packing boxes, fuel wood, kindling as well as resin production [14].

The tree species was introduced in Malawi in 1950's for timber and it is now grown in most parts of the country at different altitudes. There is a general agreement in literature that growth of *Pinus oocarpa* varies according to location [14, 15]. Despite this, no information is available on the growth performance of *Pinus oocarpa* at different altitudes in Malawi. Therefore, the purpose of this study was to investigate the influence of altitude on the growth performance of *Pinus oocarpa* in Malawi. The information on growth performance of *Pinus oocarpa* at different altitude would be incorporated into decision support system in Malawi to assist the forestry industry in planning and establishment of best seed sources sites with high genetic growth parameters.

2. Methods

2.1. Study Site

This study was conducted in Malawi near the tropical savannah region in Southern Africa at Dedza Mountain Forest Plantation (Figure 1). Dedza Mountain lies on latitude 14°20'S and longitude 34°20'E and it rises up to about 2200m above the sea level. It receives 1200 mm to 1800 mm rainfall per annum, with annual temperature ranging from 7 °C to 25 °C (Figure 2). The soils are high in ferralsols, acrisols and nitosols. It is situated about 85 km southeast of Lilongwe, the capital.

2.2. Forest Stands, Plots and Data Collection

Seedlings of *Pinus oocarpa* at the age of six month after pricking out were planted on nine compartments of 1.2 ha each at three different altitudes (1500m, 1700m, and 1900m above the sea level) in 1994. Each altitude had three compartments, totalling to 10.8 ha area planted. The mean height and root collar diameter of the seedlings were 26.8 cm and 5.1mm respectively. Hole planting was done at a spacing of 2.75m x 2.75m (1320 stems ha⁻) in all the compartments. Low thinning was carried out at the age of 9 and 15 years by removing 35% and 28% of the trees respectively. At the ages of 4, 9 and 15 years pruning was carried out in all the compartments. Trees were pruned to half of the stem height.

At the age of 18 years, 250 trees per compartment were systematically selected and measured for total height, diameter breast height (dbh) (1.3m above the ground) and

volume. Suunto clinometer and calliper were used to measured total height and dbh respectively, while tree volume was calculated from the tree diameter and height using a tree volume function [18]. The longest fresh needle found on ground for the measured trees were also measured by a 30 cm ruler. Eight circular sample plots of 0.05 ha each were systematically constructed in each compartment and total number of trees were counted. This was done in order to determine the stocking per hectare for each altitude so that the natural mortality rate at each altitude is known. The stocking per hectare and natural mortality rate was calculated using formula's as expressed by [18]. Prior to tree measurement soil analysis was done. Site characteristics information is presented in Table 1.

Table 1. *Site description.*

Altitude (m asl)	1500	1700	1900
Longitude;	34⁰20'E;	34⁰19'E;	34⁰19'E;
Latitude	14⁰22'S	14⁰22'S	14⁰21'S
Slope (%)	6	10	10
Aspect	East	East	East
Soil pH(H₂O)	6.4	6.3	6.3
P (mg kg⁻)	14.7	10.6	6.50
Zn (mg kg⁻)	0.59	0.48	0.41
Fe (mg kg⁻)	5.77	5.71	5.62
Mn (mg kg⁻)	5.98	5.73	5.46

2.3. Statistical Analysis

Data obtained were tested for normality and homogeneity with Kolmogorov-Smirnov D and normal probability plot tests using Statistical Analysis of Systems software version 9.1.3 [19]. After the two criteria were met the data were subjected to analysis of variance (ANOVA) using the same Statistical Analysis of Systems software and means were separated with Fischer's least significant difference (LSD) at the 0.05 level.

3. Results and Discussion

Mean height, dbh and volume growth are presented in Table 2. The results shows that there were significant ($P<0.001$) differences in mean height, dbh and volume of *Pinus oocarpa* among the altitudes. The altitude of 1500 m above the sea level had higher values of mean height, dbh and volume than the altitudes of 1700m and 1900m above the sea level. Mean height, dbh and volume decreased with an increase in altitude. However, diameter increment declined less, 6.5% and 15% from 1500m to 1700m and from 1500m to 1900, with increasing elevation than height, 11% and 20% respectively. There was a similar trend for dbh growth and volume growth. The reason behind this is that dbh is highly positively correlated with volume [20].

The present results are in agreement to those in literature [1, 3-5, 9-12]. Growth rates may decline with increased altitude due to reduced air and soil temperature (an adiabatic effect), increased exposure to wind, and reduced supply of nutrients [4]. This may be true for the present study since nutrients decreased with an increase of altitude (Table 1). Coomes and Allen [4] reported that phosphorus availability is known to decline with an increasing altitude and this may contribute to the slow growth at high altitudes. On the other hand, [6-8] did not find a statistically significant effect of change in altitude on

tree diameter. These differences may be attributed to the fact that various sites factors influence diameter growth [12].

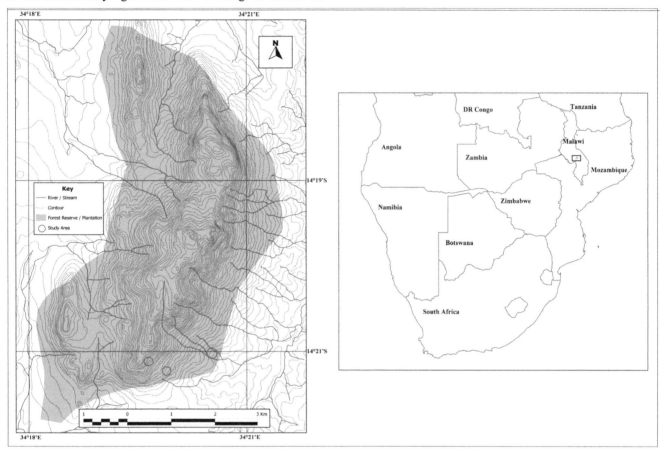

Figure 1. *Location of Dedza Mountain in Southern Africa*

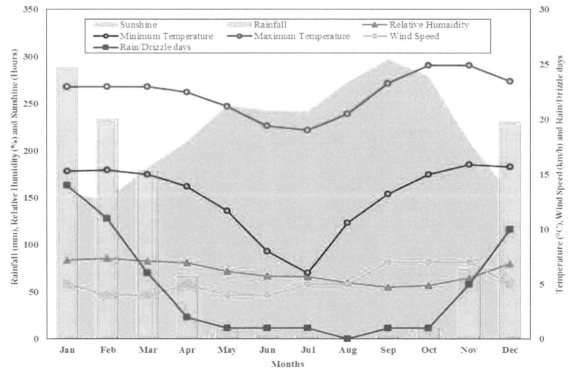

Figure 2. *Weather for the study area*

Table 2. Mean height, diameter at breast height (dbh) and volume of Pinus oocarpa at the age of 18 years for different altitudes.

Altitude (m asl)	Height (m)	dbh (cm)	Volume (m³)
1900	15.4±0.6[c]	20.8±0.2[c]	0.243±0.005[c]
1700	17.1±0.5[b]	22.9±0.2[b]	0.322±0.004[b]
1500	19.2±0.5[a]	24.5±0.3[a]	0.417±0.005[a]
CV (%)	7.9	6.3	19.6

Note: [a,b,c]means with different subscript within a column significantly differ ($P<0.001$)

Table 3 shows that there were significant ($P<0.001$) differences in estimated stocking per hectare among the different altitudes and these lead to the significant differences in natural mortality rate per year among the different altitudes. Natural mortality rate increased with an increase in altitude. Looking at this one would expect that trees grown at 1900 m above the sea level would have higher values of mean height and dbh than those grown at 1700m and 1500 m above the sea level. This is so because trees grown at 1900 m above the sea level have more space, hence the competition for light is less, which could result in high rate of photosynthesis than those grown at 1700m and 1500m above the sea level. However, the results were opposite. This was attributed to the length of the needles. The length of needles decreased with an increase of altitude (Table 3). The present results are in agreement to those in literature [21, 22]. Grace [21] reported that morphological characteristics of needles change with altitude, thus at higher altitudes needles were shorter as a result they demonstrated a lower specific leaf area. This implies a diminished importance of any trend towards increased surface conductances, hence low rate of photosynthesis.

The length of needles at higher altitudes decreases as a result of an increase in wind [21]. As wind increases with an increase in altitude, a rapid increase in damage to leaves (needles) caused by mutual abrasions of the plants parts and blown soil particles is also expected [23], such damage may not only increase transpiration, but it may disrupt the turgor relations of the epidermis and thus prevent proper stomatal closure [22]. Therefore, the decrease in length of needles is attributed to stomatal dysfunction caused by mechanical damage to the leaf and also by damage to the epidermis [21]. The present study has shown that the growth of *Pinus oocarpa* decreases with an increase in altitude. It is therefore, recommended that seed sources stands for *Pinus oocarpa* in Malawi and the surrounding region should be established at 1500m to 1600m above the sea level for better genetic growth parameters.

Table 3. Expected and estimated stocking per hectare, natural mortality rate per year and mean needle length for Pinus oocarpa at different altitudes.

Altitude (m asl)	Expected stocking (stems ha⁻)	Estimated stocking (stems ha⁻)	Natural mortality rate per year (%)	Mean Needle length (cm)
1900	620±3	559±4[c]	0.55±0.02[c]	15.4±0.8[c]
1700	620±3	572±3[b]	0.43±0.01[b]	18.7±0.6[b]
1500	620±3	598±4[a]	0.20±0.02[a]	21.2±0.9[a]
CV (%)		8.2	8.1	5.3

Note: [a,b,c]means with different subscript within a column significantly differ ($P<0.001$)

4. Conclusion

The present study has revealed that altitude has a clear effect on growth height, diameter at breast height and volume of *Pinus oocarpa*. Total height, dbh and volume decreases with an increase in altitude. It is therefore, recommended that seed sources stands for *Pinus oocarpa* in Malawi and the surrounding region should be established at 1500m to 1600m above the sea level for better genetic growth parameters.

Acknowledgements

The authors are grateful to the staff at Dedza District Forestry Office for allowing the study to be conducted at Dedza Mountain Forest Plantation and also for the assistance provided during data collection.

References

[1] Yang Y., Watanabe M., Li F., Zhang J., Zhang W., and Zhai J., 2006, Factors affecting forest growth and possible effects of climate change in the Taihang Mountains, northern China. Forestry, 79(1), 135 – 147.

[2] Vitasse Y., Delzon S., Bresson C.C., Michalet R., and Kremer A., 2009, Altitudinal differentiation in growth and phenology among populations of temperate-zone tree species growing in a common garden. Canadian Journal of Forest Research, 39, 1259 – 1269.

[3] Pacalaj M., Longauer R., Krajmerova D., and Gomory D., 2002, Effect of site altitude on the growth and survival of Norway spruce (*Picea abies* L.) provenances on the Slovak plots of IUFRO experiment 1972. Journal of Forest Science, 48(1), 16 – 26.

[4] Coomes D.A., and Allen R.B., 2007 Effects of size, competition and altitude on tree growth. Journal of Ecology, 95, 1084 – 1097.

[5] King G.M., Gugerli F., Fonti P., and Frank D.C., 2013, Tree growth response along an elevational gradient: climate or genetics? Oecologia, doi: 10.1007/s00442-013-2696-6.

[6] Li M.H., Yang J., and Kräuchi N., 2003, Growth responses of *Picea abies* and *Larix decidua* to elevation in subalpine areas of Tyrol, Austria. Canadian Journal of Forest Research, 33, 653 – 662.

[7] Norton D.A., 1985, A dendrochronological study of *Nothofagus solandri* tree growth along an elevational gradient, South Island, New Zealand. Eidg. Anst. Forstl. Versuchswes. Ber., 270, 159 – 171.

[8] Weber U.M., 1997, Dendroecological reconstruction and interpretation of larch budmonth (*Zeiraphera diniana*) outbreaks in two central alpine Valleys of Switzerland from 1470–1990. Trees, 11, 277 – 290.

[9] Körner C., 1998, A re-assessment of high elevation tree line positions and their explanation. Oecologia, 115, 445 – 459.

[10] Körner C., 1999, Alpine plant life. Springer-Verlag: Berlin, Germany.

[11] Paulsen J., Weber U.M., and Körner C., 2000, Tree Growth near Tree line: Abrupt or Gradual Reduction with Altitude? Arctic, Antarctic and Alpine Research, 32, 14 – 20.

[12] Tranquillini W., 1979, Physiological Ecology of the Alpine Timberline: Tree existence at high altitudes with special references to the European Alps. Ecological Studies (Analysis and Synthesis), Vol. 31, Springer-Verlag: Berlin, Heidelberg, New York.

[13] Zenni R.D., and Ziller S.R., 2011, An overview of invasive plants in Brazil. Brazilian Journal of Botany, 34, 431 – 446.

[14] Braga E.P., Zenni R.D., and Hay J.D., 2014, A new invasive species in South America: *Pinus oocarpa* Schiede ex Schltdl. BioInvasions Records, 3(3), 207–211.

[15] Dvorak W.D., 2005, *Pinus oocarpa* Schiede ex Schltdl. CAMCORE, 2, 628 – 631.

[16] Dvorak W., 2002, *Pinus oocarpa* Schiede ex Schltdl. In: Vozzo J (ed), Tropical Tree Seed Manual. U.S. Department of Agriculture, Forest Service: United States.

[17] Arce M., and Isaza N., 1996, Producción de semillas por cono en cuatro especies del género *Pinus* en Colombia. Informe de investigación No. 173. Investigación Forestal. Smurfit Cartón de Colombia.

[18] Ingram C.L., and Chipompha N.W.S., 1987, The Silvicultural Guide Book of Malawi, 2nd edition. FRIM: Malawi.

[19] SAS 9.1.3., 2004, Qualification Tools User's Guide. SAS Institute Inc., Cary, NC, USA.

[20] Missanjo E., Kamanga-Thole G., and Manda V., 2013, Estimation of Genetic and Phenotypic Parameters for Growth Traits in a Clonal Seed Orchard of *Pinus kesiya* in Malawi. ISRN Forestry, Volume 2013, Article ID 346982, 6 pages, doi:10.1155/2013/346982.

[21] Grace J., 1990, Cuticular water loss unlikely to explain tree-line in Scotland. Oecologia, 84, 64 – 68.

[22] Pitcairn C.E.R., Jeffree C.E., and Grace J., 1986, Influence of polishing and abrasion on the diffusive conductance of leaf surface of *Festuea arundinacea* Schreb. Plant Cell Environment, 9,191 – 196.

[23] Hadley J.L., and Smith W.K., 1986, Wind effects on needles of timberline conifers: seasonal influence on mortality. Ecology, 67, 12 – 19.

25

Preferences of ICT Tools by the Upazila Agriculture Officers (UAOs) for the Information Exchange in Bangladesh

Khondokar Humayun Kabir[*]**, Debashis Roy**

Department of Agricultural Extension Education, Bangladesh Agricultural University, Mymensingh, Bangladesh

Email address:

Kabirag09@bau.edu.bd (K. H. Kabir), droyagext@bau.edu.bd (D. Roy)

Abstract: The purpose of the study was to investigate the preferences of ICT tools by the Upazila Agricultural Officers, Bangladesh for the exchange of information. Data were collected using distributed questionnaires among the respondents. The findings showed that majority of the respondents (93.8%) had highly favorable attitude towards ICTs while 6.3% percent had moderately favorable attitude and there was no respondent had slightly favorable attitude towards ICTs. It also found that the highly preferred ICT tool by the UAOs is cell phone (1.76) and the second highly preferred tool is tab with the mean value 1.74, and on the other hand, internet (1.31) is the least preferred tools by the respondents. Correlation showed that age, job duration, personality characteristics, ambition and access to ICT tools showed significant relationship with the preferences of ICT tools by the UAOs. Challenges in using ICTs revealed that load shedding problem (2.21), lack of training facilities (2.19) and indifferences of farmers to get information through ICT (2.19) are the major challenges faced by the Upazila Agriculture Officers. Thus, it can be recommended that more ICT tools should be made available to the respondents with properly addressing the challenges so that they will be able to choose from various alternatives and also be able to gather and disseminate useful information to the farmers.

Keywords: ICT Tools, Preference, Upazila Agriculture Officer and Information Exchange

1. Introduction

1.1. Background

Agricultural extension service delivery all over the world has been concerned with communicating research findings and improved agricultural practices to farmers ([1]. The efficiency with which these information and practices are conveyed to farmers to a large extent would determine the level of agricultural productivity. The delivered information and practices may be of various types. The relevant information during the before- planting period may be crop management or scheduling of crop activities [12, 23] improved seedlings [9] input price and availability [18, 23] and soil fertility ([2]. During the growing season, other types of useful information may play crucial roles in improving the amount and the quality of products. This may include weather information [18, 23], fertilizer supply [2], fertilizer use in terms of amount and timing [12], pest surveillance and management [2, 12, 18, 23] type and dosage of pesticides [12], weed control ([2] and disease management [12, 23]. Following the harvest, information about market

opportunities [8], financial planning and market prices may be required [8, 9 and 23]. Extension organizations have been concerned with what should be the appropriate means and approaches in getting the right agricultural information to the end-users (farmers). In recent times however, there has been revolution with regards to Information and Communication Technology (ICT) in agriculture and particularly in extension service delivery of Bangladesh. This revolution is an intervention with the potential to ensure that knowledge and information on important agricultural technologies, methods and practices are put into right use by farmers [3].

Information and Communication Technologies (ICTs) are all technologies used for the widespread transfer and sharing of information. ICT tools include computer, internet, phone, television, radio, and other offline and online communication devices [6]. ICTs are rapidly consolidating global communication networks and international trade with applications for people in developing countries [14]. ICTs in agriculture promote and distribute new and existing farming information and knowledge which is communicated within the agricultural sector since information is essential for facilitating agricultural and rural development as well as bringing about social and economic changes [22]. There is growing recognition that farmers and members of rural communities have needs for information and appropriate learning methods that are not being met [4, 13]. The agriculture sector could leverage the Information and Communication Technologies (ICT) to disseminate the right information at the right time and right place. The cost factor in traditional information dissemination and the difficulties in reaching the target audiences have necessitated the introduction of ICT in agriculture [7].

1.2. Problem Statement

Agriculture is the important sector and key contributors to the national GDP of Bangladesh. Around 20.60% of the total GDP of the country comes from the agricultural sector [11].

Although Bangladesh is relatively late in entrant to e-Agriculture, the foundation for leveraging ICTs for agricultural capacity building and marketing existed for long time. It is only recently that the [15] Government of Bangladesh (GoB) through the Department of Agricultural Extension and the Department of Agricultural Marketing has started harnessing ICTs for effectively to deliver information and services to the farmers [19]. In Bangladesh, the Department of Agricultural Extension (DAE), the principle extension agency of the Government, remains the largest public agency with the representative at national, divisional, district, upazila and village levels [5]. DAE and other non-government organizations operating in rural areas of the country are providing agricultural extension services to farmers [10].

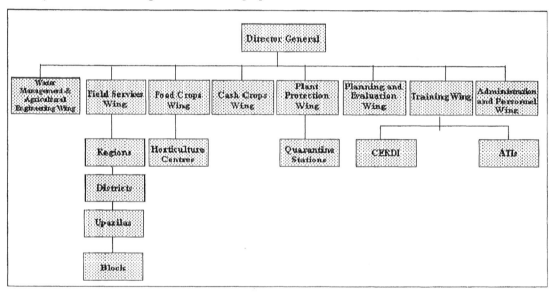

Figure 1. *Organizational structure of Department of Agricultural Extension, Bangladesh.*

Source: DAE, Bangladesh

Upazila Agriculture Officers (UAOs) under field service wing, DAE are mainly responsible for dissemination of necessary agricultural information to the farmers in the field level. They usually identify farmers' need and provide necessary solutions to them through different traditional extension teaching methods and to some extent through modern information and communication technologies. But, presently, the extension personnel are exploring to use different ICT tools in case of agricultural information/technology dissemination to the farmers. Because, the ratio of the farm families to the extension agent is 1000:1, this is really very little [16]. On the other hand, although the appointed Village Local Workers (VLWs) disseminate the information, they hardly accept any accountability. The cost factor in face-to-face information dissemination at the right time, and the difficulties in reaching the target audiences, has also created the urgency to introduce ICT [11]. Thus, the present study is an attempt to investigate the preferences of different ICT tools by the Upazila Agricultural Officers.

1.3. Specific Objectives of the Study

The following specific objectives have been taken into consideration to complete the study. These are:
a) To determine the attitude of the UAOs towards ICTs;
b) To investigate the preferences of ICT tools by the UAOs; and
c) To explore the challenges faced by the UAOs in using ICT tools.

2. Methodology

The study was carried out in Bangladesh Agricultural University, Mymensingh, Bangladesh. The participants in the workshop on 'Biotechnology with special emphasis on Bt. Brinjal' organized by Department of Agricultural Extension Education, BAU were the respondents of this study. A sample of about 80 was selected randomly from the total participants (208), covering 38.46% of the participants of workshop and representing the all agro-ecological zones of Bangladesh.

Data were collected through distributed questionnaire method and a structured questionnaire was designed based on related literature and objectives of the study. Validity of the instrument was ensured through a panel of experts in the Department of Agricultural Extension Education and extension professionals from the Department of Agricultural Extension (DAE), Bangladesh.

Measurement of personality: Personality of the respondents was measured by using a four-point rating scale. Some personality characteristics viz. knowledgeable, ability to initiate, flexible, fearless, industrious, sincere and persevere were included in the scale. Weights of responses were: 3 for highly, 2 for moderately, 1 for low and 0 for not at all. The personality of a respondent was, therefore, determined by adding the total responses against 7 characteristics. The personality score could vary from 0 to 21, 0 indicating 'no personality' and 21 indicating 'high personality'.

Collected data were tabulated, coded and assigned scores, wherever required in order to make them meaningful. Coded/scored data were analyzed using Statistical Package for Social Sciences (SPSS). Descriptive statistics such as frequency distribution, percentage analysis, mean, standard deviation, rank order (whenever necessary) and Pearson product-moment correlation analysis was used to test the strength of association among variables.

3. Findings and Discussion

3.1. Socio-Economic Profile

Socio-economic characteristics of the Upazila Agriculture Officers were analyzed and presented in Table 1.

From Table 1, the mean age of the UAOs was 44.6 years with majority of them (71.3%) had Master of Science degree. The average job duration of the UAOs 19.27 years and a mean of 10.40 years as using ICT tools as first experience.

Table 1. Salient Features of the Selected Characteristics of the Upazila Agriculture Officers.

Socio-economic variables	Respondents (n=80)	
	Mean (Std. deviation)	Percentage
Personal data		
Age (years)	40.65 (7.10)	-
Young (up to 30)	-	6.2
Middle Aged (31-50)	-	90.0
Old Aged (above 50)	-	3.8
Level of Education	-	-
B. Sc. Ag. (Hons.)	-	22.5
Master of Science (MS)	-	71.3
Doctor of Philosophy (PhD)	-	6.2
Total Duration of Job (years)	19.27 (7.92)	-
First Experience of ICT tools (years)	10.40 (4.43)	-
Level of Job Satisfaction		
Low	-	12.5
Medium	-	67.5
High	-	20.0
Personal Character	16.41 (2.29)	-
Low Personality (up to 7)	-	0
Medium Personality (8-14)	-	27.5
High Personality (above 14)	-	72.5
Ambition	8.55 (1.80)	
Low (up to 3)	-	2.5
Medium (4-6)	-	11.2
High (above 6)	-	86.3
Do you think ICT can change our technology dissemination system?	-	-
0=No	-	7.5
1= Yes	-	92.5
Extension Contact (days)	19.54 (7.93)	-
Possible Range (0-51)		
Low Contact (up to 17)	-	40.0
Medium Contact (18-34)	-	58.9
High Contact (above 34)	-	1.1
Training Received (In country)		
0=No	-	46.3
1= Yes	-	53.8
Foreign Training Exposure		
0=No	-	70.0
1= Yes	-	30.0

Source: Field survey, 2014

More than two-third of the UAOs (67.5%) had medium job satisfaction and majority of them (72.5%) possessed high personality. More than four-fifth of the UAOs (86.3%) was highly ambitious, while, majority of them (92.5%) thought that ICTs can change technology dissemination system. The mean extension contact of the UAOs was 19.54 and about half of them (53.8%) had in country training exposure, while, majority of them (70.0%) had no foreign training exposure.

3.2. Access to ICT tools

This section examines the level of access and nature of access to selected ICT tools by the respondents. The findings have been shown in Table 2.

Table 2. Access to ICT tools.

ICT tools	Level of Access				Mean	Rank Order
	Adequate access	Moderate access	Limited access	No Access		
Cell Phone	63	11	6	0	2.71	1
Internet	33	36	9	2	2.25	2.5
Digital Camera	35	23	21	1	2.25	2.5
Print Media	29	28	18	5	2.01	4
Multimedia Projector	29	21	27	3	1.95	5
Video Chatting	7	18	37	18	1.18	7
Tab	4	8	36	32	0.80	8

The results of the analysis on the access to ICT tools indicate that majority of the respondents (2.71) had the access to cell phone for transferring agricultural information. Williams and Agbo [24] found similar findings. The second majority of the respondents had access to internet (2.25) and digital camera (2.25) where some respondents had access to print media (2.01), multimedia projector (1.95) and video chatting (1.18). The findings indicate that only few of the respondents (0.80) had access to tab and hence it got the lowermost position in the rank order.

3.3. Use of ICT Tools

Use of common ICT tools was evaluated by inquiring the availability or accessibility to the cell phone, tab, internet, video chatting, multimedia projector, digital camera and print media. The findings have been presented in Table 3.

As shown in Table 3, more than four-fifth of the UAOs (88.8%) uses cell phone while the access to tab was 41.3%. This is in line with research by Seepersad [21] who reported that cell phones are fairly common among extension employees in Trinidad and Tobago. Samansiri and Wanigasundera [20] found similar results. Majority of the respondents (87.5%) uses internet to transfer necessary information while 52.5% and 75.0% of the respondents uses

video chatting and multimedia projector for transferring agricultural information, respectively.

Table 3. Use of ICT tools by Upazila Agriculture Officers.

ICT tools	Yes		No	
	Frequency	Percentage	Frequency	Percentage
Cell Phone	71	88.8	9	11.3
Tab	33	41.3	47	58.8
Internet	70	87.5	10	12.5
Video Chatting	42	52.5	38	47.5
Multimedia Projector	60	75.0	20	25.0
Digital Camera	70	87.5	10	12.5
Print Media	61	76.3	19	23.8

More than four-fifth of the respondents (87.5%) uses digital camera while 76.3% of them uses print media to disseminate information to the farmers.

3.4. Purpose of Using ICT Tools

The purpose of using ICT tools formulated five major purposes such as retrieve information, video chatting, downloading videos, communication and collection of different training materials. The findings have been presented in Table 4.

Table 4. Purpose of using ICT tools.

Purposes	Yes		No	
	Frequency	Percentage	Frequency	Percentage
Retrieve Information	73	91.3	7	8.8
For Video Chatting	51	63.8	29	36.3
For Downloading different Videos	60	75.0	20	25.0
For Communication	68	85.0	12	15.0
For Collection of different Training Materials	71	88.8	9	11.3

Table 4 shows that majority of the respondents use ICT to retrieve different information while 63.80% of them use for video chatting. Three-fourth of the respondents use ICT for downloading different videos and more than four-fifth of them communicate through ICTs, where 88.8% of the respondents use ICT for collecting different training materials.

3.5. Attitude towards ICTs

The possible attitude score of the respondents towards ICTs could range from 8-24, 8 indicating highly unfavorable attitude and 24 indicating highly favorable attitude. The computed attitude scores of the respondents ranged from 13-22 where, mean being 18.11 with a standard deviation of 1.98.

Table 5. Classification of the respondents according to their attitude towards ICTs.

Possible Score	Observed Score	Categories of Farmers	Number	Percentage	Mean	SD
8-24	13-22	Slightly favorable (up to 7)	-	-	18.11	1.98
		Moderately favorable (8-14)	5	6.3		
		Highly favorable (above 14)	75	93.8		
		Total	80	100		

Data indicated that almost all of the respondents (93.8%) had highly favorable attitude towards ICTs while 6.3% percent had moderately favorable attitude and there was no respondent had slightly favorable attitude towards ICTs. The finding indicates that the respondents were enthusiastic to adopt different ICT tools to disseminate required agricultural information to the farmers.

3.6. Preferences of different ICT Tools

To understand the preferences of ICT tools by the respondents a rank order of the commonly used ICT tools was made. The rank order is based on the mean value of each ICT tools and the mean value of individual tool ranges from 1.31 to 1.76.

Table 6. Preferences of different ICT tools.

ICT tools	Degree of Preference			Mean	Rank Order
	Highly prefer	Moderately Prefer	Not at all		
Cell Phone	63	15	2	1.76	1
Tab	62	15	3	1.74	2
Video Chatting	60	17	3	1.71	3.5
Multimedia Projector	58	21	1	1.71	3.5
Digital Camera	57	18	5	1.65	5
Print Media	50	24	6	1.55	6
Internet	36	33	11	1.31	7

Table 6 reveals that the highly preferred ICT tool by the UAOs is cell phone (1.76) and thus stood first in the rank order. Mabe and Oladele [14] stated that prominent information communication technologies among extension officers were mobile phones. The finding is also in line with research by Seepersad [21]. The second highly preferred tool is tab with the mean value 1.74, and video chatting (1.71) and multimedia projector (1.71) are equally preferred by the respondents. It also indicates that internet (1.31) is the least preferred ICT tool by the respondents. This finding may be due to there was insufficiency of internet facilities and may also be the difficulties in operating internet.

3.7. Relationship between Independent Variables and Focus Variable

This subsection is carried out an assessment of the relationship between the socioeconomic characteristics of the respondents and their preferences of ICT tools. The results have been shown in table 7.

Table 7. Correlation between the selected characteristics of the UAOs and their preferences on using ICT tools.

Independent Variables	Correlation of Coefficients (r)
Age	-0.251*
Duration of first ICT tools	0.200
Total duration of training	0.151
Total duration of job	-0.301**
Personal characteristics	0.249*
Ambition	0.295**
Access to ICT tools	0.638**

*Significant at the 0.05 level of probability (2-tailed)
** Significant at the 0.01 level of probability (2-tailed)

Table 7 reveals that out on seven characteristics five viz. age, duration of job, personal character, ambition and access to ICT tools found to be related with the focus variables. Personal characters, ambition and access to ICT tools showed positive relationship with the focus variable. It is found the personality characteristics shows significant and positive relationship with the focus variable and it can be implied that the higher personality, more preferences to ICT tools. The person who posses high personality characteristics usually willing to make them updated with the modern technologies and are likely to accept any change rapidly. It is also found that ambition of the respondents showed significant and positive relationship with the focus variable. Thus, it can be implied that the higher ambition of the officers, higher preferences to ICT tools. The result may be due to the ambitious people are more positive to any kind of change and are likely to make their betterment by using modern facilities. Access to ICT tools showed significant and positive relation with the preferences of ICT tools. It can be implied that more access to ICT tools, higher the preferences of ICT tools. Access to ICT tools provide the range of different tools where the respondents have facilities to use different tools and hence the result seems to be like this. On the other hand, age and job duration showed negative relationship i. e. older and more job experienced officers are likely resistant to use ICT tools for the exchange of agricultural information rather they use traditional methods to do so. Older people are likely to have put in more years of work experience and be more resistant to change and innovations [17]. And hence, age and job duration showed significant and negative relationship with the focus variable.

3.8. Challenges Faced in Using Different ICT Tools

This section examines the challenges to ICT usage among the respondents. The challenges in using ICT tools among the respondents are as presented in Table 8.

Table 8. Challenges faced in using different ICT tools.

Challenges	Extent of Challenges					
	High	Medium	Low	Not at all	Mean	Rank Order
Load shedding problem	32	37	7	4	2.21	1
Lack of training facilities	35	30	10	5	2.19	2.5
Indifferences of farmers to get information through ICT	33	34	8	5	2.19	2.5
The price of different ICT tools have increased sharply	16	52	6	6	1.98	4
Most of the ICT tools are sophisticated to use	10	40	24	6	1.68	5
Gender Issues	8	38	18	16	1.48	6

Table 8 reveals that load shedding problem (2.21), lack of training facilities (2.19) and indifferences of farmers to get information through ICT (2.19) are the major challenges faced by the Upazila Agriculture Officers. Omotesho [17] found that load shedding is the second major challenge to extension agents in Nigeria. The increased prices of the ICT tools (1.98) and sophistication of using ICT (1.68) are the challenges faced by the respondents in using ICT tools. Furthermore, gender issue (1.48) is another challenge but it bears no significance and hence, got the lowest position in the rank order.

4. Conclusions

For easy, fast and cost effective means it is necessary to incorporate ICT tools in extension services for the exchange of useful information with the farmers. As the findings demonstrate favorable attitude towards ICT, it would be very much effective to disseminate useful agricultural information through different ICT tools and the extension personnel as well as the farmers could harvest maximum benefits out of it. The findings also indicate that cell phone and tab are the most preferred tools and thus, availability of these tools should be ensured. On the other hand, internet facilities should be available so that the respective stakeholders could make its better utilization in agricultural development activities. Load shedding problem, lack of training facilities, indifferences of farmers to get information through ICT are the major challenges faced in using ICT tools. Necessary efforts to overcome these challenges would ensure increased use of ICTs for agricultural activities. So, it can be mentioned that the authority could take necessary steps to provide infrastructural facilities such as constant electricity, ensure periodic training for both farmers and extension agents on the necessity and operation of ICTs, subsidize ICTs facilities and make available all networks in collaboration with the network providers.

Acknowledgements

The authors convey deepest thanks to Upazila Agriculture Officers (UAOs) for their cooperation during data collection. Moreover, authors convey their gratefulness to the Department of Agricultural Extension Education, Bangladesh Agricultural University, Mymensingh for arranging such workshop from where authors could collect data easily.

References

[1] Agbamu, J. U. (2007). Essentials of Agricultural Communication in Nigeria. Malthouse Press Limited Lagos. pp. 20-24, 91-93.

[2] Ekoja, I.I. (2004). Sensitising users for increased information use: the case of Nigerian farmers. African Journal of Library. *Archives and Information Science*, 14(2):193-204.

[3] Ezeh Ann N, (2013). Extension Agents access and utilization of Information and Communication Technology (ICT) in Extension Service Delivery in South East Nigeria. *Journal of Agricultural Extension and Rural Development*, 5(11):266-276.

[4] Greenidge, C. B. (2003). ICTs and the rural sector in ACP State: Mirage or Marriage? Address delivered at the CTA's ICT observatory U.S.A.

[5] Haque, J. T. (n.d.). Agrarian Transition and Livelihoods of the Rural Poor: Agriculture Extension Services. Unnayan Onneshan, Bangladesh.

[6] Hasan, M. R. & Shariff, A.R.M. (2007). Study on the Telecenter Movement in Bangladesh: Obstacles and Opportunities; Rural ICT Development Conference, University Utara Malaysia, Kedah, Malaysia (November).

[7] Hasan, R., Islam, S., Rahman, M. S. & Jewel, K. N. A. (2009). Farmers Access to Information for Agricultural Development in Bangladesh. *Bangladesh Research Publication Journal*, 2(1): 319-331. Retrieve form http://www.bdresearchpublications.com/admin/journal/upload/08043/08043.pdf

[8] International Institute for Communication and Development [IICD]. (2006). ICTs for agricultural livelihoods: Impact and lessons learned from IICD supported activities. The Hague: The Netherlands.

[9] Irivwieri, J.W. (2007). Information needs of illiterate female farmers in Ethiope East local government area of Delta State. *Library Hi Tech News*, 9(10): 38-42.

[10] Karim, Z., Bakar, M. A. and Islam, M. N. (2009). Study of the Implementation Status and Effectiveness of New Agricultural Extension Policy for Technology Adoption. Center for Agriresearch and Sustainable Environment and Entrepreneurship (CASEED). Final Report.

[11] Kashem, M. A., Farouque, M. A. A., Ahmed, G. M. F. and Bilkis, S. E. (2010). The complementary roles of information and Communication technology in Bangladesh agriculture, *Journal Science Foundation*, 8(1&2): 161-169.

[12] Krishna Reddy, P. & Ankaiah, R. (2005). A framework of information technology-based agriculture information dissemination system to improve crop productivity. *Current Science*, 88(12): 1905 –1913.

[13] Lightfoot, C. (2003) Demand –driven extension: some challenges for policy makers and managers. *ICTs – Transforming Agricultural Extension*? The 6th Consultative Expert Meeting of CTA's Observatory on ICTs, Wageningen, 23- 25 September, 2003.

[14] Mabe L. K. & Oladele O.I. (2012). Awareness level of use of Information Communication Technologies tools among Extension officers in the North- West Province, South Africa. *Life Science Journal*, 9(3): 440-444.

[15] Ministry of Agriculture. (1999). Agricultural Extension Manual, Department of Agricultural Extension, Bangladesh

[16] NAEP. (1996). New Agricultural extension Policy, Govt. of the People's Republic of Bangladesh, Ministry of Agriculture. p-8.

[17] Omotesho, K. F., Ogunlade, I. O. & Lawal, M. (2012). Assessment of Access to Information and Communication Technology among Agricultural Extension Officers in Kwara State, Nigeria. *Asian Journal of Agriculture and Rural Development*, 2(2): 220-225.

[18] Rao, S. S. (2004). Role of ICTs in India's rural community information systems. *Info*, 6(4): 261 –269.

[19] Rubaiya, A. (2010). *Bangladesh*. In R. Saravanan (ed). ICTs for Agricultural Extension: Global Experiments, Innovations and Experiences (pp. 1-43). New India Publishing Agency, Pitam pura, New Delhi.

[20] Samansiri, B. A. D. and Wanigasundera, W. A. D. P. (2014). Use of Information and Communication Technology (ICT) by Extension Officers of the Tea Small Holdings Development Authority of Sri Lanka. *Tropical Agricultural Research*, 25 (4): 360 – 375.

[21] Seepersad, J. (2003). Case study in ICTs in Agricultural Extension in Trinidad and Tobago. In CTA *ICTs – Transforming Agricultural Extension?* The 6th Consultative Expert Meeting of CTA's Observatory on ICTs, Wageningen, 23- 25 September, 2003.

[22] Swanson, B, E. & Rajalahti, R. (2010). *Strengthening Agricultural Extension and Advisory Systems: Procedures for Assessing, Transforming and Evaluating Extension Systems*, The International Bank for Reconstruction and Development/The World Bank, Washington, pp. 98 – 127.

[23] Tiwari, S. P. (2008). Information and communication technology initiatives for knowledge sharing in agriculture. *Indian Journal of Agricultural Sciences*, 78(9): 737 –747.

[24] Williams, E. E. & Agbo, I.S. (2013). Evaluation of the Use of ICT in Agricultural Technology Delivery to Farmers in Ebonyi State, Nigeria. *Journal of Information Engineering and Applications*, 3 (10):18-26.

26

Change in diversity and abundance of nematode destroying fungi in land use under irrigation in selected small scale irrigation schemes in Kenya

Wachira P. M.[1,*], **Kimenju J. W.**[2], **Otipa M.**[3]

[1]School of Biological Sciences, University of Nairobi, Nairobi, Kenya
[2]Department of Plant Science and Crop Protection, University of Nairobi, Nairobi, Kenya
[3]Kenya Agricultural and Livestock Research Organization, Nairobi, Kenya

Email address:

pwachira@uonbi.ac.ke (Wachira P. M.), wkimenju@yahoo.com (Kimenju J. W.), otipamj@gmail.com (Otipa M.)

Abstract: Intensity of land cultivation is usually associated with increase in crop production and loss of soil biodiversity or its function. This study was conducted to determine the effect of intensity of land use under irrigation on the occurrence, abundance and diversity of nematode destroying fungi in selected small scale irrigation systems in Kenya. The study was conducted in four spatially separated irrigation schemes namely Kabaa and Kauti in Machakos and Kathiga Gacheru and Mbogooni) in Embu. The study areas were stratified according to land use, which included the irrigated land, rain-fed cultivated land and undisturbed land under fallow. The period of cultivation also differed with the oldest cultivated irrigation system, having been opened in 1960, while the youngest having been opened in 2011.Soil samples were collected from the study site for isolation of nematode destroying fungi. The soil sprinkle and culture technique was used to isolate soil nematode destroying fungi from the soil samples. A total of 216 fungal isolates were identified as nematode destroying fungi belonging to six genera namely *Acrostalagmus, Arthrobotrys, Haptoglossa, Harposporium* and *Monacrosporium*. All the isolates were identified resultingto nine species. 49.5% of all the fungi were isolated from irrigated land while, the rain-fed and the undisturbed land uses accounted for 29.7 and 20.8% of the isolates, respectively. The oldest irrigation systems had the least diversity (0.110) of nematode destroying fungi compared to the youngest which had a diversity index of 1.311.The species *Arthrobotrys oligospora* was the most frequently isolated fungus followed by *Monacrosporium cionapagum* with occurrence frequencies of 57 and 53%, respectively. The least frequently isolated species was *Nematoctonus leiospora* with an occurrence frequency of 2.3%. Of the total identified species, only *Nematoctonus leiosporus* and *Arthrobotrys dactyloides* were not affected by the irrigation activities.). From the study, it is evident that land use intensity under irrigation system and the duration of cultivation impacts on occurrence and diversity of nematode destroying fungi in the soil.

Keywords: *Arthrobotrys oligospora*, Bio -Control, *Monacrosporium cionopagum*, Plant Parasitic Nematodes, Soil Biodiversity

1. Introduction

Agriculture is the main occupation and source of income for the majority of Kenya population accounting for one third of the Gross Domestic Product and employs more than two thirds of the country's labor force (Republic of Kenya, 2005). With three quarters of the country's land being uncultivable, coupled with rapid increase in population, sustainability of food production becomes a key area of focus. In an effort to increase food production for the increasing population in

Kenya, continuous fragmentation of arable land and illegal forest clearance have only compounded the problem. This has resulted in intensive agricultural production through irrigation systems in many parts of Kenya. These agricultural production systems are characterized by a low fallow ratio and high use of pesticides and chemical fertilizers relative to land area with the aim of increasing farmer's income and reducing poverty (Wu and Li, 2013). The main characteristics of land under irrigation include frequent watering and regular turning of the soil, addition of inorganic fertilizers and pesticides and continuous cultivation of the land all geared towards increased crop production. The examples of crops grown in these schemes are water-intensive and high-value crops such as French bean, green maize/corn, snow peas, onion, tomato, spinach, cabbage, kale, and watermelon among other horticultural crops. Products with these schemes are used to meet subsistence demands as well as domestic and export markets (Neubert, 2007). Irrigation therefore plays an important role in national economic development by increasing crop diversity and yield. Concomitantly there is a raise in food security, an increased income and empowerment of the people in this sector. This leads to improve quality of life to smallholder farmers, government scheme households and persons employed in commercial farms.

Due to this, there is persistent pressure on the soils under irrigated production systems to produce more and more harvests. This pressure has resulted in disruption of the soil ecological balance (Li et al., 2013). It has been reported that addition of inorganic fertilizer into the soil has led to increased soil pests and diseases and reduction of beneficial microorganism (Singh, 2000). In addition, continuous and prolonged cultivation of the soil affects the soil quality by changing the soil structure that has been associated with reduced soil biodiversity (Muya et al., 2009). Other negative effects of intensive cultivation include buildup of pests and diseases (Maina et al, 2009), siltation, low irrigation water use efficiency, high cost of production and declining soil fertility. For example, intensive agriculture has been associated with increase in soil pathogenic fungi especially *Fusarium* spp. (Luque et al., 2005). Some of the farmer practices in the irrigation schemes such as monoculture, excessive tillage, and pesticide use, disrupt the natural regulatory mechanisms in the soil leading to build up of soil pests (Altieri and Nicholls, 2003).

The major crop pests in irrigation systems are the plant parasitic nematodes (Jones et al., 2011.) They are known to cause mechanical damage and malfunctions of the plant roots leading to alteration of the plant growth and development, resulting in poor growth and reduced yield (Bridge et al., 2005). Damage by nematodes is often associated with retarded growth and chlorosis due to inability of the roots to deliver water and nutrients and thus may be confused with similar symptoms resulting from poor soil conditions and nutrient deficiency. Plant parasitic nematodes are very important pests especially in irrigated systems due to the susceptibility of the crops grown, diversity of the host range and also favorable moisture, which is provided by the regular watering. In particular, the root knot nematodes cause severe yield losses by reducing the quantity and quality of the harvestable products. They have been reported to cause more than a third of the total yield. In addition, they are known to open up avenues in the roots for other pathogens like fungi and bacteria. They have also been recorded as the responsible organisms for the huge crop loss in tomato for smallholder growers in Kenya (Oruko and Ndungu, 2001).In this regard,

the management of plant parasitic nematodes in agricultural crops has attracted a lot of attention (Garcia et al., 2004) with chemical nematicides being regular inputs (Akhtar and Malik, 2000). Although chemical nematicides are efficient and fast acting, they are currently being reappraised with respect to the environmental hazards associated with them and unaffordability to many small-scale farmers (Wachira et al., 2009).This has created pressure on farmers to adopt nematode management strategies that are environmentally friendly and affordable. As a result, the crave to biological control of plant parasitic nematodes as a viable practice in modern agriculture and horticulture has increased dramatically (Mashela et al., 2008). This has led to the exploration of the nematodes destroying fungi, which are natural enemies of plant parasitic nematodes as the possible candidates for development. So far more than 160 fungal species that live on nematodes, partially or entirely, have been reported (Elshafieet.al. 2006). The fungi use specialized structures and toxins to capture and destroy nematodes (Luo et al., 2004; Yang et al., 2007).Consequently, this group of fungi has drawn much attention because of their potential as biological control agents of nematodes that are parasitic on plants (Masoomeh et al., 2004; Yan et al., 2005).

The aim of this study was therefore to determine the effect of irrigation soil for agricultural production on the occurrence and diversity of nematode destroying fungi with the ultimate goal of utilizing these fungi in the management of plant parasitic nematodes in the smallholder irrigation schemes in Kenya.

2. Materials and Methods

2.1. Characterization of the Target Irrigation Schemes

The study was conducted within small scale irrigation schemes in Machakos, Embu, and Meru counties of Kenya: The criteria for selection of the sites were soil type, cultivation period and crops cultivated. The schemes selected for this study were Kauti and Kabaa irrigation schemes in Machakos county, Kathiga-Gacheru in Embu county and Mboogoni in Meru. The schemes were characterized in terms of crops grown, water source and production constraints (Table 1). All the irrigation schemes under study had different crop production constraints with pests and diseases being common in all of them. Other production constraints identified included, siltation, low irrigation water use efficiency, inappropriate management skills, and high cost of production, lack of organized marketing, poor seed quality and declining soil fertility. The two irrigation schemes in Machakos (Kabaa and Kauti-Kathiani) obtained irrigation water from dams while those in Embu and Meru obtained water from the rivers. A wide range of crops is grown in the irrigation schemes with tomato production being common to all of them. Kabaa irrigation scheme was the largest scheme with 240 ha being under irrigation while Kauti-Kathiani was the smallest scheme with only66 ha, being under irrigation. Mboogoni irrigation scheme had 100 ha under irrigation while only 80 ha were under irrigation in Kathiga- Gacheru irrigation scheme. Kabaa irrigation scheme would be considered the oldest irrigation scheme among the four schemes, having been established in 1960. The irrigation canal was rehabilitated and the area under irrigation was expanded in 2011. Kathiga- Gacheru irrigation scheme was fully established in 1984. The other two irrigation schemes (Mboogoni and Kauti – Kathiani) were established in 2011.

Table 1. *Characteristics of four selected smallholder irrigation schemes selected for a study of nematode destroying fungi in Kenya.*

Scheme	Location	Acreage	Year Established	Main crops	Water source	Main crop production constraint
Kabaa,	Machakos	240 ha,	1960	French bean, tomato, onion, banana, kale, cabbage and passion fruit.	Dam	Pests and diseases, siltation, low irrigation water use efficiency, inadequate management skills, high cost of production and declining soil fertility.
Kauti-Kathiani	Machakos	66 ha	2011	French bean, kale, tomato, maize and coffee	Dam	Pests and diseases, inadequate water supply, low soil fertility, lack of skilled labourand high post-harvesting losses
Mboogoni	Meru South	100 ha	2011	Banana, tomato, green maize, watermelon and mango.	River	Pests and diseases, inadequate water, no organized marketing group and poor seed germination
Kathiga-Gacheru	Mbeere	80 ha	1984	Pawpaw, tomato, butternut, watermelon, kale and onion.	River	Pests and diseases, poor cropping patterns

2.2. Soil Sampling

Each of the study area was stratified into three main land uses, the irrigated land, rain-fed cultivated land and the undisturbed/ natural land. The soil sampling method was adopted from Moreira *et al,* 2008 with some modification. From each land use type, a total of ten farms were randomly identified for soil sampling. From each farm/sampling point, a central position was determined and marked. From the center, four diagonals of six meters long were drawn and the soil sampled at the three and six meter lengths including the center. Soil sampling was done using a soil auger, which was sterilized between sampling points to avoid cross contamination. A total of 120 soil samples were collected for this study. The soil samples were placed in a cool box and transported to the laboratory for isolation of nematode destroying fungi. All the laboratory work was done at the University of Nairobi, Mycology laboratory.

2.3. Isolation of Nematode Destroying Fungi

Isolation of the fungi was done using the soil sprinkle technique as described by Jaffee *et al.*, (1996). One gram of soil from each soil sample was transferred to a previously prepared sterile solid potato dextrose agar (PDA) medium in a Petridish and spread evenly. A suspension of approximately 500 juveniles *Meladogyne incognita* was added as bait in each petri dish and incubated at room temperature. Observations on fungal growth were conducted every week after the third week of incubation for three weeks. Observations on dead nematodes and the mycelia growth in the petridish were conducted under the dissecting microscope and then under a compound microscope at a magnification of x40. Identification was based on the type and size of conidia, the habit of the conidiophore and the type of nematode destruction structure produced.

2.4. Data Analysis

Generalized linear models were fitted to test the effect of

land intensity on the occurrence of nematode destroying fungi. Frequency of occurrence, evenness, Renyi profiles and the Shannon diversity index were also calculated (Kindt and Coe, 2005*)*.

3. Results

Land use intensification under irrigation has a significant ($P = 0.05$) effect on the occurrence of nematode destroying fungi. Overall, 49.5 % of all the isolated nematode destroying fungi was obtained from the irrigated land while, 29.7 and 20.8% were recovered from the rain fed and uncultivated land respectively (Fig. 1). Similar trend of occurrence of nematode destroying fungi was observed in all the irrigation schemes with the irrigated land uses having higher numbers of occurrence of nematode destroying fungi followed by the rain fed land uses and the least being the uncultivated land uses.

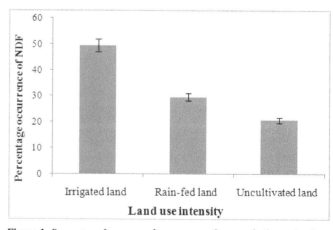

Figure 1. *Percentage frequency of occurrence of nematode destroying fungi in land under different levels of disturbance.*

From the results, it was evident that the old irrigation systems had the least number of nematode destroying fungi compared to the recently established irrigation systems (Fig 2). Kauti irrigation scheme had the highest percentage record

(45.8%) of nematode destroying fungi. Kabaaand Mboogoni had a record of 20.8 while 12.5% was recorded for Kathiga-Gacheru.)

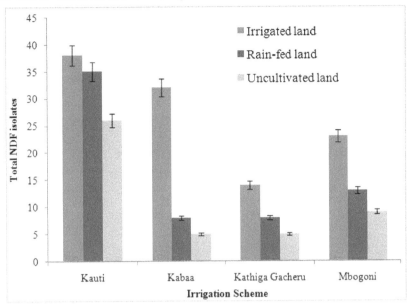

Figure 2. Nematode destroying fungi recorded in irrigation scheme.

The diversity of nematode destroying fungi varied among the irrigation scheme (Table 2). Kauti irrigation scheme was the most diverse with a mean Shannon of 1.311, followed by Kauti rain fed systems with 1.218 while the third was Kabaa irrigated systems with 1.078. The rest had mean Shannon of less than 1. The land uses with the least number of nematode destroying fungi were in Kathiga Gacheru irrigation schemes with mean mean richess of 8, 5 and 4 in the rain-fed, uncultivated and irrigated lands, respectively.

A total of 216 fungal isolates were positively identified as those of nematode destroying fungi. The fungi were grouped into six genera and nine species. The six genera were, *Arthrobotrys, Acrostalagmus, Harposporium, Haptoglossa, Monacrosporium* and *Nemtoctonus*. The genus *Arthrobotrys* was represented by three species namely, *A.dactyloides, A.longispora* and *A. superba*. All the other genera were represented by one species. With the exception of *A. dactyloides* and *Nematoctonus leiosporus,* all the other species of nematode destroying fungi were significantly affected by land use intensity (Table 2).

Arthrobotrys oligospora was the most frequently isolated nematode destroying fungal species with 26.1 % occurrence, which was closely followed by *Monacrosporium cionapagum* with isolation frequency of 24.5%. The least frequently isolated species were *Harposporium aungullilae* and *Nematoctonus leiosporus* with percentage frequencies of 4.2 and 2.3%, respectively (Table 3)

The species cumulative curve indicated that the 120 samples collected in this study were sufficient to detect all the available nematode destroying fungi in the study site. The curve had already flattened at 80 samples (Figure 3)

Table 2. Richness, Shannon indices and abundance of nematode destroying fungi in various land use systems in selected irrigation schemes in Kenya.

Irrigation Scheme/land use	Mean richness	Shannon index	Abundance
Kathiga Gacheru Irrigated	0.4	0.277	4
Kathiga Gacheru rain-fed	0.8	0.139	8
Kathiga Gacheruuncultivated	0.5	0.110	5
Kauti- Kathiani Irrigated	3.8	1.311	38
Kauti- Kathiani rain-fed	3.5	1.218	35
Kauti- Kathianiuncultivated	2.6	0.855	26
Mbogoni Irrigated	2.3	0.717	23
Mbogoni rain-fed	1.3	0.277	13
Mbogoni uncultivated	0.9	0.139	9
Kabaa Irrigated land	3.2	1.078	32
Kabaa Rain-fed production	0.8	0.179	8
Kabaa uncultivated	0.5	0.139	5

Table 3. *Effect of land use intensity on occurrence of nematode destroying fungal species in selected irrigation schemes in Kenya.*

Land use intensity	Nematode destroying fungal species								
	Arthrobotrys dactyloides	*Arthrobotrys longispora*	*Arthrobotrys oligospora*	*Arthrobotrys supreba*	*Acrostagmas obatus*	*Harposporium aungullilae*	*Haptoglosa heterospora*	*Monacrosporium cionopagum*	*Nematoctonus leiosporus*
KathigaGacheru Irrigated	0.0	0.1	0.5	0.1	0.0	0.4	0.0	0.3	0.0
KathigaGacheru rain-fed	0.1	0.0	0.3	0.1	0.0	0.0	0.0	0.3	0.0
KathigaGacheru uncultivated	0.0	0.1	0.2	0.1	0.0	0.0	0.0	0.1	0.0
Kauti Irrigated	0.1	0.4	0.8	0.8	0.2	0.0	0.4	1.0	0.1
Kauti rain-fed	0.1	0.4	0.9	0.8	0.3	0.0	0.3	0.6	0.1
Kauti uncultivated	0.1	0.4	0.4	0.7	0.2	0.1	0.1	0.6	0.0
Mbogoni Irrigated	0.3	0.4	0.6	0.0	0.0	0.0	0.0	0.9	0.1
Mbogoni rain-fed	0.0	0.2	0.6	0.1	0.0	0.0	0.0	0.3	0.1
Mbogoni uncultivated	0.0	0.0	0.3	0.1	0.0	0.1	0.0	0.3	0.1
Kabaa Irrigated	0.2	0.4	0.8	0.5	0.1	0.2	0.5	0.5	0.0
Kabaa rain-fed	0.1	0.0	0.2	0.1	0.1	0.0	0.0	0.3	0.0
Kabaa uncultivated	0.1	0.0	0.1	0.1	0.0	0.1	0.0	0.1	0.0
P value	0.4853	9.806×10^{-3}	3.545×10^{-04}	2.205×10^{-09}	8.417×10^{-2}	0.01114	1.059×10^{-05}	2.201×10^{-05}	0.7941

Table 4. *Rank, abundance and proportion of nematode destroying fungi isolated from four irrigation schemes in Kenya.*

Species	Rank	Abundance	Proportion	Accumulative frequency
Arthrobotrys oligospora	1	57	26.1	26.4
Monacrosporium cionapagum	2	53	24.5	50.9
Arthrobotrys superba	3	35	16.2	67.1
Arthrobotrys longispora	4	24	11.1	78.2
Haptoglossa heterospora	5	13	6.0	84.3
Arthrobotrys dactyloides	6	11	5.1	89.4
Acrostalagmus obatus	7	9	4.2	93.5
Harposporium aungullilae	8	9	4.2	97.7
Nematoctonus leiosporus	9	5	2.3	100.0

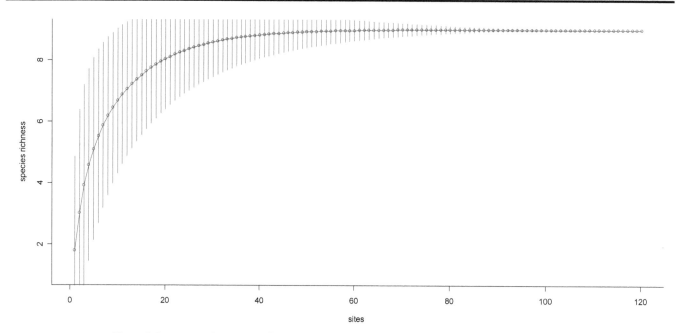

Figure 3. *Species cumulative curve of nematode destroying fungi in the selected irrigation systems in Kenya.*

4. Discussion

Agricultural intensification under irrigation is one of the key practices for increased crop production world allover. In Kenya the total area under irrigation is about 80,000 hectares with public and private small-scale irrigation being less than 50,000 ha. This is still very small compared to the estimated potential of more than 300,000 ha, with water availability and quantity being one of the challenges. From this study, dams and rivers were identified as the main sources of irrigation water although other sources like wells, lakes and seas have been reported in other countries. From the study area, the farms were characterized by heavy usage of pesticides, and chemical fertilizers in order to increase productivity. It was noted that the irrigation systems were inserting great pressure on the environment by degradation and depletion of the natural resource, like soil, water and natural plants as identified in the production constrains from the four irrigation schemes, similar to those reported by Burney et al., 2010 and Moeskops, et al., 2010.

The main production constrains identified in the four studied irrigation schemes were poor soil fertility and pests and disease. This was thought to arise from reduction of soil organic matter through removal of organic materials and soil erosion. In all the four irrigation schemes, removal of crop materials is highly practiced in order to reduce disease inoculum from the soil. However, this practice has been associated with reduced soil fertility (Su et al., 2006; Lou et al., 2011). It has been proposed that one of the strategies to improve soil fertility is to manage soil organic matter, which, other than maintaining soil fertility, it is important in sustaining the productivity of agro ecosystems (Lou et al., 2011).

Despite the fact that application of chemical fertilizers and chemical pesticides has improved crop production, this has fundamentally changed farming practices world over leading to loss of mixed agriculture with farms becoming increasingly specialized. Arable farming (field crops) and traditional crop rotation has been abandoned leading to a population decline of many species living on farmland (Boatman et al., 2007), including the non-target species, e.g. the birds, (Walker, et al, 2008). Pesticides therefore are a major factor affecting biological diversity, along with habitat loss and climate change. Terrestrial and aquatic biodiversity have also declined rapidly due to excessive use of fertilizers, pesticides, tillage and even crop rotation (Tilman et al., 2002; Tilman et al., 2006). Furthermore, microbial biomass has been reported to correlate positively with yield in organic farming compared to conventional farming systems (Tu et al., 2006).

From the current study, soils under irrigation harbored more nematode destroying fungi compared to the rain-fed and the non-cultivated soils. This could be attributed to the frequent tillage of land, which scatters the fungi mycelia in the farm increasing the chances of detection (Wachira et al., 2009). Irrigated land receives frequent tillage compared to the rain fed land use. Irrigated land was characterized with a diversity of crops, which are grown on short cycles increasing microbial host range. On the other hand, the rain-fed land use experience high organic matter especially after crop harvest. Although the crop debris is important in promoting microbial activity, the land use is challenged with reduced moisture which is important for microbial activity hence the reduced occurrence of nematode destroying fungi. The undisturbed land use had the least abundance and diversity of nematode destroying fungi in all the irrigation systems probably due to reduced soil moisture and organic matter compared to the other two land uses systems. Microbial communities in undisturbed ecosystems achieve ecological stability and therefore occupy specific niches and are more resilience. This reduces their chances of detection and hence the low records of nematode destroying.

Continuous cultivation of land for a long period lead to reduced detection of nematode destroying fungi. This was evidenced among the four irrigation schemes in this study. This was demonstrated by the fact that Kauti irrigation scheme had the highest number of nematode destroying fungi and was established most recently (year 2011) compared to Kathiga – Gacheru that was established in 1985 and had the least number of nematode destroying fungi. In agreement with this observation, Powlson et al., (2013) reported that prolonged cultivation breaks down the soil structure that negatively affects microbial population. This then could explain the low frequency of detection of nematode destroying fungi in the older irrigation systems. A similar trend was observed between Kabaa and Mboogoni where Kabaa was established in 1960 and only rehabilitated in 2011 while Mboogoni was established in 2011.

It can be concluded that nematode destroying fungi are spread in all land uses and forms part of the soil microbial community and that their occurrence and diversity is influenced by soil management system. It was also clear that prolonged cultivation of soil reduces the diversity of soil microorganism, hence reducing the chances of isolation and detection of the nematode destroying fungi. It is proposed that the high number of nematode destroying fungi could be utilized for the management of soil pests and diseases especially the management of plant parasitic nematodes in the cultivated land.

Acknowledgment

The authors would like to acknowledge the following institutions for facilitating this study, The Ministry of Agriculture, The African Development Bank, Kenya Agricultural and Livestock Research Organization and the University of Nairobi.

References

[1]　Akhtar, A., and A. Malik, 2000. Roles of organic soil amendments and soil organisms in the biological control of plant parasitic nematodes: a review. *Bioresource Technology* 74: 35 – 47.

[2]   Altieri,M.A and Nicholls, C. I., 2003. Soil fertility management and insect pests: harmonizing soil and plant health in agro ecosystems. Soil and Tillage Research 72: 203–211.

[3]   Boatman ND, *et al*, Impacts of agricultural change on farmland biodiversity in the UK, In: Hester RE, and Harrison RM (eds), Biodiversity under threat, RSC Publishing, Cambridge, UK 2007, pp. 1-32.

[4]   Burney, J.A., Davis, S.J., Lobell, D.B., 2010. Greenhouse gas mitigation by agricultural intensification. PNAS 107, 12052-12057.

[5]   Elshafie, A.E., Al-Mueini, R., Al- Bahry, Akindi, A., Mohmoud, I., and Al- Rawahi, S., 2006. Diversity and trapping efficiency of nematophagous fungi from Oman.*PhytopathologiaMediterranea.*45: 266 – 270.

[6]   García-Álvarez, A., Arias, M., M. A. Díez-Rojo and A. Bello, 2004. Effect of agricultural management on soil nematode trophic structure in a Mediterranean cereal system. *Applied Soil Ecology* 27:197-210

[7]   Jones J. et al. (eds.), 2011. *Genomics and Molecular Genetics of Plant-Nematode Interactions.* Published by Springer Science and Business Media B.V. 2011.Pages 21 – 43.

[8]   Kindt R and R. Coe, 2005.Tree diversity analysis. A manual and software for common statistical methods for ecological and biodiversity studies. Nairobi: World Agro-forestry Center (ICRAF).

[9]   Li, Man, JunJie Wu, Xiangzheng Deng, 2013. Identifying drivers of land use change in China: A spatial multinomial logit model analysis. Land Economics 89: 632-654.

[10] Lou, Y., Wang, J., Liang, W., 2011. Impacts of 22-year organic and inorganic N managements on soil organic C fractions in a maize field, northeast China. Catena. Vol. 87 (3) 386–390.

[11] Luo, H., Mo M, Huang X,Li X., and Zhang, K., 2004. *Coprinuscomatus*: A basidiomycete fungus forms novel spiny structures and infects nematode. Mycologia 96: 1218-1224.

[12] Mashela, P.W., H.A. Shimelis and Mudau F. N., 2008. Comparison of the efficacy of ground wild cucumber fruits, aldicarb and fenamiphos on suppression of *Meloidogyne incognita* in tomato. *Phytopathology* 156: 264 -267.

[13] Masoomeh, S.G., Mehdi, R.A., Shahrokh, R.B., Ali, E., Rasoul, Z. and Majid E., 2004. Screening of Soil and Sheep Faecal Samples for Predacious Fungi: Isolation and Characterization of the Nematode-Trapping Fungus *Arthrobotrysoligospora. Iranian Biomedical Journal8: 135-142.*

[14] Moeskopsa, B., Sukristiyonubowob, Buchana, D., Sleutel, S., Herawatyb, L., Husenb, E., Saraswati, R., Setyorini, D., Nevea, S., 2010. Soil microbial communities and activities under intensive organic and conventional vegetable farming in West Java, Indonesia. Applied Soil Ecology 45,112–120.

[15] Moreira, F. M. S.; Huising, E. J.; Bignell, D. E. (2008). A Handbook of Tropical Soil Biology: Sampling and Characterization of Below Ground Biodiversity. Earthscan, UK 218 pp.

[16] Oruko, L. and Ndungu, B., 2001. Final social – economic report for the Peri-Urban Vegetable IPM Cluster.CABI/KARI/HRI/NRI/ University of Reading/ IACR Rothamsted Collaborative Project.

[17] Powlson, D.S.; Gregory, P.J.; Whalley, W.R.; Quinton, J.N.; Hopkins, D.W.; Whitmore, A.P.; Hirsch, P.R.; Goulding, K.W.T . 2013. Soil Management In Relation To Sustainable Agriculture and Ecosystem Services. *Food Policy*36 (1): 572–587.

[18] Republic of Kenya (2005). Economic Survey 2005. Government Printer, Nairobi.

[19] Singh, R.B., 2000. Environmental consequences of agricultural development: a case study from the Green Revolution state of Haryana, India. Agriculture, Ecosystems and Environment 82: 97–103.

[20] Su, Y.Z., Wang, F., Suo, D.R., Zhang, Z.H., Du, M.W., 2006. Long-term effect of fertilizer and manure application on soil-carbon sequestration and soil fertility under the wheat–maize cropping system in northeast China. Nutrient Cycling in Agro ecosystems 75, 285–295.

[21] Tu, C., Ristaino, J.B., Hu. S., 2006. Soil microbial biomass and activity in organic tomato farming systems: Effects of organic inputs and straw mulching. Soil Biology and Biochemistry 38, 247-255.

[22] Wachira,P. M., J.W. Kimenju, S.A. Okothand R. K. Mibey, 2009. Stimulation of nematode –destroying fungi by organic amendments applied in management of plant parasitic nematode. *Asian Journal Plant Sciences.* Volume 8: (2) 153 – 159.

[23] Walker et al., 2008. The Predatory Bird Monitoring Scheme: Identifying chemical risks to top predators in Britain, Ambio 37(6): 466-471, 2008)

[24] WuJunJie and Li Man, 2013.Land Use Change and Agricultural Intensification: Key Research Questions and Innovative Modeling Approaches.A Background Paper Submitted to The International Food Policy Research Institute.Final Report, November 2013.

[25] Yan, L.K., Hyde, R., Jeewon, L., CaiD., and Vijaykrishnak, Z., 2005. Phylogenetics and evolution of nematode-trapping fungi (Orbiliales) estimated from nuclear and protein coding genes. Mycologia 97: 1034 -1046.

[26] Yang, Y., Yang E., An Z. and Liu X, 2007:Evolution of nematode-trapping cells of predatory fungi of the Orbiliaceae based on evidence from rRNA-encoding DNA and multiprotein sequences. *PNAS* 104: 8379 – 8384.

27

The ecological status and uses of *Ricinodendron heudelotii* (Baill.) Pierre and *Gnetum* species around the Lobeke National Park in Cameroon

Roseline Gusua Caspa[1, 3, *], **Isaac Roger Tchouamo**[2], **Jean-Pierre Mate Mweru**[1], **Joseph Mbang Amang**[3]

[1]Regional Postgraduate School of Integrated Tropical Forest and Landscape Management (ERAIFT), University of Kinshasa, P.O. Box 15373 Kinshasa, Democratic Republic of Congo
[2]Department of Economics and Rural Sociology, University of Dschang, Dschang, Cameroon
[3]Institute of Agricultural Research for Development (IRAD), P.O. Box 2123 Yaounde, Cameroon

Email address:

rcgusua@gmail.com (R. G. Caspa), rogetchouam@yahoo.fr (I. R. Tchouamo), jpmatemweru@gmail.com (J. P. Mate Mweru), mbang4@yahoo.com (J. M. Amang)

Abstract: *Ricinodendron heudelotii* (Baill.) Pierre (Euphorbiaceae), *Gnetum buchholzianum* Engl. and *Gnetum africanum* Welw. (Gnetaceae) are among the major species of non timber forest products (NTFPs) of the Lobeke National Park (LNP). The growing demand for these products has led to an increase in exploitation; but no information exists on the status of the resource base. There is equally very little information on the importance of these resources to the local population. A survey was conducted in 152 households to get the perception of locals on the availability and use of these resources in the Lobeke landscape. An inventory was also conducted to determine the abundance of the species in three forest systems including protected area, production forest and agroforest. Productivity of *Ricinodendron heudelotii* was assessed by counting fruits from fifteen trees in each forest system. 88 % of respondents collected *Ricinodendron heudelotii*, mostly for sale and earned between 69300 and 1002000 FCFA per year. All respondents collected *Gnetum* leaves, and up to 35 % of collectors solely for consumption. Sale of *Gnetum* leaves ranged from 200 to 9200 FCFA per week. 99 % of collectors thought that the quantity of *Gnetum* in the wild has reduced over the years while *Ricinodendron heudelotii* is still abundant. There were more trees of *Ricinodendron heudelotii* in the production forest and agroforest than in the protected area. No significant difference was observed in mean diameter at breast height (DBH) of trees in all forest systems. Mean number of fruits was significantly higher for trees in the agroforest than for those in the production forest and protected area. A significantly higher number of *Gnetum buchholzianum* vines was observed in the production forest than in the agroforest and protected area while the growth of *Gnetum buchholzianum* seedlings were significantly lower in the agroforest than in the protected area and production forest which showed no significant difference. The presence of productive individuals of *Ricinodendron heudelotii* is an indication that the resource will be available for some time but the near absence of younger individuals is unfavorable for the perpetuity of the species. It is recommended that the local population be sensitized on the importance of domesticating these species and trained on appropriate techniques to propagate and incorporate them into suitable agro systems.

Keywords: Non-Timber Forest Products, Resource Base, Forest System, Baka Pygmies, Bangando

1. Introduction

The Congo basin forest is the second largest contiguous area of tropical rainforests in the world [1]. These forests extend over 228 million ha, storing 36.815 billion tons of carbon and providing habitats for over 11000 species of plants (many of which are endemic to the area) and associated fauna, including endangered species such as gorillas and chimpanzees [2]. Cameroon has a tropical dense forest of about 19 million hectares which forms part of the Congo basin forest [1]. In 2003 there were almost 80 million inhabitants in the Congo basin with 62 % of them living in rural areas and deriving their livelihoods from natural resources [3]. The international community has recognized that the conservation of biodiversity and the reduction of greenhouse gas emission are global environmental issues that affect the future of the entire human society; thus the conservation of tropical rainforests has attracted global concern [4]. Non timber forest products (NTFPs) are a major source of livelihoods for forest communities. They are sources of food, income, medicine, construction materials and are also of cultural and spiritual value to forest dwellers [5, 6, 7]. About 80 % of the population in developing countries uses forest products on a daily basis and about 75 % of poor people that live in rural areas depend on forests for subsistence [8, 9]. These forest products are available during times of scarcity of staples and are often used as safety nets where the rural community depends on the resources to bridge the hunger gap [10]. NTFPs provided employment for an estimated 283,000 people in Cameroon and generated a market value of US$ 54 million in the 2007-2008 seasons [6]. According to [11], the management of NTFPs has been recognized as a way of ensuring forest conservation and as an alternative to conversion of forests into agricultural or other land use. NTFP harvest can affect ecological processes at many levels, from the genes, individual and population to community and ecosystem, all of which have important consequences [12, 13]. The fragile ecological balance maintained in a tropical forest is easily disrupted by human intervention and extractive activities that at first glance appear very benign but can later have a severe impact on the structure and dynamics of forest tree populations [14]. The NTFP sector in Cameroon is also negatively affected by inconsistent legal and institutional policies that do not favor effective management and commercialization of the resources [15, 16]. Conflicting and overlapping customary and formal regulations make management difficult and recognition of positive customary regulations could aid sustainable management [17]. *Ricinodendron heudelotii* and *Gnetum* species (*Gnetum buchholzianum* and *Gnetum africanum)* were identified among priority NTFPs in the study area because they were major sources of food and income [6].

Ricinodendron heudelotii (Baill.) Pierre (Euphorbiaceae) is a fast-growing late secondary forest tree found in the Guinea-Congolean humid forests of West and Central Africa, reaching a height of up to 50 m and girth of 2.7 m [18]. It occurs typically as a species of the fringing, deciduous and secondary forests, common throughout the semi-dry, wooded savannah zone, where it is scattered in gaps at forest edges and in secondary scrub and thickets [19]. *Ricinodendron heudelotii* trees first produce fruits after about 4 years of existence [20]. On the other hand, [21] and [22] reveal that trees first produce fruits at 6-7 years and 8-10 years of age respectively. *Ricinodendron heudelotii* trees fruit once in every two to three years between September and October, although some can fruit every year [23]. [24], however observed ripe fruits of the species from late August to end of

December in the humid forest zone of Cameroon.

According to [25], *Ricinodendron heudelotii* has four major use categories including consumption, medicinal, sociocultural and soil fertility improvement. It is valued for its distinctively flavored seeds, commonly called 'njansang' which are dried and used as flavoring and thickening agent in food. *Ricinodendron heudelotii* bark is used to treat yellow fever, anemia, malaria, stomach pain and disease of infants. It is also used as aphrodisiac in parts of Cameroon and in pregnancy to ease delivery [25]. The decoction of *Ricinodendron heudelotii* bark is used to treat hernia and abdominal pain in the Lobeke area [26]. It is grown by farmers for soil fertility improvement, forage, shade, poles and light wood work; and is a major source of income [18, 23]. About 71585 metric tons of *Ricinodendron heudelotii* kernels were produced and traded in the humid forest zone of Cameroon between 2003 and 2010; and valued at about equivalents to US$ 708770 [27].

Gnetum buchholzianum Engl. and *Gnetum africanum* Welw. are gymnosperms of the family Gnetaceae. They are understorey lianas found in the humid tropical forest from Nigeria through Cameroon, Central African Republic, Gabon, and Democratic Republic of Congo to Angola [28]. In Cameroon, these *Gnetum* species are often found in fallow farmlands, secondary forests and closed forests. The vines climb on trees, saplings and shrubs for support in the complex tropical humid forest where they grow luxuriantly and produce great quantities of leaf biomass [29]. *Gnetum* plays a significant nutritional and social role across the sub-region where the plant is found and is consumed by people of all social strata [30]. In Cameroon, leaves of *Gnetum* species are harvested on a daily basis for sale in local and regional markets. They are evergreen and therefore available throughout the year [29]. However, there are increasingly reports of destructive and unsustainable harvesting through cutting and removal of the entire plant and/or felling of support trees [31]. [32] point out that the vulnerability of some major traded species is exacerbated by the lack of knowledge about sustainable harvest techniques such as in Cameroon where 40 % of *Gnetum* species are harvested using unsustainable techniques. The high volume of trade in *Gnetum* species combined with low levels of domestication results in unsustainable harvests [33, 18]. *Gnetum* species were declared endangered by the Ministry of Environment and Forestry (MINEF) in Cameroon in 1995 [34]. However, no action has been taken to protect the species. Although it is listed as a special forestry product, no particular protection measures other than a quota system which is not based on any inventory has been put in place [32]. Likewise, both species of *Gnetum* are Red List classified as near threatened [35, 36]. In Cameroon, harvesters of *Gnetum* species can earn US$ 98 to 110 per month, which is higher than the guaranteed minimum wage [37]. According to [38], about 607900 metric tons of *Gnetum* leaves were traded between 2002 and 2008, and valued at about 631,167,345 F (CFA) which is equivalent to US$1,262,334. Incomes from sales of *Gnetum* leaves allow families to pay for food, healthcare and children's education [7]. *Gnetum* species exist in a wide range of habitats and are harvested in both fallow farm areas and closed canopy forests. These species contribute to food security, for harvesters who consume them directly and for households that buy leaves in the markets, and income diversification. In the South west, coastal and central areas of Cameroon, the annual harvest is estimated at 4,180 tons. The *Gnetum* species sector directly involves at least 1,885 people in Cameroon. It represents a valuable trade that is estimated at US$ 3.8 million per year in South west Cameroon [7]. However, the growing importance of *Gnetum* species both for the purposes of nutrition and for earning income increases the pressure placed on the resource. Improved harvesting and domestication techniques are therefore required to ensure the sustainability of this non-timber forest product [7]. A survey in southern Cameroon found that 86 % of 200 households surveyed ate *Gnetum* more than 3 times a week [18]. In addition to being a major source of food and income, *Gnetum* species also have a number of medicinal properties. In Cameroon it is recorded as being used in the South west to ease childbirth [39] and the leaves are used as a disinfectant for wounds, to treat hemorrhoid and as an anti-hangover agent when the fresh leaves are crushed and used to neutralize the effects of alcohol [32]. The Bulu ethnic group uses the leaves to treat colds, increase blood production and to treat spleen problems [40]. The leaves are used to treat nausea and act as an antidote to certain types of poison [7]. The leaves also have very high nutritional value as their cellulose can extend digestion periods and reduce cholesterol levels [42].

These species feature in most lists of priority NTFPs in Cameroon [43, 44, 18, 45]. [46] identified *Ricinodendron heudelotii* among trees preferred in cocoa agroforests whereas [47] proposed these species as appropriate for inclusion into multi strata agroforests and for domestication. Surveys carried out by ICRAF (World Agroforestry Center) and partners to find farmers' preferences for multipurpose trees ranked *Ricinodendron heudelotii* third among species considered useful for domestication in the West African humid lowlands [48], and were ranked fourth in other such surveys in Onne, Nigeria and Mbalmayo, Cameroon [49]. *Gnetum* species are ranked among the main NTFPs in terms of their high economic importance in Central and West Africa [6, 7]. The high economic value of these species to the local population is the cause of high exploitation pressure hence; this study was aimed at determining the status of their resource base, the possible impact of harvesting on these species and their uses within the Lobeke landscape.

2. Materials and Methods

2.1. Study Site

This study was carried out at the Lobeke National Park in Cameroon. The LNP is situated between latitudes 2°05' to 2°30'N and longitudes 15°33' to 16°11'E. It has a surface area of about 217 854 hectares and found in the Moloundou Sub Division of the Boumba and Ngoko Division of the East Region of Cameroon. The LNP forms a part of the Sangha Trinational Landscape (28000 km²), a forest conservation area established by three African countries, including Cameroon, the Central African Republic and the Republic of Congo, with the objective of protecting the native tropical rainforests with their diverse flora and fauna [50] . It is rich in forest resources and habitats of diverse and unique wildlife on which generations of indigenous communities have depended for millennia [50] and was recently designated a UNESCO world heritage site. The climate of the area is typically equatorial with four seasons - two rainy and two dry seasons. The long rainy season extends from September to November with its peak in October while the short rainy season extends from late March to June with its peak in April. The maximum precipitation is about 1500 mm per year. The long dry season extends from December to February while the short dry season is between July and August [51]. The mean monthly temperature varies between 23 and 25°C with a mean annual temperature of 24°C. The relative air humidity also varies from 60 to 90 % [52]. This area was chosen for the study because of the increasing importance of NTFPs to the local population in the Lobeke landscape.

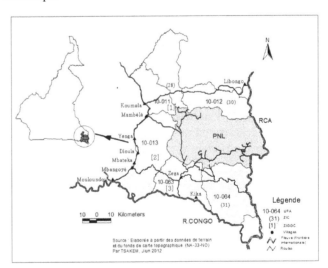

Figure 1. *Map of the Lobeke National Park (LNP)*

2.2. Survey

Nine villages (Koumela, Mambele, Yenga, Dioula, Mbateka, Mbangoye 2, Zega, Kika and Libongo) were chosen around the LNP by virtue of their close proximity to the park for administration of questionnaires. This was to ensure that villagers had a high possibility of collecting NTFPs within the Lobeke landscape. Semi structured interviews were conducted between June and August 2012 in 152 households which had been involved in the exploitation of *Ricinodendron heudelotii and/or Gnetum* species for at least five years, with the objective of getting the perceptions and attitudes of respondents towards the management and use of these resources.

2.3. Ecological Assessment

The materials used for this assessment included a 100 m measuring tape for setting up the plots for inventory, a diameter tape for measuring the girth of trees, cutlasses for clearing and opening up paths for movement and delineating plots, a meter ruler for measuring the heights of seedlings nylon ropes and wooden poles to set up plot limits and calipers for measuring seedling diameter. An inventory was carried out to assess the structure, abundance, distribution and regeneration of *Ricinodendron heudelotii* and *Gnetum buchholzianum* in the study area. (Inventory was carried out on *Gnetum buchholzianum* because locals indicated that they preferred harvesting it to *Gnetum africanum*). Six plots of 100 m x 100 m were systematically set, separated from each other by a distance of 200 m in three systems including protected area, production forest (community forest) and agroforest (old fallow/active agricultural land), making a total of eighteen (18) plots [53]. A community forest in Cameroon is a forest forming part of the non-permanent forest estate, which is covered by a management agreement between a village community and the forestry administration. Management of such forest – which should not exceed 5,000 ha – is the responsibility of the village community concerned, with the help or technical assistance of the forestry administration.

All individuals of *Ricinodendron heudelotii* with at least 10 cm diameter at breast height (DBH)) were tagged, counted and measured for DBH in each plot. Each plot was divided into sub plots of 5 x 100 m. Younger trees of 5 – 9 cm DBH and saplings (young plants of *Ricinodendron heudelotii* greater than 1 m height but less than 5 cm DBH) were counted and measured in ten alternating sub plots per plot. *Gnetum buchholzianum* vines were also counted in ten alternating sub plots per plot. Regeneration plots of 2 x 2 m were set at 50 m into each sub plot for enumeration of seedlings of both species [53]. Twenty five (25) seedlings of *Ricinodendron heudelotii* were selected and tagged under five widely separated productive trees (within and out of sample plots depending on availability) in all three forest systems making a total of three hundred and seventy five (375) seedlings. Seedling survival and growth were determined by counting number of surviving seedlings, number of leaves per seedling and measuring seedling height and collar diameter every three months for eighteen months. On the other hand, one hundred (100) seedlings of *Gnetum buchholzianum* were identified and tagged in all forest systems making a total of three hundred (300) seedlings (both within and outside sample plots) and assessed for survival and growth for eighteen months (March 2013 to September 2014). Fruit production from *Ricinodendron heudelotii* was assessed on a weekly basis by counting fruits throughout the production period. Fifteen producing trees were identified and tagged in each forest system, making a total of forty five trees. Fifteen 1m x 1m plots were demarcated and cleared under each fruiting tree, from which fallen fruits were collected, counted and removed on a weekly basis for a period of 2 months (September and October).

2.4. Data Analysis

Data was analyzed using Excel and the SAS statistical package Version 9.0, with the General Linear Model Procedure (GLM) and the Student-Newman-Keuls test to separate the means.

The abundance was determined by counting the total number of individuals of *Ricinodendron. heudelotii* and *Gnetum buchholzianum* in the site while the density was estimated by determining the total number of individuals of each species per hectare.

3. Results and Discussion

3.1. Demographic Information

3.1.1. Ethnic Origin, Age Range, Sex, Marital Status and Level of Education of Collectors of Ricinodendron Heudelotii and Gnetum Species

Collectors of *Ricinodendron heudelotii* and *Gnetum* species came from two major ethnic groups Bangando (53 %) and Baka (29 %) who are indigenes of the area. A small proportion (18 %) of collectors came from 6 other groups in the East region of Cameroon. These have probably migrated from their villages of origin and settled in the area in search of jobs in the logging and mining companies found there.

The survey revealed that 49 % of collectors belonged to the 40 to 55 age group exclusive, while 35 % belonged to the 25 to 40 age group. 10 % of producers were 25 years old or younger, while a minority 6 % was 55 years old or more. This indicates that mostly individuals of the middle aged group are involved in NTFP collection in the Lobeke landscape. An almost equal proportion of men and women collect NTFPs in the study area, with 51 % of women and 49 % of men involved. A distribution of NTFP collectors according to marital status indicated that 87 % of respondents were married while 10 % were single. A small proportion (2 %) was divorced while a minute 1% was widowed. This indicates a stable society in which the development of NTFP enterprises can be successful. It was observed that 24 % of NTFP collectors had no formal education, while 43 % had been to primary school. 33 % of respondents indicated that they have been to secondary school. This is an indication that new ideas or innovations can to an extent be adopted in the community at least at individual level. This study revealed that 48 % of collector households had 6 to 9 members, while 22 % of households had 10 to 14 members. 17 % of households were made of more than 15 individuals while a small 13 % was made of less than 5 members. Family sizes are generally medium to large and this is possibly as a result of the high level of polygamy observed in the zone. This also explains why most households depend on family members as a source of labor for their NTFP collection and agricultural activities.

It was observed from this study that respondents' major

sources of income include collection of NTFPs (43 %) and crop production (39 %). This result is in line with [66] who found that 44.44 % of income of people in villages around the LNP is made of forest income while 18.34% is made of income from agriculture. Fishing also makes a modest contribution to household incomes with 7 % of respondents involved in this activity. The near absence of development, with mainly logging and mining companies; and no other industries in the area leaves only exploitation of NTFPs and agriculture as the mainstay of the people. The fact that different species of NTFPs produce at different periods of the year, and usually when staples are scarce also explains why the populace is highly involved in NTFP collection. Most NTFPs in the study area, unlike in the past now have market (monetary) value thereby attracting many people into their exploitation. [26] observed that some collectors of NTFPs take their produce to the border towns of Moloundou and Kika for sale as prices are higher than in their villages.

3.1.2. Collection of Ricinodendron Heudelotii and Gnetum Species (Gnetum Buchholzianum and Gnetum Africanum)

All respondents in this study indicated that they had free access into the forest to collect both *Ricinodendron heudelotii* fruits and *Gnetum* leaves. *Ricinodendron heudelotii* fruits are collected by picking from the forest floor. The fact that fruit collection does not harm the parent plant is a positive indication for the sustainability of the species in the zone. Only 88 % of respondents indicated that they collected this product in marketable quantities while 12 % did collect in small quantities for household consumption. The later explained that their reticence towards collecting this product is because the extraction of its kernels from the hard shell is very tedious and time consuming.

Gnetum species were also collected by all respondents in this study, either just for food or for both food and trade. Collectors indicated that they harvest *Gnetum* leaves mostly in the secondary (production) forest (112 citations), followed by fallows (70 citations) and farms (50 citations). The most common methods of harvesting *Gnetum* leaves were felling support trees (95 citations) and pulling vines from support trees (84 citations) which are very destructive and detrimental to the sustainability of the species and also those of support tree species. Some respondents indicated that they climb up big support trees (67 citations) by use of larger vines on such trees to collect *Gnetum* leaves. These were mostly the Baka pygmies who have lived in the forest for centuries. Others indicated that they collect mature leaves from accessible vines without destroying the vines (52 citations). This seemed to be the most sustainable harvest method cited by respondents, but nearly all respondents indicated that they also used the destructive harvest method indicated above. This result agrees with [31, 32, 33] who observed destructive harvest of *Gnetum* species in the South West region of Cameroon. In this study, 89 % of respondents indicated that they collect *Gnetum* leaves 2 to 5 times per week while 11 % of respondents indicated that collection is

done on a daily basis. 98 % of collectors indicated that although they collected both *Gnetum* species, they preferred *Gnetum buchholzianum* to *Gnetum africanum*. Their reasons were that *Gnetum buchholzianum* has larger leaves, is easier to harvest and slice; and also has a better taste than *Gnetum africanum* which has smaller leaves, is more difficult to collect and slice; and has a slight bitter taste when cooked.

Collectors' views regarding the availability of *Gnetum* species in the Lobeke landscape showed that 99 % of respondents observed a reduction in the availability of the resource while 1 % declared that the resource was still abundant as in the past. Respondents identified agriculture (79 citations) and increased collection (60 citations) as the most possible reasons for a reduction in the stock of *Gnetum* species in the Lobeke landscape. Deforestation (43 citations), establishment of cocoa plantations and use of chemical inputs (23 citations) and prohibition on hunting (20 citations) were other reasons mentioned for the dwindling stock of *Gnetum* species. The need to feed the ever growing rural population and increased cash cropping (cocoa) for financial gains, establishment of the protected area which has led to a restricted access to forest resources and reduction in peoples' extractive activities have all resulted in increased agriculture and consequently destruction of some of the species' habitat. Respondents pointed out that *Gnetum* is one of the vegetables that can be prepared without meat, so prohibition on hunting certain species of wildlife and the increased market value of this resource are possible reasons for increased collection. This agrees with [26] who indicates that *Gnetum* leaves are considered a meat substitute in the area and that frequency of collection is higher during the dry season when wild meat is scarce.

Problems encountered by collectors of *Gnetum* species in the Lobeke landscape include walking long distances into the forest where the resource abounds, (103 citations). This was followed by threats from wild animals such as gorillas, elephants, and chimpanzees, (78 citations) that are said to also consume this resource. Difficulty climbing large support trees (39 citations), limited markets (8 citations) and high perishability of product (5 citations) were to a smaller extent identified as hindrances to the collection of *Gnetum* leaves.

3.1.3. Conservation of Ricinodendron Heudelotii and Gnetum Species by Collectors

All collectors of *Ricinodendron heudelotii* indicated that they conserve the species mostly by protecting wildlings on their farms. They also protect mature and producing trees during farm creation by not burning around them. Up to 59 % of respondents had less than 10 trees of *Ricinodendron heudelotii* on their farms while 37 % of respondents indicated they had between 10 and 20 trees of this species on their farms. A minute 3 % had between 21 and 30 trees while 1 % had more than 30 trees on their farms. Those with more than 20 trees in their farms are probably those who also own cocoa farms where these trees also serve for shade provision and soil fertility improvement [25]. This reveals that respondents have a poor attitude towards planting the species, probably because they think it is abundant in the forest. On the other hand, *Gnetum* species were conserved by collectors only through the protection of wildlings.

3.1.4. Uses of Ricinodendron Heudelotii and Gnetum Species in the Lobeke Landscape

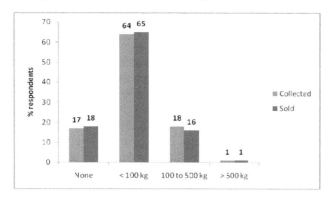

Figure 2. *Estimates of Ricinodendron heudelotii seeds collected and sold around the LNP*

Figure 2 shows that 67 % of collectors ranked *Ricinodendron heudelotii* as the second major source of income while 11 % ranked it third among four major NTFPs including *Irvingia gabonensis*, *Gnetum* species and *Aframomum pruinosum*. Up to 65 % of collectors sold less than 100 kg of *Ricinodendron heudelotii* kernels meaning that they earned at most from 69300 FCFA to 198000 FCFA for at most 99 kg of product, considering that the price of a kg ranged between 700 FCFA when the product is abundant and 2000 FCFA when the product is scarce. Only 16 % of collectors sold between 100 kg and 500 kg for at least 70000 FCFA and at most 1000000 FCFA while a minute 1 % earned between 350700 FCFA and 1002000 FCFA for at least 501 kg of the product. Figure 2 also reveals that this product is mostly collected in small quantities. Collectors indicated that this is because the collection, cleaning and cracking to get the kernels is very tedious and time consuming. Almost all of the *Ricinodendron heudelotii* kernels collected are sold. This can be explained by the fact that indigenes of the area consume this product is very small quantities, except the Baka pygmies who use it as a major condiment in some meals. Those who collected in appreciable quantities indicated that they do this by drying the nuts after having boiled, and then crack gradually almost throughout the year. Income earned from the sale of *Ricinodendron heudelotii* is used for buying household needs, education of children through payment of tuition fees and other school requirements, provision of health care, buying of clothing, farm tools and payment of hired labor on-farm. This result is consistent with those of [7 and 67] who found that incomes from forest products play roles in buying agricultural implements and purchasing basic household needs as well as sending children to school.

Figure 3 reveals that 76 % of collectors harvested less than 15 kg of *Gnetum* leaves per week while 17 % harvested between 15 to 30 kg of the product per week. Only 2 % of collectors harvested between 30.5 and 45 kg while 5 % of

collectors harvested more than 45 kg of this product per week. Up to 35 % of collectors indicated they did not sell but consumed the product whereas 51 % sold less than 15 kg per week, earning between 200 FCFA and 2800 FCFA for at least 1 kg and at most 14 kg of the product respectively per week, being that 1 kg costs 200 FCFA. 9 % of collectors sold between 15 and 30 kg of *Gnetum* leaves between 3000 FCFA and 6000 FCFA per week, 1 % sold between 30.5 and 45 kg between 6100 FCFA and 9000 FCFA per week while 4 % of respondents sold at least 46 kg at 9200 FCFA per week. *Gnetum* is also highly consumed in households in the area on a year round basis. It is not sold in large quantities because of the perishability of the product and poor state of roads to major markets in the country where this product has very high demand.

Exchange rate:
1 €=Approximately 650 FCFA; 1 $US = Approximately 500 FCFA.

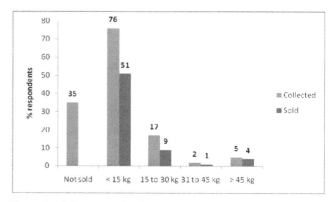

Figure 3. *Collection and sale of Gnetum leaves around the Lobeke landscape*

Up to 61 % of NTFP collectors indicated that *Ricinodendron heudelotii* had some medicinal values and used variously for the treatment different ailments notably the decoction of bark for anemia. On the other hand, 12 % of respondents indicated that *Gnetum* species have some cultural values especially during the "jengi" religious festival and initiation rites of Baka pygmies. 7 % of respondents indicated that *Gnetum* leaves are used as charm to influence peoples' decisions while 25 % of respondents indicated that *Gnetum* leaves have medicinal values and used particularly to speed up walking in slow children.

Respondents indicated that they preferred consuming NTFPs to other conventional food stuffs such as groundnut and cassava leaves which are used in the absence of *Ricinodendron heudelotii* and *Gnetum* leaves respectively because NTFPs are freely collected from nature; and also because they form part of their traditional meals. This is supported by [50], who observed that dependence on forest products is not related to individual household incomes, as wealthier households remain dependent on forest-derived products. This indicates that the traditional link to forests cannot be easily changed through subsidies and compensations.

3.2. Inventory

3.2.1. Size of Ricinodendron Heudelotii in Three Forest Systems

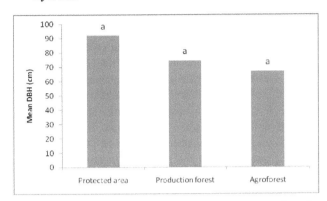

Figure 4. Mean DBH of Ricinodendron heudelotii trees in three forest systems around the LNP. Means with the same letter are not significantly different

There was no significant difference in mean tree DBH of *Ricinodendron heudelotii* in all forest systems, although trees within the protected area were larger than those in the production forest and agroforest (Figure 4). This is probably because only the fruits of this tree species are mostly exploited and not the wood, such that even in zones of frequent human intervention like the agroforest and production forest, people have the tendency to protect it for its fruits.

More trees of *Ricinodendron heudelotii* were observed in the production forest and agroforest than in the protected area (Figure 5). Most trees recorded in this inventory had a DBH of at least 10 cm, meaning that mainly productive individuals of this species are found in the forest in the study area. This is possibly due to younger plants being hindered at some point in their growth process by some natural or biological factors such as pests from attaining the larger size classes. Very few seedlings were recorded in this inventory. This could be as a result of the high gathering pressure by the local population, and the fact that rodents and other small mammals found in the area also feed on the seeds. *Ricinodendron heudelotii* was also observed to suffer from attack of the psyllid *Diclidophlebia xuani* which led to the death of many seedlings within two months. This is in line with [54] who stated that this species suffers severe attack from this psyllid. According to [14], the extraction of commercial quantities of fruits and seeds can cause notable changes in the structure and dynamics of a tree population, which are typically precipitated by a reduction in seedling establishment due to over-harvesting. Human intervention harvesting fruits of species such as *Ricinodendron heudelotii* from the forest may have an impact on the species' population structure. [55] have found some evidence that over-gathering of fruits, combined with the length of time needed for the seed to reach germination point, is having an impact on natural regeneration rates. According to [56], the stony endocarp may lie for more than two years after the

pulp has rotted if it does not germinate or get carried away by rodents.

In this study, more individuals of *Ricinodendron heudelotii* were observed in the agroforest and production forest than in the protected area. This result confirms the observation by [23] that *Ricinodendron heudelotii* is mostly found as a pioneer species in the secondary forest. According to [57], this species occurs in fringing deciduous and secondary forests, and is found throughout the South West of Cameroon in forest, fallow, cocoa and other farms [18]; environments which meet the light requirement of the species. *Ricinodendron heudelotii* also possibly enjoys some protection within the agroforest and production forest by the local population who frequent these areas more, because of its economic value. The density of *Ricinodendron heudelotii* which was considered as the number of individuals of this tree species per hectare was 3.4 trees per hectare overall. On the contrary, [58] reported a density of 10.3 stems/ha for *Ricinodendron heudelotii* around the Boumba-Bek National Park in Southeastern Cameroon. In the South Province of Cameroon, an average density of 2.1 stems/ha was observed in secondary forest, with a maximum density of 4.1 stems/ha [59] while in the Mbalmayo Forest Reserve in Cameroon, a higher density of 5 individuals/ha was recorded [60]. Likewise, [18] reported densities of 2.4 stems/ha for trees with DBH greater than 10 cm and 0.5 stems/ha for trees of DBH greater than 70 cm at an altitude of 1650 m in the Dja Fauna Reserve in East Cameroon.

3.2.2. Abundance of Ricinodendron Heudelotii Trees

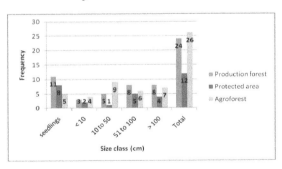

Figure 5. Size class distribution of Ricinodendron heudelotii trees in three forest systems around the LNP

The survival and growth assessment of *Ricinodendron heudelotii* seedlings ended within two months as seedlings were attacked by a psyllid (*Diclidoplebia xuani*) and died out completely within this period as seen in plate 1.

Plate 1a. Healthy seedlings of Ricinodendron heudelotii .

Plate 1b. Infected seedlings of Ricinodendron heudelotii

3.2.3. Fruit Production of Ricinodendron Heudelotii

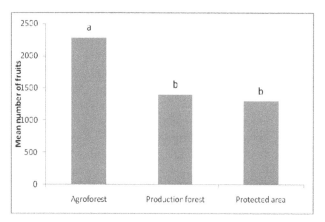

Figure 6. *Fruit production by Ricinodendron heudelotii trees in three forest systems around the LNP. Means with the same letters are not significantly different.*

The mean number of fruits from *Ricinodendron heudelotii* trees within the agroforest (approximately 2300 fruits) was significantly higher than for trees in the production forest and protected area (approximately 1500 fruits) between which there was no significant difference (Figure 6). [18] point out that fruits of *Ricinodendron heudelotii* are produced in large quantities, while [23] and [21] observed that an individual tree can produce up to 900 fruits in a fruiting year while kernel yield varies between provenances. On the other hand, [24] observed more than 4000 fruits (172 kg) per tree in the fallows of the humid forest zone of Cameroon in line with [62]. The fruits are eaten by animals [20] and are dispersed by bats, hornbills and rodents [18].

3.2.4. Abundance of Gnetum Buchholzianum Vines

Gnetum vines were observed to be significantly higher in number in the production forest than in the agroforest and protected area (Figure 7). The low vine population in the agroforest could be as a result of overharvesting of the resource over the years and its accessibility to the population, in line with [41] who notes that in Gabon, the Central African Republic and the Democratic Republic of Congo, overharvesting is shrinking wild populations of *Gnetum*. The very reduced *Gnetum* vine population in the protected is probably due to the presence of an important population of large mammals such as gorillas, elephants and chimpanzees

who also feed on the leaves of *Gnetum* species, in line with the observations of [62] who observed that the leaves of *Gnetum africanum* and *Gnetum buchholzianum* are consumed by gorillas in the Nouabali-Ndoki forest of the Republic of Congo. [41] also notes that *Gnetum* leaves also play a role in the diet of other forest dwellers such as chimpanzees and gorillas. Very few seedlings were recorded in all forest systems during the inventory. This could be as a result of the fact that ripe fruits of this species are eaten by birds or due to poor germination as the seeds take too long to germinate [33]. [31] reports observing birds, squirrels and other rodents eating the fruit as they ripen, reducing the availability of seeds for regeneration. However, [41] suggests that it may be necessary for the seeds of *Gnetum* species to pass through the intestines of a bird, fruit bat or other animals before they can readily germinate. *Gnetum* vines were however observed to regenerate mainly by the production of new shoots and root suckers. This is confirmed by [63] who indicated that *Gnetum* species are capable of generating root suckers, as offshoots of lateral adventitious roots. [64] report production of suckers from *Gnetum* provenances growing in the *Gnetum* gene bank of the Limbe Botanic Garden in Cameroon. This suckering can be quite prolific in the wild, suggesting that vegetative regeneration is important [65]. However, *Gnetum* species may be favored by forest disturbance, a probable explanation for its abundance in degraded forest, bush fallow and crop fields [34].

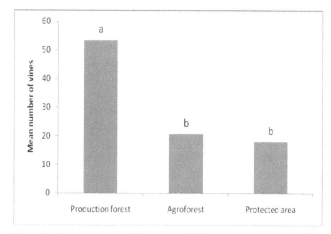

Figure 7. *Mean number of Gnetum buchholzianum vine in three forest systems around the LNP. Means with the same letter are not significantly different*

3.2.5. Growth of Gnetum Buchholzianum

3.2.5.1. Vine Length

There was a continuous increase in the vine length of *Gnetum* seedlings in all three forest systems with time. No significant difference was observed in mean *Gnetum* vine lengths in the production forest and protected area, which were each significantly higher than mean *Gnetum* vine length in the agroforest. This is possibly due to the higher shade levels in the production forest and protected area, unlike in the agroforest with higher sunlight; as *Gnetum* species are known to grow better in shaded environments[30] .

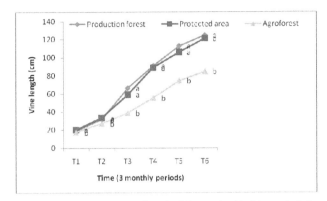

Figure 8. Mean increase in vine length of Gnetum buchholzianum in 3 forest systems around the LNP. Means with the same letter are not signifcantly different

3.2.5.2. Number of Leaves

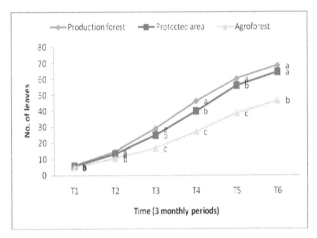

Figure 9. Mean increase in number of leaves of Gnetum buchholzianum in 3 forest systems around the LNP. Means with the same letter are not significantly different

According to Figure 9, there was a continuous increase in the the mean number of *Gnetum* leaves in all three forest systems, but a significantly higher mean number of *Gnetum* leaves was observed in the protected area than in the production forest at 9, 12 and 15 months of observation. It is possible that side shoots developed at vine nodes at these periods and led to big increases in leaf number. There was also a significantly higher mean number of *Gnetum* leaves in the production forest and protected area than in the agroforest. *Gnetum* vines in these forest systems were longer and therefore had many nodes for leaf development.

3.2.6. Survival of Gnetum Buchholzianum Seedlings

According to Figure 10, a general decrease in survival was observed for *Gnetum* seedlings in all three forest systems, with the protected area having the least survival rate (64 %) and the production forest having the highest survival rate of 82 % after 18 months of observation. The high incidence of large mammals in the protected area could lead to trampling of seedlings while the high humidity in this environment can provide suitable conditions for microbial attack, which may lead to death of some seedlings.

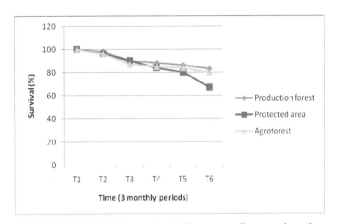

Figure 10. Survival of Gnetum buchholzianum seedlings in three forest systems around the LNP.

3.3. Conclusion

The result of this study shows that the abundance of these species is possibly witnessing a reduction. This is because these products which in the past were mainly exploited for home consumption currently have market (money) value, and are one of the major sources of income in the locality. The increase in demand will lead to an increase in the number of people collecting the products and also an increase in the quantity collected per person. It is recommended to educate the local population on the importance of domesticating these species of NTFPs, different methods of producing improved planting materials and appropriate agro systems in which they can be planted.

Acknowledgment

We are grateful to the Congo Basin Forest Support Program (PACEBCo) and Wallonie Bruxelles International for providing funds for this research

References

[1]   de Wasseige C., de Marcken P., Bayol N., Hiol Hiol F., Mayaux Ph., Desclée B., Nasi R., Billand A., Defourny P. and Eba'a Atyi R. 2012 The forests of the Congo basin-state of the forest 2010. Publications Office of the European Union. Luxembourg. 276 p. ISBN: 978-92-79-22716-5, doi:10.2788/47210

[2]   FAO, 2007. State of the World's Forests 2007. Rome. ISBN: 978-92-5-105586-1.144pp.

[3]   FAO, 2005. State of the World's Forests 2005. FAO-Rome. ISBN: 92-5-105187-9. 153 pp.

[4]   Ichikawa, M. 2012. Central African forests as hunter-gatherers' living environment: An approach to historical ecology. *African Study Monographs Supplementary Issue*, 43: 3−14.

[5]   Hoare, A.L. 2007. The use of non timber forest products in the Congo basin, Constraints and opportunities. http://www.rainforestfoundationuk.org/files/Forest%20Produc ts%20Low%20PDF.pdf

[6]   Ingram, V. and Schure, J. (2010). Review of Non Timber Forest Prouducts (NTFPs) in Central Africa, Cameroon. Center for International Forestry Research.

[7]   Ingram, V., O. Ndoye, D.M. Iponga, J. C. Tieguhong, R. Nasi, 2012. Non-timber forest products: contribution to national economy and strategies for sustainable management In: de Wasseige C., de Marcken P., Bayol N., Hiol Hiol F., Mayaux Ph., Desclée B., Nasi R., Billand A., Defourny P. and Eba'a Atyi R. 2012 The forests of the Congo basin-state of the forest 2010. Publications Office of the European Union. Luxembourg.   276   p.   ISBN: 978-92-79-22716-5, doi:10.2788/47210

[8]   IFAD 2004. Commerce et developpement rural: enjeux et perspectives pour les ruraux pauvres. 31pp.

[9]   Noubissie E.T., Chupezi, J., Ndoye, O. 2008. Studies on the Socio-Economic Analysis of non-timber forest products (NTFPs) in Central Africa. Synthesis of reports of studies in the Project GCP/RAF/398/GER. Fao. Yaounde, Cameroon, FAO GCP/RAF/398/GER Enhancing Food Security in Central Africa through the management and sustainable use of NWFP: p. 43.

[10] Neumann, R. P., Hirsch, E. 2000. Commercialization of non-timber forest products: review and analysis of research. Bogor: Centre for International Forestry Research.

[11] Ankila, J.H. 2004. The Ecological Consequences of Managing Forests for Non-Timber Products. *Conservation & Society* 2 (2) 211-216

[12] Ticktin, T. 2004. The ecological implications of harvesting non-timber forest products. *Journal of Applied Ecology*, 41, 11–21

[13] Brites, A. D., Morsello, C. 2012. The Ecological Effects of Harvesting Non-Timber Forest Products from Natural Forests: a Review of the Evidence. VI Encontro Nacional da Anppas 18 a 21 de setembro de 2012, Belém - PA – Brasil

[14] Peters, C.M. 1994. Sustainable harvest of non-timber plant resources in tropical moist forest: An ecological primer. Washington, Biodiversity Support Program.

[15] Ndoye O., and Awono A., 2009. Regulatory policies and *Gnetum* spp. trade in Cameroon: Overcoming Constraints that Reduce Benefits and Discourage Sustainability. In: Wild Product Governance: Finding Policies that Work for Non-Timber Forest Products, edited by S.A. Laird, R. McLain and R.P. Wynberg. London: EarthScan.

[16] Tieguhong J.C., Ndoye O., Vantomme P., Zwolinski J., and Masuch J., 2010. Coping with crisis in Central Africa: enhanced role for non-wood forest products. Unasylva 233 (60):49-54.

[17] Laird S., Ingram V., Awono A., Ndoye O., Sunderland T., Lisinge E., and Nkuinkeu R., 2010. Integrating customary and statutory systems: The struggle to develop a legal and policy framework for NTFPs in Cameroon. In: Laird S.A., McLain R., and Wynberg R.P., 2010. Wild Product Governance: Finding Policies that Work for Non-Timber Forest Products, London: Earthscan.

[18] Clark, L. E., Sunderland, T. C. H. 2004. The Key Non-Timber Forest Products of Central Africa: State of the Knowledge. Technical Paper No. 122, SD Publication Series

[19] Orwa, C., Mutua, A., Kindt, R., Jamnadass, R., Anthony, R. 2009. Agroforestry Database: a tree reference and selection guide version   4.0 (http://www.worldagroforestry.org/sites/treedbs/treedatabases.asp)

[20] Eyog-Matig. O, O. Ndoye, J. Kengue, A. Awono, (eds). Les fruitierscomestibles   du   Cameroun.   International   Plant resources Institute.

[21] SCUC, 2006. Ndjanssang: Ricinodendron heudelotii, Field manual for extension workers and farmers. Southhampton, UK;   Southhampton   Center   for   Underutilized   crops, University of Southhampton.

[22] Lemmens, R.H.M.J, Louppe, D and Oteng-Amoaka, A.A. 2012 (eds.).Plant resources of tropical Africa 7 (2). Timbers 2. PROTA Foundation/CTA, Wageningen, Netherlands.

[23] Tchoundjeu, Z., and Atangana, A.R. (2006). *Ricinodendron heudelotii (Baill.)* Southampton Centre for Underutilized Crops. University of Southampton: Southampton, UK; 74p.

[24] Ngobo, M.P., Weise, S.F., Macdonald, M.A. 2003. Non-wood forest products in short duration fallow lands of Southern Cameroon. Paper presented at the XII world forestry congress, 2003, Quebec, Canada.

[25] Fondoun,   J.M.   Tiki   Manga,   T.   Kengue,   J.   1999. *Ricinodendron heudelotii* (Djansang): ethnobotany and importance for forest dwellers in southern Cameroon. *Plant Genetic Resources Newsletter* 118: 1–6

[26] Jell, B. 1998. Utilisation des produits secondaires par les Baka et les Bangando dans la region de Lobeke au Sud Est Cameroun. Etude de cas. Report submitted to GTZ on Projet de conservation des forets naturelles au sud-est Cameroun (PROFORNAT). September 1998.

[27] Nkwatoh, A.F., Labode, P., Ebobenow, J., Nkwatoh, F.W., Ndumbe, N.L., and Ewane, M.E. (2011). Gathering Processing and Marketing of *Ricinodendron species* (Bail) in the humid forest zone of Cameroon. *Agric. Sci. Res. Journal 1(9).* 213 – 221.

[28] Mialoundama, F. 1980. Corrélations intervenant dans la croissance des rameaux chez *Gnetum africanum* Welw. Compte Rendus des Sceances Academie Scientifique serie D, 291, 509-512.

[29] Shiembo,   P.N.,   Newton,   A.C.,   Leakey,   R.R.B.   1996. Vegetative propagation of *Gnetum africanum* Welw., a leafy vegetable from West Africa. *Journal of Horticultural Science* 71(1): 149–55.

[30] Mialoundama, F. 1993. Nutritional and socio-economic value of *Gnetum* leaves in Central African forest. In: *Tropical forests, people and food: biocultural interactions and applications to development*. Carnforth, UK: Parthenon Publishing Group.

[31] Shiembo, P.N. (1997). Domestication of *Gnetum* spp.by vegetative propagation techniques. In R.R.Schippers and L. Budd (eds.) *African indigenous vegetables*. Workshop proceedings. January 13–18, 1997, Limbe, Cameroon.IPGRI and NRI.

[32] Ndumbe L., Ingram V., and Awono A., 2009. Baseline study on *Gnetum* spp. in the South West and Littoral Regions of Cameroon, edited by CIFOR. Yaoundé, Cameroon: FAO CIFOR-SNV-World Agro-forestry Center- COMIFAC.

[33] Shiembo, P.N. 1999. The sustainability of eru (*Gnetum africanum* and *G. buchholzianum*): an overexploited non-wood forest product from the forests of Central Africa. In: Sunderland, T.C.H. Clark, L.E., Vantomme, P (eds.) *.Non-wood forest products of Central Africa: current research issues and prospects for conservation and development*. Rome: Food and Agriculture Organization, pp. 61–66.

[34] Fondoun, J. M., Tiki-Manga, T. 2000. Farmers indigenous practices for conserving *Garcinia kola* and *Gnetum africanum* in Southern Cameroon." *Agroforestry Systems* 48: 289-302.

[35] Lakeman Fraser, P., Bachman, S. 2008. *Gnetum africanum*. In: IUCN 2011. IUCN Red List of Threatened Species. Version 2011.1. http://www.iucnredlist.org. Accessed 20 February 2014.

[36] Baloch, E. 2009. *Gnetum buchholzianum*. In: IUCN 2011. IUCN Red List of Threatened Species. Version 2011.1. http://www.iucnredlist.org. Accessed 20 February 2014

[37] Awono A., Manirakiza D. and Ingram V., 2009. Étude de base de la filière *Gnetum* spp. (*Fumbwa*) dans les Provinces de L'Équateur et de Kinshasa, RDC. FAO-CIFOR-SNVWorld Agro-forestry Center-COMIFAC, February 2009. Web site: http://www.fao.org/forestry/ enterprises/45716/en/.

[38] Nkwatoh, A.F., Labode, P.,Iyassa, S.M., Nkwatoh, F.W., Ndumbe, N.L., Ewane M.E. 2010. Harvesting and marketing of *Gnetum* species (Engl) in Cameroon and Nigeria. *Journal of Ecology and the Natural Environment* 2(9) 187-193.

[39] Jiofack, T., C. Fokunang, V. Kemeuze, E. Fongnzossie, N. Tsabang, R. Nkuinkeu, P. M. Mapongmetsem and B. A. Nkongmeneck (2008). "Ethnobotany and phytopharmacopoea of the South-West ethnoecological region of Cameroon." Journal of Medicinal Plants Research 2(8): 197-206.

[40] CIFOR (2008). *Gnetum* spp (Okok or Eru) Fact Sheet. CIFOR. Yaounde, CIFOR: 2

[41] Schippers, R.R. (2000). *African indigenous vegetables: an overview of the cultivated species*. U.K.: Natural Resources Institute/ACP/EU Technical Centre for Agriculture and Rural Cooperation.

[42] Toirambe B., 2002. Utilisation des feuilles de *Gnetum* spp. Dans la lutte contre la pauvreté et l'insécurité alimentaire dans le Bassin du Congo, cas de la RDC. Kinshasa, FAO: 32.

[43] Ayuk, E.T., Duguma B., Franzel, S., Kengue, J., Mollet, M., Tiki-Manga, T., Zenkeng, P. 1999. *Forest Ecology and Management* 113. 1-9.

[44] ICRAF, 1999. Diversification of Tree crops: Domestication of companion crops for poverty reduction and environmental services.

[45] Egbe, E.A., Tabot,P.T., Fonge, B.A. 2012. Ethnobotany and prioritization of some selected tree species in south-western Cameroon. *Ethnobotany Research and Applications* 10:235-246

[46] Asare, R. 2005. Cocoa agroforests in West Africa. A look at activities on preferred trees in the farming system. *Forest and Landscape*. Working Paper No. 6, 2005.

[47] Leaky, R.R.B. 1998. Agroforestry in the humid lowlands of West Africa: some reflections on future directions for research. *Agroforestry Systems* 40: 253–262.

[48] Leakey, R. R. B., Tomich, T.P. 1999. Domestication of Tropical Trees: from Biology to Economics and Policy. Agroforestry In: *Sustainable Agricultural Systems*. L. E. Buck, J. P. Lassoie and E. C. M. Fernandes. Nairobi, ICRAF: 319-338.

[49] Adeola, A. O. 1995. The process of multipurpose tree species prioritization for agroforestry research. Forestry and the small scale farmer, Kaduna, Kaduna State, Nigeria, Forestry Association of Nigeria

[50] Tieguhong, J.C. and Zwolinski, J. (2008). Unrevealed economic benefits from the forests in Cameroon; IUFRO 4.05.00: emerging needs of society from forest ecosystems, University of Ljubljana, Slovenia, May 22-24, 2008

[51] Ekobo, A. 1995. Conservation of the African forest elephant (*Loxodonta africana cyclotis*) in Lobéké, Southeast Cameroon. PhD. Thesis, University of Kent, 151.

[52] WWF, 2006. Plan d'Amenagement du Parc National de Lobeke et de sa zone peripherique. Période d'exécution : 2006 – 2010. WWF.

[53] Wong, J.L.G. 2001. The biometrics of Non-Timber Forest Product resource assessment: A review of current methodology.

[54] Alene, D .C., Djieto-Lordon, C., Burckhardt, D., Messi, J., 2008. Population dynamics of the Ricinodendron psyllid, *Diclidophlebia xuani*, and its predators in Southern Cameroon. *Mitteilungen Der Schweizerischen Entomologischen Gesellschaft* 1 (2) 87-103

[55] Sunderland, T. C. H. and P. Tchouto (1999). A Participatory Survey and Inventory of Timber and Non-Timber Forest Products of the Mokoko River Forest Reserve, SW Province, Cameroon.

[56] Shiembo, P. (1994). Domestication of multipurpose tropical plants with particular reference to *Irvingia gabonensis* Baill., *Ricinodendron heudelotii* (Baill.) Pierre ex Pax and *Gnetum africanum* Welw. Edinburgh.

[57] Burkill, H.M. (2000). The useful plants of west tropical Africa, vol 5. London: Royal Botanic Gardens, Kew.

[58] Fongzossie, F.E., Ngansop, T.M., Zapfack, L., Kameuze, V.A., Sonwa, D.J., Nguenang, G.M., Nkongmeneck, B.A. 2014. Density and natural regeneration potential of selected non-timber forest products species in the semi deciduous rainforest of southeastern Cameroon. African study monographs, suppl. 49:69-90

[59] van Dijk, J.F.W. 1999. An assessment of non-wood forest product resources for the development of sustainable commercial extraction. In T.C.H. Sunderland, L.E. Clark and P. Vantomme (eds.). Non-wood forest products of Central Africa: current research issues and prospects for conservation and development. Rome: Food and Agriculture Organisation, pp. 37–50.

[60] Musoko, M., Last, F.T. and Mason, A. (1994). Populations of spores of vesicular-arbuscular mycorrhizal fungi in undisturbed soils of secondary semideciduous moist tropical forest in Cameroon. *Forest Ecology and Management* 63 (2–3): 359–77.

[61] ICUC, 2004. Fruits for the future. Ndjanssang. International Centre for Underutilized Crops. Factsheet No. 10. April 2004. 1-2pp.

[62] Moutsamboté, J.M. 1994. Végétation et plantes alimentaires de la région de la Sangha (Nord-Congo). In L.J.G. van der Maesen, X.M. van der Burgt & J.M. van Medenbach de Rooy (eds.). *The biodiversity of African plants:proceedings of XIVth AETFAT Congress.* Kluwer Academic Publishers, pp. 754-756.

[63] Halle, F., Oldemann, R.A.A. and Tomlinson, P.B. (1978). Tropical trees and forests. An architectural analysis. Springer-Verlag Berlin Heidelberg, New York, 242-244,441.

[64] Nkefor, J., Ndam, N., Blackmore, P , Engange, F. and Monono, C. (2000). Transfer of eru *(Gnetum africanum* Welw. and *G. buchholzianum* Engl.*)* domestication model to village-based farmers on and around Mount Cameroon. (Unpublished.) Report for CARPE.

[65] Sunderland, T.C.H.(2001). *Cross River State Community Forest Project: non-timber forest products advisor report.* Department for International Development/Environmental Resources Management, Scott Wilson Kirkpatrick & Co Ltd.

[66] Tieguhong, J.C. 2008. Ecotourism for sustainable development: Economic valuation of recreational potentials of protected areas in the Congo Basin. Unpublished PhD Thesis. University of KwaZulu Natal.

[67] Tieguhong J.C. and Ndoye O. 2006. Transforming subsistence products to propellers of sustainable rural development: Non-timber forest products (NTFPs) production and trade in Cameroon. *Africa-Escaping the Primary Commodities Dilemma. African Development Perspective Yearbook Vol. 11.* Unit 1. VERLAG Berlin. Pp. 107-137. ISBN 3-8258-7842-2

28

Economic Assessment of New Herbicides Used to Fight the Weeds in Wheat

Vasko Nikolov Koprivlenski[1], Maya Dincheva Dimitrova[2], Ivan Stoyanov Jalnov[2], Ilian Dimitrov Zheliazkov[2], Plamen Ivanov Zorovski[2]

[1]Department of Management end Marketing, Faculty of Economics, Agricultural University, Plovdiv, Bulgaria
[2]Department of Foundation of Agriculture, Faculty of Agronomy, Agricultural University, Plovdiv, Bulgaria

Email address:

koprivlenski@au-plovdiv.bg (V. Koprivlenski)

Abstract: Within the period 2011-2014, at the experimental base of the Agricultural University – Plovdiv we conducted field experiments with new herbicides in wheat. Based on the obtained results, we made a summarized economic assessment of the chemical fight against weeds. We established the amount of production expenses needed for growing wheat in 10 tested variants including 9 treated with herbicidal preparations. It was found the critical levels of the yield for each variant as well as the factors determining them. The economic effectiveness of production has been analyzed using a system of various indicators: the value of the permanent and variable costs per unit of area; the level of the average yield; the value of the total revenue per unit of area; the amount of the profit from 1 ha; the prime cost of the production and the profitability rate. The highest average yield was obtained from the variants treated with herbicides: Axial 1 plus 050 EK (5180.60 kg/ha); Sekator OD (5200.00 kg/ha) and Pasifica VG (5210, 46 kg/ha). Upon the application of these herbicides, the rate of increase of the additional production from unit of area exceeds the rates of the investments made. This makes the use of these herbicides economically substantiated. With the exception of the untreated control sample, all other herbicides give good economic results from the production of wheat, a sufficient profit rate from unit of area and profitability ranging from 20% to 22%.

Keywords: Wheat, Herbicides, Economic

1. Introduction

The high level of weed presence directly affects the growth, the development and the yield obtained from grain crops. This necessitates exercising control over weeds by using various selective herbicides for the purpose of restricting their negative influence on the production results.

Scientist from Pakistan (Mushtaq Ali et al., 2004 and Ashiq H. Sangi et al. 2012) have tested the influence of various herbicidal preparations on the final economic indices related to the production of wheat, comparing them to the untreated control sample. The results obtained from their experiments show that the use of herbicides increases the number of grains per one ear and the yield increases from 13,98% to 32,08%. As a result, the economic indices – net returns from 1 ha and profit rate significantly exceed those of the control sample. Similar results are presented by Cheema, M. and Akhtar, M., 2005; Qazi, M. A. and al.. 2002; Ahmad, S., and al. 1995;

Ashiq H. S., and al. 2012; Jabbar, A., and al., 1999; Khan, M. and N. Haq, 1994; Sharar, M.S.,and al., 1994.

M. Sayili, and al. (2006) have made a comparative analysis regarding the use of herbicides in wheat under production conditions. They have studied 78 farms around Turkey, 70 of which use herbicides and 8 do not apply herbicidal preparations. The results from the comparative analysis show that the average yield of wheat treated with herbicides increases by 29,09% in case of irrigation and 10,53% if no irrigation has been applied, compared to the untreated control sample. When conducting irrigation, the amount of the profit from unit area for the farmers using herbicides is 181,47 dollars per ha and those not using herbicides obtain 28,06 dollars/ha, which is 6.5 times more. The profit threshold (the critical average yield) for which the wheat production ends without any profits or losses is 244,4 kg/ha for the irrigated wheat and 176,40 kg/ha for the non-irrigated wheat. Similar studies regarding the effectiveness of the chemical fight against weeds in wheat have also been presented by Obst, A.,1981 and van Heems, T., 1985.

The area of the fields sown with wheat in Bulgaria varies from 1.1 million ha to 1.2 million ha. Every year, the fight against weeds is conducted using different herbicides (Aleksiev A. and al., 2003; Tityanov M. and al., 2009) but there is insufficient data about the economic effectiveness of the new herbicidal preparations.

The purpose of this study was to make aggregated economic assessment of the experimental results of the application of the herbicides to control weeds in wheat production. Implementation of the set objective passes through solving the following four interrelated tasks:
- Creation of Methodological Tools for Development and Assessment of Technological and Economic Estimates for the Test Culture;
- Accurate amount of the cost of production elements and technological units;

- Splitting the cost of fixed and variable based on carefully selected evaluation criteria.
- Determination of the critical level of the average yield and economic efficiency of the production of wheat.

2. Materials and Methods

2.1. Conducting the Experiment

Within the period 2011-2014 in the experimental field of the Agricultural University, Plovdiv, we made field experiments using new herbicides applied to the leaves during the vegetation period of the wheat, variety Diamand. The sowing during the three years of the experiment was performed on October 20-30. The experiments were made using the block method over an area of 21 m^2 in four repetitions (As shown in Table 1).

Table 1. Variants of experiments.

Variants	Active substans	Dose
1. Derby super	150,2 g/kg florasulam + 300,5 g/kg aminopiralid	33 g/ha
2. Arat	500 g/kg dicamba + 250 g/kg tritosulfuron	100 g/ha
3. Laren 20 CG	200 g/kg metsulfuron-methyl	30 g/ha
4. Secator OD	106 g/l amidosulfuron + 25 g/l iodosulfuron	100 cm^3/ha
5. Lintur 70 WG	41 g/kg triasulfuron + 659 g/kg dicamba	150 g/ha
6. Axial 1 plus 050 EK	45 g/l pinoxaden + 5 g/l florasulam	1000 cm^3/ha
7. Axial 050 EK	50 g/l pinoxaden	600 cm^3/ha
8. Pallas 75 WG	75 g/kg piroxulam	200 g/ha
9. Pacifica WG	30 g/kg mezosulfuron + 10 g/kg iodosulfuron	350 g/ha
10. Control	untreated	

The agrotechnical activities were conducted in accordance with the commonly used technology for wheat (processing of the soil, fertilization, sowing, rolling). The herbicides were applied using a knapsack sprayer and a solution of 300-400 l/ha.

2.2. Methodological Tools for Development and Assessment of the Technological and Economic Estimates

The amount of the production costs for growing wheat has been calculated on the grounds of a technological chart containing information about the physical volume and value of all necessary material and labour costs for growing and gathering the crop. The expenses for raw materials and other materials (seeds, fertilizers, preparations, fuel and lubricants, electricity and others) have been calculated on the grounds of their volume and unit price, on average for the period 2011-2013. The expenses for labour, mechanized activities (ploughing, cultivating, harrowing, transporting, fertilizing, spraying, gathering and others), additional costs, machine and tractor and transport expenses have been calculated on the grounds of the estimates valid for the Agricultural University – Plovdiv.

The value of the finished produce – the maize grain, has

been calculated based on the average price for the period of the experiment 2011-2013.

2.3. Splitting the Cost of Fixed and Variable Based on Carefully Selected Evaluation Criteria

The fixed costs are notable for the following major characteristics:
- The manager can not change them in a short run;
- They do not depend on the size of production;
- With increasing the size of the produce they reduce in a unit of it;
- In a particular production it is difficult to calculate the fixed costs, because of which they are called yet indirect and are usually allocated as total economic costs.

The fixed costs include: the depreciation cost, costs of interest payments, maintenance charges as well as the overhaul, tax, tariff and insurance expenses.

The fixed costs include also the land rent, preliminarily specified rents of using the fixed assets, rents for different services and the labor cost (after signing the labor agreements).

The variable costs are investments that are directly used and transformed into final produce. Therefore, they are known as direct production cost. The main criteria to determine these costs as variable ones are:

- They can be managed or their size depends directly on the manager's decision;
- They directly influence the amount of the produce or are incurred only in case of carrying out production (herbicides, ets.);
- Their a per unit value remains relatively constant quantity under unchanged production conditions;
- They can be easily related to the production of one particular product (wheat, etc.)

2.4. Methodological Tools for Determining the Critical Levels of the Yield and the Economic Thresholds of Efficiency in Variants of the Experiment

The determination of the critical levels of the yield and the thresholds of efficiency in the production of wheat has been made using various indices shown in tables 3 and 4 (Koprivlenski, 2011).

The thresholds of efficiency, expressed through the critical level of the average yield for the separate variants, have been calculated using the formula:

$$Q_{BEP} = \frac{FC}{p - VC_1}$$

where:

Q_{BEP} – the critical level of the average yield;
FC – total value of the stable costs;
p – average costs of the implementation;
VC_1 – variable costs for unit of production.

The critical level of the average yield determines the amount of the yield for which the wheat production results in no profits or losses, after which each kilogram of the finished produce above the critical yield shall secure profits.

3. Results and Discussion

The amount and the structure of the production costs for growing wheat vary within a relatively narrow range for all the variants included in the experiment (As shown in Table 2).

***Table 2.** Production costs for growing wheat, lv/ha.*

Variants	Average yield, kg/ha	Material costs /A/	Expenses for tractor and transport /B/	Expenses for manual labour /C/	Direct costs I /A+B+C/	Additional costs II	Total costs I+II
1	5082	678.1	366.4	7.0	1051.5	218.7	1270.2
2	4828	674.9	366.4	7.0	1048.3	218.2	1266.5
3	4665	674.9	36.6.4	7.1	1048.4	218.1	1266.5
4	5200	679.9	366.4	7.1	1053.4	219.1	1272.5
5	4935	677.9	366.4	7.0	1051.3	218.7	1270.0
6	5186	751.9	366.4	7.2	1125.5	231.3	1356.8
7	4067	698.9	366.4	7.0	1072.3	222.3	1294.6
8	5153	696.4	366.4	7.0	1069.8	221.9	1291.7
9	5214	727.9	366.4	7.0	1101.3	227.2	1328.5
10. Control	3522	663.9	366.4	6.6	1036.9	216.3	1253.2

* The direct and indirect costs have been calculated based on the estimates for the Agricultural University

The most important factor determining this is the price of the used preparations and the doses applied to the separate variants. In relation to this, the direct production costs have a greater influence as the difference between the untreated control sample and the most material-consuming variant 6 is 8.54%. Other factors affecting the amount of the costs include the selected technological variant and the volume of the finished product. As the volume of the finished product increases, the costs related to machines, tractors and transport also rise.

The drafted technological and economic calculations show that the variants with the highest level of production costs are those treated with the herbicides Axial 1 plus 050 EK and Pasifica WG. The total amount of material costs together with the additional charges for the two variants is 751.9 lv/ha and 727.9 lv/ha, respectively. For the other variants, these values are 8-12% lower.

The structure of the general production costs is typical of the crops with a high level of mechanization of the production processes, such as wheat.

The highest average yield is obtained from the variants treated with the herbicides Axial 1 plus 050 EK (5186.0 kg/ha), Secator OD (5200.0 kg/ha) and Pacifica WG (5214.0 kg/ha). The largest increase of the general production costs – 6.0% is registered between the control sample and the variant treated with the herbicide Pacifica WG. These two variants show the greatest difference in yield, reaching 48%, which indicates that the growth rate of the additional production from unit area outstrips the investments made (As shown in Table 2).

The main results from the economic assessment of the tested herbicides used in the fight against weeds in wheat have been summarized in table 3.

The amount of the production costs and the level of the average yield are the main factors affecting the economic effectiveness in the production of wheat between the separate variants of the experiment. The data in tables 2 and 3 indicates that the control sample has the lowest economic effectiveness.

Table 3. Economic efficiency and critical levels of the average yield in the production of wheat.

Variants	Average yield, kg/ha	Critical level of the average yield, kg/ha	Value of the total produce, lv/ha	Net profit from 1 dka, lv/kg	Prime cost of the produce, lv/ha	Profitability rate, %
1	5082	2430.0	1524.6	0.25	254.4	20.03
2	4828	2727.5	1448.4	0.2.6	181.9	14.36
3	4665	2726.2	1399.5	0.27	133.0	10.50
4	5200	2191.0	1560.0	0.24	287.5	22.59
5	4935	2430.0	1480.5	0.25	210.5	16.57
6	5186	2891.2	1555.8	0.26	199.0	14.67
7	4067	5557.5	1220.1	0.31	- 74.5	- 5.75
8	5153	2465.6	1545.9	0.25	254.2	19.68
9	5214	2524.4	1564.2	0.25	235.7	17.74
10. Control	3522	21630.0	1056.6	0.36	-196.6	-15.69

The low level of the yield from the untreated control sample as well as the comparatively high production costs are the reasons for the unsatisfactory value of the total production obtained from unit area – 1056.6 lv/ha (As shown in Table 2). With these levels, the overall production cannot provide net returns, in which case the production of this variant will lead to a loss and the profitability rate will be a negative value. A similar trend can be observed with variant 7 (Axial 050 EK), where the average yield is low and the level of the investments per unit area has been increased. The reason for this is the spectrum of activity of the herbicide, which destroys the annual monocotyledon weeds and the latter are not dominant among the wheat plants.

The cost price of production for variants 7 and 10 exceeds the purchase price of wheat and leads to negative economic indices for the aforementioned variants. For the remaining variants, the high yield provides normal returns of the investments and sufficient amount of the profit from 1 ha.

The highest economic effectiveness of the production was registered for variant 4 (Secator OD - 22,5%), variant 1 (Derby super - 20,0%) profitability and variant 8 (Pallas 75 WG - 19,68%).

4. Conclusion

The use of herbicides in the fight against weeds is economically expedient considering the fact that they provide outpacing growth rates of the overall revenue from the production. In other words, the use of herbicides is justified when they cause a faster rate of increase in additional revenue, resulting in total revenues are higher or at least equal to variable costs.

For a short period of time to cover fixed costs can be "delayed" because in the short term they are not affected by the volume of production. Furthermore, we found out that the value of fixed costs per unit of output fell by any excess product. Therefore, for the fixed costs, in the short term, growth in the volume of production is always justified.

Therefore, the criterion for assessing the appropriateness of wheat production is the ratio between total revenue and total variable costs. The difference between them must be positive or zero, to justify the production of wheat.

All the variants treated with herbicides provide high economic results in the production of wheat, sufficient amount of the profit from unit area and profitability rate ranging from 10, 50 to 22,59%. The critical level of the yield from these variants varies from 42 to 58% of the production actually obtained from unit area.

The specified thresholds of effectiveness for all variants treated with herbicides provide high competitiveness and profitability of the crops, which allows for normal reproduction.

Acknowledgements

This study was funded from the Research Fund of Ministry of Education and Science, Bulgaria: Project number DDVU 02/82 "Agro-ecological Assessment of New chemical Products for Plant Protection in Modern Agriculture".

References

[1] Ahmad S, Sarwar M, Tanveerand A, Khaliq A, (1995). Efficacy of some weedicides in controlling P. minor Retz. In: Wheat. Proc. 4th All Pakistan WeedSci. Corf., Faisalabad. 26-27 March, 1994, 89-94.

[2] Aleksiev A., Tonev T, Tityanov M, (2003). Economik Analysis of the Effekt of the Chemical Weed Control in Wheat Crops. Agricultural Economiks and Management, 4, 76-83.

[3] Sangi AH, Aslam M. Javed S, Khalid L, (2012). Efficacy and Economics of Mixing Different Herbicides for Controlling Broad and Narrow Leaved Weeds in Wheat. J. Agric. Res., 50 (1), 79-87.

[4] Ashraf MY, Baig NA, Khan MA, (1989). Chemical weed control in wheat (Triticum aestivum L). Pakistan J. Agri., Agri. Engg. and Vet. Sci., 5, 21- 4.

[5] Cheema MS, Akhtar M, (2005). Efficacy of different post emergence herbicides and their application methods in controlling weeds in wheat. Pak. J. Weed Sci. Res.,11(9-12), 23-29.

[6] Jabbar A, Saeed M, Ghaffar A, (1999). Agro-chemical weed management in wheat. Pakistan J. Agri. Sci., 36, 33-8.

[7] Khan M, Haq N, (1994). Effect of post emergence herbicides on weed control and wheat yield. J. Agric. Res., 32, 253-9.

[8] Koprivlenski V, (2011). Production costs and thresholds of efficiency in plant-growing, Academic publishing house of the Agricultural University, Plovdiv (Bg), 17 - 36.

[9] Sayili M, Akca H, Önen H, (2006). Economic analysis of herbicide usage in wheat fields. Journal of Plant Diseases and Protection, 755-760.

[10] Mushtaq A, Sabir S, Din QM, Ali MA, (2004). Efficacy and Economics of Different Herbicides Against Narrow Leaved Weeds in Wheat. International Journal of Agriculture & Biology, 647-651.

[11] Mushtaq A, Shahzad S, Din QM, Ali MA, 2004. Efficacy and economics of different herbicides against narrow leaved weeds in wheat. Int. J. Agri. Biol. 6(4): 647-651.

[12] Obst A, (1981). Chemical control in intensive wheat cultivation. In: Kommedahl, T. (ed): Proceedings of Symposia IX International Congress of Plant Protection, 5-11 August 1979, Entomological Society of America, 448-451.

[13] Qazi, MA, Ullah AS, Ali A, (2002). Weed management-hand weeding vs chemical weed control in wheat. Balochistan J. Agric. Sci., 2, 39-42.

[14] Sharar MS, Sharif M, Shah SH, Tanveer A, (1994). Efficacy of some herbicides in controlling weeds in wheat (Triticum aestivum L.). Proc. 4th All Pakistan Weed Sci. Conf., Faisalabad. 26-27 March, 18.

[15] Tityanov M, Tonev T, Mitkov A, (2009). New Opportunities for Efficient Chemical Control of Weeds in Wheat. Plant Science, vol. XLVI, 154-160.

[16] Tmo, Statistical data (www.tmo.gov.tr), 2005

[17] Heemst DJ, (1985). The influence of weed competition on crop yield. Agricultural Systems, 18, 81-93.

29

Effectiveness of Communication Strategies used in Creating Awareness and Uptake of Food Quality and Safety Standards in the Informal Market Outlets of Camel *Suusa* and *Nyirinyiri*

Madete S. K. Pauline[1], Bebe O. Bockline[2], Matofari W. Joseph[1], Muliro S. Patrick[1]

[1]Department of Dairy and Food Science and Technology, Egerton University, Nakuru, Kenya
[2]Department of Animal Sciences, Egerton University, Nakuru, Kenya

Email address:
mdtpauline@yahoo.com (M. Pauline)

Abstract: The *Nyirinyiri* and *Suusa* products from camel meat and milk processed by pastoral women using indigenous knowledge and traded in the informal markets presents opportunities to enhance household food security and income and also health benefits to consumers. However, safety and quality concerns by consumers are market barriers, especially acceptability beyond the traditional camel eating communities and in urban niche markets. It is possible to break this market barrier with effective communication of the food safety and quality standards but there exist knowledge gaps on the extent to which use of seminars and trainings, media briefs, radios, television and manuals increase awareness and uptake of the food standards and benefits to actors in the informal food markets. This study therefore identified the effectiveness of communication strategies used in promoting awareness and uptake of food quality and safety standards in the informal market outlet. Survey, Focus Group Discussion and Participatory appraisal of actors along the value chain were the methods used in data collection. The results showed that communication strategies in place were meant for the formal market hence the camel *Suusa* and *Nyirinyiri* chain actors gave the perceived effectiveness of the communication strategies if they were to be for the informal market outlet for promote acceptance and access for *Suusa* and *Nyirinyiri* in the high value markets.

Keywords: Food Quality and Safety, Informal Markets, Communication, Consumer Concern

1. Introduction

Safer food can generate both health and wealth for the poor. Wanyoike et al., (2008) and Kaitibei et al. (2010) noted that products such as *Suusa* and *Nyirinyiri* do not exploit higher market value due to their inability to meet safety and quality standards. Food quality and safety are the totality of characteristics of the food products that bear on their ability to satisfy all legal, customer and consumer requirements (Will and Guenther, 2007). Quality includes all product attributes that influence its value to consumers, whereas safety includes all measures intended to protect human health (Nelson, 2005).

Women play an important role in activities dealing with livestock management, transformation and marketing (IFAD, 2009). Pastoral women's human capital includes their knowledge and skills on animal health and husbandry, livestock management, natural resource management and environmental conservation, much of which is unrecognized both by outsiders and sometimes within their own societies. Their human capital may however be limited by their lack of knowledge of the market economy and their rights under national law. In spite therefore of their indigenous knowledge, strong capacities and resilience, pastoral women begin life with less human capital than their male counterparts and rarely get the opportunity to redress this balance (Watson, 2010). In Kenya food quality and safety standards are generated by two main regulators; Kenya Bureau of Standards (KEBS) and Kenya Dairy Board (KDB). Thereafter, the generated standards are disseminated using different communication strategies to the formal and informal market outlets.

The information shared by the regulators are on hygienic and sanitation, functional food properties, packaging, health attributes to consumers, shelf life, aroma, flavour, texture (FAO/ WHO, 2005). Although food safety and quality standards information is available from various sources, awareness is likely to be low because the information dissemination is not demand driven. The general packaging and medium of communication are unlikely to be appropriate to the needs of the users in the informal market because of barriers with access to the channels of choice by the regulators. There is however knowledge gaps about the extent to which these information sharing strategies can influence awareness and uptake and benefits of food quality and safety actions in the informal markets where pastoral women process and trade Suusa and Nyirinyiri.

2. Methodology

This study used survey, Focus Group Discussions (FGD) and participatory rapid appraisal approaches to sample actors along the camel *Suusa* and *Nyirinyiri* value chains in Isiolo, Marsabit and Nairobi (Eastleigh) where consumption of camel products is predominant in Kenya. Sampling targeted the value chain regulators, supporters and operators to identify communication strategies which they use in creating awareness on food safety and quality standards. Data collected was subjected to cross tabulation, Chi-square test statistics and

post hoc Anova.

3. Results and Discussion

Survey results could not identify any communication strategy specifically targeted to promoting awareness and uptake of food safety and quality standards in the informal market outlets of camel *Suusa* and *Nyirinyiri*. The communication strategies mentioned in Figure 5 are best suited to communicating formal food quality and safety standards as part of risk communication in formal market outlets. This calls for an in-depth knowledge of target audiences, which include their level of awareness and understanding of food safety issues; their attitudes to food in general and food safety in particular; the possible impact of communications on behaviour; and the appropriate channels for effective dissemination of messages (FAO/WHO, 2002). This is because the camel Suusa and Nyirinyiri is produced using indigenous knowledge.

Table 1 highlights the magnitude of effectiveness of communication strategies as rated by value chain actors of camel *Suusa* and *Nyirinyiri* The rating suggest that all actors consider mass media equally effective but they differ on the effectives of print and electronic media and direct methods. These points to mass media as being an acceptable communication strategy among all actors for communicating food safety and quality standards.

Table 1. Mean ± SD rating (1=High; 5=low) of effectiveness of communication strategies in the informal market outlet of camel Suusa and Nyirinyiri.

Chain actors	Sample (n)	Mass media	Print media	Electronic media	Direct methods
Producers	12	1.58±0.79	4.17±1.12[b]	4.17±1.11[ab]	1.50±0.80[ab]
Processors	10	1.80±0.78	2.60±1.17[a]	4.40±0.52[b]	2.40±1.27[b]
Transporters	8	2.13±1.60	3.25±1.17[ab]	4.25±0.71[b]	1.13±0.35[a]
Marketers	10	2.90±1.25	3.40±0.52[ab]	3.20±0.63[a]	1.70±0.68[ab]
Consumers	30	2.37±1.89	2.17±1.29[a]	3.70±0.95[ab]	2.63±1.19[b]
F		2.276	7.492	3.409	5.540
Sig		0.070	0.00	0.014	0.001

Table 2 summarizes the proportional responses by actors when actors asked of which communication strategy would (Table II) be the most effective in promoting uptake of food quality and safety standards. Differences were noticeable among producers and consumers (p< 0.05) but not among processors, transporters or marketers (p >0.05). Among

producers, majority (66.7%) indicated direct method was the most effective strategy while mass media was the most effective strategy for majority (56.7%) of the consumers. Results points to need for use of diverse communication strategies to reach wider audience of the actors, for they express different preferences.

Table 2. Most effective communication strategies to promote awareness and uptake of food safety and quality standards among actors of the informal market outlet (% response).

Communication strategies	Producers	Processors	Transporters	Marketers	Consumer
Mass media	25	40	25	80	56.7
Print media	8.3	10	12.5	0.0	0.0
Electronic media	0.0	0.0	0.0	0.0	6.7
Direct methods	66.7	50	62.5	20	36.7
Sample size (n)	12	10	8	10	30
Chi- square value	6.500	2.600	3.2500	3.600	11.400
P value	0.039*	0.273	0.197	0.058	0.003*

Figure 1 elaborates the effectiveness of communication strategies per the respondent feedback (n=70) indicating that about a third (35.7%) of the actors found mass media to be more effective in communication of food quality and safety

standards in the informal market outlet. Direct methods had about three quarters (75.7%) of the respondents citing it as most effective compared to print media and electronic media (5.7%) respectively, hence, direct methods having the overall effect as the most effective communication strategy for promoting awareness and uptake of food quality and safety standards in the informal market outlet of camel *Suusa* and *Nyirinyiri*.

Actors along the camel *Suusa* and *Nyirinyiri* value chain preferred direct methods (Figure 1) because it allows direct communication between actors and information service providers. Direct methods are most effective among the actors as it is easy to seek their audience. Direct methods alone are not enough in emphasizing on adoption of food quality and safety standards in the informal market outlet yet, in order to reach high value market producers and processors are under high pressure to meet the required standards with an aim of building consumer (pastoralist and non- pastoralist) confidence and ensure safety yet they lack relevant and crucial information on food safety and quality standards (APCAEM, 2008). This calls for strengthening of other communication strategies by the regulators and information service providers in creating awareness and uptake of food quality and safety standards.

Mass media was more effective among camel *Suusa* and *Nyirinyiri* owing to the fact that radios are in many households and are portable. Mass media (radio) remains the most powerful, favorite and yet cheapest tool because information is conveyed in the simplest form yet the information is for the formal market. Mass media comes in handy as the pastoral women processors can listen to it even while doing other household chores if broadcast is in local language.

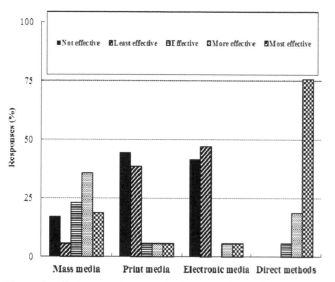

Figure 1. *Effectiveness of communication strategies used in promoting awareness and uptake of food quality and safety standards in the informal market outlet of camel Suusa and Nyirinyiri.*

Electronic media includes the use of ICT and the actors have found mobile phones very important in getting information on market prices and customer information, delivery and payment of goods. The analysis implies that the use of ICT is yet to be embraced in communicating food quality and safety standards in the informal market of Suusa and Nyirinyiri. The ICT industry is rapidly growing hence

the need to embrace technology and promote production of safe and quality food for consumer and pastoral women benefit. Electronic media is being used by pastoral women to communicate with herders, transporters, cooperatives and customers in Nairobi. This is done to ensure that the product reaches the market in good time. Regulators thus need to incorporate the use of mobile phones in communicating food quality and safety standards in the informal market of camel Suusa and Nyirinyiri. Use of radio among the pastoralists is not common as they move up and down and don't have time to listen to it. In the informal market of Suusa and Nyirinyiri use of radio is not very effective as many of them don't listen to radio due to communication barrier and limited frequency of disseminating information.

4. Conclusion and Recommendation

The effectiveness of communication strategies varied among the chain actors with direct methods being most preferred followed by mass media. Print media and electronic media ranked least effective communication strategy among the chain actors as it was not preferred because of complex language and being non informative. Appropriate ways of strengthening communication of food quality and safety standards are known but policy remains a key issue because the standards in place are for international markets and not national and informal markets. Policy review was majorly suggested by regulators so as to develop standards for informal market outlet, use of vernacular FM radios and increased frequency of information on food quality and safety standards were the major options given by the actors of *Suusa* and *Nyirinyiri* value chain.

There is therefore need for reviewing formal communication strategies with potential to reaching more actors in the informal markets of Indigenous Knowledge Food Products (IKFP). With increased outreach there is a perceived benefit for both pastoral women processors and consumers. This calls for participatory risk analysis, risk management and risk communication so that the right standards are disseminated and linked to the most effective communication strategies with ICT adoption being key in creating awareness and promoting uptake of food quality and safety standards in the informal market outlets.

Acknowledgements

This work was supported by a grant from RUFORUM which the authors are grateful to. The authors would also like to thank Egerton University and Pastoral women in ASAL areas of Isiolo and Marsabit Kenya.

References

[1]   APCAEM. (2008). Food safety issues in Agriculture trade, policy brief, issue no.3. www.unapcaem.org/publication/epolicybrief.pdf. Retrieved on 12/08/2012.

[2]   FAO/WHO. (2002). Post- harvest technology and food quality. Part 6.Science and technology for sustainable development FAO/WHO. Joint FAO/WHO Food Standards Programme Report of the Sixteenth Session of FAO/WHO Coordinating Committee for Africa, Codex Alimentarius Commission FAO. Rome. 25 – 28 January 2005.

[3]   FAO/WHO. (2005). Assuring food safety and quality. Guidelines For strengthening national control.system.http://www.who.int/foodsafety/publications/capacity/en/Englsih_Guidelines_Food_control.Retrieved 0n 26/11/2012.

[4]   IFAD. (2009). Rota, A. and Sperandini, S. "Value chains, linking producers to the markets", in Livestock Thematic Papers: Tools for project design. Rome: International Fund for Agricultural Development (IFAD).

[5]   Kaitibie S., Wanyoike F. , Kuria S. ,Brustel A. , Thendue I. N., Mwangi D. M., Omore A. (2010). Consumers' preference and willingness to pay for improved quality and safety: Case of fresh camel milk and camel meat (nyirinyiri) in Kenya.ILRI Research report No 24.

[6]   KEBS. (2009). The Benchmark. The official magazine of Kenya Bureau of Standards. www.kebs.org. Retrieved on 03/03/2013

[7]   Nelson M. B. (2005). International Rules, Food Safety and the Poor Developing Country Livestock Producer, Pro-Poor Livestock Policy Initiative Working Paper No. 5. FAO, Rome. 20th July 2005.

[8]   Wanyoike F., Kaitibei S., Omore A. (2008). Can small scale actors earn higher returns from improved quality and safety animal products? www.ilri.org/publication.Retrieved on 12/11/2012.

[9]   Watson Cathy. (2010). Gender issues and pastoral economic growth and economic development in Ethiopia. Pgs. 2-4

[10]   Will M and D Guenther. (2007). Food Quality and Safety Standards as required by EU Law and the Private Industry with special Reference to MEDA Countries' Exports.

30

Economic Assessment of Integrated Fish Farming (Fish-Rice-Piggery) in Sierra Leone

Olapade Olufemi Julius[1, *], Alimamy Turay[1], Momoh Rashid Raymond[2]

[1]Department of Aquaculture and Fisheries Management - School of Forestry and Horticulture, Njala University, Njala, Sierra Leone
[2]Department of Extension and Rural Sociology School of Agriculture, Njala University, Njala, Sierra Leone

Email address:

fem66@hotmail.com (O. O. Julius), ojulius@njala.edu.sl (O. O. Julius), 6532rashid@gmail.com (M. R. Raymond),
raymondmomoh2006@yahoo.com (M. R. Raymond)

Abstract: The present study evaluates the profitability and environmental effect of integrated fish cum rice and piggery production at Njala University, Sierra Leone. The research carried out between June and November, 2014 consists of the pigsty (2.5m × 11m), a maggoty and integrated pond (395.2m^2) sown with 0.94kg NERICA 19 rice at spacing of 20cm inter - rows and 5cm intra - rows. The pond was stocked with *Clarias gariepinus* juveniles (mean weight 25.6± 1.78g) at the density of 4.8fish per m^2. Water quality parameters viz., temperature, pH, Dissolved oxygen, NO_3-N, NH_3/NH_4, hardness and alkalinity were determined in the pond and were found to be within the recommended range for the culture of tropical fish. Economic analysis of the adaptive research gave a negative incremental benefit and Net Present Value (NPV) in both the first and second year of production. The sensitivity analysis evaluation shows that the enterprise is fairly sensitive to price fluctuation (-46.02%) and highly sensitive to survival rate of the fish and the scale of production of the pigs. Increasing scale of production and selling at market price will greatly enhance profitability and short term payback of costs.

Keywords: Profitability, Integrated Aquaculture, Water Quality, Njala University, Sierra Leone

1. Introduction

Hunger and malnutrition are among the foremost problems facing the majority of the world`s poor and needy, and this continue to dominate the health of the world`s poorest nations [1]. About 30% of humanity, including infants, children, adults and elderly within the developing world, are known to be currently suffering from one or more of the multiple forms of malnutrition [1]. Aquaculture (the farming of fish, other aquatic animals and plants) is considered as one of the world`s global food production systems that can effectively be used to tackle the problems of malnutrition and poverty particularly in the developing countries [2]. Coastal and inland fisheries are stagnating or declining in the sub-region, which presents a real concern in terms of fish supply and food security. The development of aquaculture appears as a possible solution for this growing supply gap in the future. Integrated aquaculture which link aquaculture to conventional farming systems will be the most appropriate weapon to fight hunger and poverty in sub Saharan Africa. The development of such systems has been driven by different needs in different parts of the world, including a desire to improve food security on small, subsistence family farms; or to minimize pollution and use valuable resources (such as water) more efficiently and effectively. Integrated agriculture aquaculture system (IAAS) practices were established long ago in many Asian countries for subsistence purposes, but are increasingly being developed for more commercial, income generating purposes in both Asia and developed "Western" countries. The practice is however quite a new way of farming in Africa. The principle behind integrated agri-aquaculture business is the maximal use of all available land and water resources. Integrated pig/fish/rice production is a promising way of making the most of a small farmer`s land and labour. Economic studies have shown that traditional production system is wasteful and unprofitable due to poor feed conversion, high mortality rates, low reproductive rates and final products [3]. It is now established that integrated fish cum livestock farming is a good strategy that can be adopted by small scale farmers in developing countries to boost farm yield and returns. Similarly, integrated farming is a way of ensuring stability in the production processes spreading the risk of production over several activities. Since livestock and fish production are not usually characterized by co-variant risks, the farmer is able to stabilize inter temporal flow of daily income. Moreover, integrated farming maintains environmental friendliness by facilitating productive use and recycling of wastes [4].The multi-stage utilization of feedstuffs and fertilizer makes it possible to supply the community with more produce and to increase the income for the fish farm as well. The importance of integrated livestock/piggery, rice and fish production system has begun to be more appreciated. This present study seeks to evaluate and document information on the interplay of components of the integrated aquaculture system especially in relation to returns on investment, sustainability issues, and utilization of resources within the system and the impact of the research innovation on the environment.

2. Materials and Methods

2.1. Description of the Study Area

The research was conducted at Njala University fish farm in Moyamba district, southern province, Sierra Leone (Fig. 1). Sierra Leone is located on the west coast of Africa, between the 7th and 10th parallels north of the equator. The country is bordered by Guinea to the north and northeast, Liberia to the south and southeast, and the Atlantic Ocean to the west. The country has total area of 71,740km^2 (27,699m^2) divided into a land area of 71,620km^2 (27,653m^2). The country has four distinct geographical regions; coastal Guinean mangroves, the wooded hill country, an upland plateau, and the eastern mountain. Eastern Sierra Leone is an interior region of large plateaus interspersed with high mountains, where mount Bintumani rises to 1,948 m (6,391 ft) [5]. The climate of Njala is mainly tropical and has distinct dry and rainy seasons. Daily mean temperature ranges from 21^{0}Cto 23^{0}Cfor the greater part of the dry season. The vegetation consists of farm bush, grassland and inland valley swamps.

Figure 1. Map of Sierra Leone Showing location of Njala University inMoyamba.

2.2. Experimental Design

The experiment lasted six months (June - November, 2014). The adaptive research platform consisted of the pigsty (2.5m × 11m in dimension) partitioned into three units with wastes discharge channels (Plate 1) and a maggoty made of bush stick and corrugated iron sheet (Plate 2). The integrated pond has a surface area of 395.2m^2 which was divided into fish refuge (208.2m^2) and rice paddy platform 10m × 18.7m (187m^2)(Plate 3).

Plate 1. The pigsty and the piggery (feeding time).

Plate 2. The maggoty.

Plate 3. The paddy field.

2.3. Cultural Practices

Lowland NERICCA 19 rice (0.94kg) was nursed for transplanting into the rice paddy at a spacing of 20cm inter rows and 5cm intra rows. The rice field was fertilized with organic manure at recommended rate (150kg fresh manure/ha/day). This was done twice in a month to avoid the problem of overloading that could lead to water chemical imbalance. Water in the integrated pond was kept low at the level of the paddy to prevent the young seedlings from logging by simply lowering the stand pipe at the outlet. Weed was manually removed twice a month (rouging) (Plate 4).

Plate 4. a and b. Cultural practices in the paddy field.

2.4. Piggery Production

One in-gilt and eight weaners were housed in the pigsty (4 weaners per unit while the in-gilt was isolated). The animals were washed every three days and fed to satiation once a day. Random weighing was done every month. The pig feed consists of palm kernel cake, salt, fish scraps, rice bran, corn meal and a little addition bone meal. The animals were also fed regularly with leafy vegetables and fruits such as *Ipomoea reptans* (gogodi) *Alternanthera braziliana*, pawpaw leaves and fruits (this contain piperazine which serves as dewormer). Mango fruits were also given to them when available. Wounds sustained through wall rubbing were treated with Gentian Violet and disinfection was carried out with Dettol and Izal.

2.5. Fish Culture

Clarias gariepinus juveniles mean weight 25.6g obtained from Magbosie village flood plains was stocked at a density of 4.8fish per m². The fish were initially acclimated for three days without food so as to adjust them to taking imported expanded pellets (45% crude protein). The fish were fed at 5% of body weight and feeding was adjusted every month to attune to weight gain. Feeding was supplemented with maggots generated from the poultry manure every one week. Extruded sinking pellets were made locally to reduce the high cost of imported diet which would make it impossible for local farmers to adopt the innovation. The culture period lasted six months.

2.6. Maggot Production

Wastes generated in the poultry house was divided and bagged in polyethylene with the second portion poured into plastic bowls as shown in the plate 5. The wastes were thereafter wetted with water and layered with handful of rice bran for ease of maggot production and were left in the open in the maggoty for the blow flies to have easy access to them. Big brown colored larvae of the blow flies which are rich in protein are produced in sufficient quantities after seven days. These are sieved, weighed and poured into the pond for consumption by the fish. Digested wastes left after maggots have been removed are used as fertilizer for the rice paddy and also for pond productivity.

Plate 5. Festering Maggots.

2.7. Water Quality Assessment

Water quality parameters measured bi-weekly include: Dissolved oxygen, water temperature, pH, water hardness, water alkalinity, ammonia, nitrate, BOD and nitrite. These parameters were determined using Jenway analytical probes and Pondlab multi-parameter kits. Water samples were collected at three different points - Point A (water inlet point); Point B (mid - pond close to the paddy platform) and Point C (outlet point). Three samples were collected in cleaned 600ml water bottles per sampling points. The water samples were preserved with one percent concentrated HNO_3 and kept in the refrigerator at 4°C prior to analysis in the laboratory.

2.8. Statistical Analysis

Measure of central tendency (mean and standard deviation) and one-way analysis of variance (ANOVA) at P=0.05 were used to analyze the results of the water quality variables measured in the adaptive research pond. Significant differences were separated using the least significant difference (LSD).

2.9. Economic Analysis

Estimation of production costs, and gross revenues, cash flow, sensitivity analysis of the project, gross merging and benefit – cost ratio were evaluated using the formulae below.

Incremental benefit = Revenue – Costs

Discounted costs = Discount factor (15%) multiply by Costs

Discounted revenue = Discount factor (15%) multiply by Revenue

Net Present Value at 15% = Discounted revenue – Discounted costs

Discounted factor at 15% was calculated as follows

$$\text{Year } 1 = 1 \div (1 + 15\%)^1$$

$$\text{Year } 2 = 1 \div (1 + 15\%)^2$$

$$\text{Year } 3 = 1 \div (1 + 15\%)^3$$

Sensitivity Analysis of the project was estimated as =
$$\frac{Total\ NPV\ (3\ years) * 100}{Total\ Discounted\ Costs\ (3\ years)}$$

Net profit = (Gross revenue) – Total operating costs)

Gross profit margin = (Gross profit) ÷ (Gross revenue)

Payback period = (Total capital cost) ÷ (Net profit)

Benefit – Cost Ratio = Discounted Revenue ÷ Discounted Costs

3. Results

3.1. Physico-chemical Parameters

The results of the water quality parameters evaluated in this study are presented in Fig2, 3 and 4.Mean surface water temperature ranged from $28 \pm 0.15 - 28.8 \pm 0.51^0$C in the integrated pond. The highest and lowest mean water temperature values were obtained in June and November respectively and were not significantly different ($p < 0.05$). pH value of 6.2 ± 0.32 obtained in November was higher than 6.1 ± 0.26 recorded in October and 5.8 ± 0.17 obtained in July. The highest value of 6.73 ± 0.25 was recorded in the month of June while July recorded the lowest pH value.

Mean dissolved oxygen values of 6.70 ± 0.20 mg/L $- 7.09 \pm 0.18$mg/L were recorded for the study. The highest value was obtained in July while the least was obtained in September. These mean values were however not below the critical level of 4mg^l for fish culture in the tropics. BOD values obtained throughout the sampling period ranged from 3.07 ± 0.12 mg/L -3.93 ± 0.06 mg/L. These values showed no significant difference ($p<0.05$) with locations and months of sampling. Alkalinity and general hardness values were constant throughout the period of study (35 mg/L and 54 mg/L respectively). Nitrate – Nitrogen and Ammonia/Ammonium (NH_3/NH_4) was not detected in the ambient water of the integrated pond which is an indication of low organic loading.

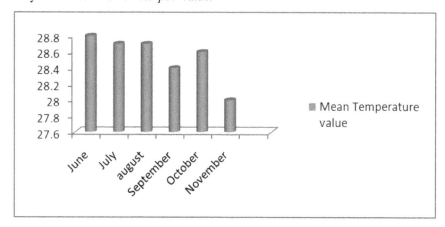

Figure 2. *Mean temperature recorded for the study.*

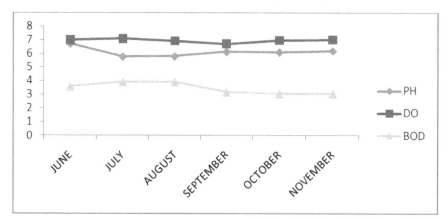

Figure 3. *pH, DO and BOD recorded for the study.*

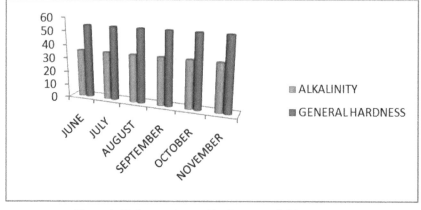

Figure 4. *Alkalinity and general hardness recorded for the study.*

3.2. Economic Performance of the Study

Table 1 presents the cost and revenue of the investment in the first year. The cash flow analysis presented in table 2. The Net Present Value (NPV) of the project is Le 4,231,351.52 (US$ 950.87) while the benefit/cost ratio is calculated as Discounted revenue/Discounted costs= 0.54. The assumption that a project is feasible when NPV is positive and benefit/cost ratio is greater than 1 holds for this projection.

3.3. Sensitivity Analysis (Uncertainty Determination)

This a measure for testing risk in investment feasibility especially the influence of unstable prices of inputs on costs of project. For this project the sensitivity is:

NPV at 15% discount rate = Le -4,231,351.52

Total Costs Present Values (CPV) at 15% discount rate = Le 9,195,431.52

$$Senstivity = \frac{-4231351.52 \times 100}{9195431.52} = -46.02\%$$

This analysis assumed that costs of this project are highly sensitive to price fluctuations and therefore the investment should be properly managed achieve a positive incremental benefit and Net Present Value (NPV).

Table 1. Budget – Costs and Revenues.

1st year of operation (1st production cycle)	
Fixed capital expenses	
1. construction of piggery house	Le 8,363,474
2. construction of maggoty house	Le 70,000
3. construction of fish pond	Le 2,221,234
4. Equipment	Le 1,280,000
Operating Expenses	
1. Stock	
Purchasing of pigs	Le 1,300,000
Fish seeds (1000 Clarias gariepinus)	Le 1,000,000
Rice seeds (NERICA 19)	Le 20,000
2. Labor	
Permanent and Hired	Le 2,400,000
3. Feed	
Piggery feeds	Le 2,500,000
Fish feed	Le 1,194,000
4. Others	
Chemical lime and disinfectants	Le50, 000
Drugs and disinfectants	Le 500,000
Summary of cost	
Fixed capital	Le 11,934,708
Operating expenses	Le 8,964,000
Total	Le 20,898,708
Revenue	
Rice	Le 70,000
Fish	Le 4,400,000
Pigs	Le 6,812,000
Total	Le 11,282,000

Table 2. Cash flow analysis for a years (1 production cycle).

Cycle	Costs	revenue	Incremental Benefit (1)	Discount Factor at 15% (2)	Net Present Value at 15% (3)	Discounted Costs (4)	Discounted Revenue (5)
1	20,898,708	11,282,000	-9,616,708	0.44	-4,231,351.52	9,195,431.52	4,964,080

4. Discussion

The physico–chemical parameters of the study site are within the recommended level for the survival of fish and other aquatic biota [6]. Physico-chemical parameters of water body serve as measure of water quality [7]. Changes in the intensity of rainfall are known to affects physico-chemical parameters of water [8]. Water quality in fish ponds is affected by the interactions of several chemical components. Carbon dioxide, pH, alkalinity and hardness are interrelated and can have profound effects on pond productivity, the level of stress and fish health, oxygen availability and the toxicity of ammonia as well as that of certain metals [9].The pH of the integrated fish pond with the exception of June that recorded 6.73 was below the recommended level for the survival and performance of tropical fish species and also below the range associated with most natural waters [10]. The low pH could be attributed to the component of the soil on which the pond was constructed. NO_3 – N and NH_3/NH_4 were not detected in the ambient water.

4.1. Economic Performance of the Study

The Net Present Value and the benefit/cost ratio calculated for the business does not conform to the assumption that a

project is feasible when NPV is positive and benefit/cost ratio is greater than 1. The NPV of this study is negative (-Le 4,231,351.52) and the benefit – cost ratio is below 1 (0.54). The sensitivity analysis of the study was negative (-46.02%). The negative NPV with the benefit – cost ratio of the enterprise that was below 1 is a clear pointer to the fact that the investment was not profitable in the first year due to so many factors. These factors include the high initial capital outlay especially in the construction of the pigsty and the paddy pond; the scale of production of both the piggery and the integrated pond which is assumed to be highly germane to the profitability of the investment. Other factors that require special attention in order for the investment to be profitable are physical factors; which include the environmental conditions of the farm area. There is the need to improve on the water quality parameters of the integrated fish pond especially the pH, dissolved oxygen, alkalinity and hardness. These factors could be limiting to the productivity of the pond. Yields form the investment can be determined by either of topographies (sources of water supply, water quality, type of soil and weather), types of fish pond, and farm sizes. Yields are known to be sensitive to physical factors, farm size, farm type and stocking rate. Besides the stocking rate, types of fish stocked also determine the returns on investment. In this study fish stocked were sourced from the wild since there was no single fish hatchery in Sierra Leone. Importing fish from abroad attract a lot of money and immigration restrictions. Consequently, the high mortality recorded as a result of the source of the fish grossly affected the revenue by reducing the overall incremental benefit of the investment. Survival rate determines the quantity produced at the end of the production period [11]. Increasing mortality rates leads to low survival rate and thus lower yields. Higher yields coupled with good prices are needed to increase revenue and thus profitability.

The average total cost of Le 20,898,708.00 (US$4,860.16) was expended in the first production cycle of the first year and gross revenue of Le 11,282,000 (US$ 2,623.72) was earned. This result explained the fact that integration of rice cum fish and piggery production is capital intensive and that the business will not recoup its cost under the circumstances in which it was carried out. Running a three years cash flow analyses and calculating the internal rate of return (IRR) of the investment will clearly throw light on the actual payback period which is presumed to be two year provided all the bottle necks are circumvented.

The findings of this study show disparity in pricing of the farm produce. In the capital city (Freetown), a pound of pork sells for Le 11,000 as against Le 7,000 that it was sold to the university community – a big gap of Le 4,000 and this could have contributed to up the revenue. In Nigeria, a kilogram of catfish sells for ₦ 500 (US$ 3.03) farm gate prices whereas it was sold at Le 10,000 (US$ 2.25) in Sierra Leone. Increasing the selling prices of both the pork and the fish to match current market price elsewhere will make the investment profitable and highly attractive to farmers. For the present study to break even there is the need to increase the stocking density of the fish to 9.8 fish/m^2. The rearing period of the pigs should equally be shortened to six months to accommodation two cycles of production for the pig per annum. This will increase the revenue base of the study without significantly affecting the variable costs especially when the final products are sold at the existing marketing price which was far higher than what was used in this study.

4.2. Conclusions

The economic benefit of integrated fish farming is enormous. It contributes immensely to the economic empowerment of families' especially in rural communities and enables the farmer to be productive all the year round provided the different components of the farming system are well utilized by the farmers to advantage. The findings of the study clearly showed that the venture will be profitable in the long run especially if the operating cost could be reduced through the supply of input locally. A range of public and private sector investments and initiatives are needed to realize the potential for the development of integrated fish farming especially in the area of supply of inputs such as fingerlings, feeds, seeds and other ancillaries. Public private partnerships offer potentially important opportunities for pro-poor agricultural development. Such collaborations have

already contributed to food security in many developing countries.

It was also discovered that the price at which the pork and table fish were sold is below what is obtained in the urban cities. Therefore, if the price of the fish could be increased from Le 10,000 (US$ 2.25) to Le 15,000 (US$ 3.37) and the pork to Le 10,000 (US$ 2.25) it will enhance quick profitability and early payback of invested cost and encourage farmers to adopt the innovation. To make the best out of the integrated fish farming investment, increasing the scale of production for pigs, rice and fish has been found to be related to increase profitability.

Acknowledgement

The authors greatly acknowledged CORAF/WECARD who through Poverty Eradication and Grassroots Empowerment through Sustainable Integrated Aquaculture Development: Fish cum rice and Piggery production (Project Number: 03PA11) supported this adaptive research at the Department of aquaculture and Fisheries Management Njala University, Sierra Leone.

References

[1]   World Health Organization, Malnutrition the global picture 2000 http://www.who.org.nut.welcome.htm

[2]   A.G.J. Tacon, A.G.J, Increasing the contribution of aquaculture for food security and poverty alleviation. In: R.P. Subashinghe, P. Bueno, M.J. Philips, C. Hough, S.E, McGladdery and J.R. Arthur, Eds. Aquaculture in the Third Millennium. Technical Proceedings of the conference on Aquaculture in the Third Millennium, Bangkok, Thailand, 20 – 25 February 2000. Pp. 63 – 72. NACA, Bangkok and FAO, Rome 2001.

[3]   A. Verhulst, Lessons from field experiences in the development of monogastric animal production. In: Mack, S. (Ed.), Strategies for sustainable animal agriculture in developing countries. Proceedings of the FAO expert consultation held in Rome, Italy, 10 – 14 December 1990. FAO Animal Production and Health Paper 107, 1993 pp. 261-271(http://www.fao.org/DOCREP/004/TO582E/TO582EOO.htm#TOC.

[4]   O. Amarasinghe, Some Economic Aspects of Integrated Livestock – Fish Farming in Sri Lanka. Integrated – Fish Production Systems: Proceedings of the FAO/IPT Workshop on Integrated Livestock – Fish Production Systems, 16 – 20 December 1991, Institute of Advanced Studies, Universities Malaya, Kuala Lumpur, Malaysia 1992 ISBN 983 – 9576 – 16 – X.

[5]   S. Levert, Sierra Leone (Cultures of the World). Cavendish Square Publishing, New York, USA. ISBN 13:978 – 0761423348, 2006 Pages: 144.

[6]   World Health Organization, www.who.int/water_sanitation.health/publicationns/facts2004/enindex.html. 2004 Downloaded February 2015.

[7] B.O. Offem, Y. Akegbejo – Samson, I.T. Omoniyi, and G.U. Ikpi, Dynamics of the limnological features and diversity of zooplankton populations of the Cross River System SE Nigeria. *Knowledge and Management of Aquatic Ecosystems*, 2008, 393, 2 – 19.

[8] A.A Adebisi, The physico-chemical hydrology of tropical seasonal river upper Ogun River. *Hydrobiologia*, 1981, 79, 157 – 165. http://dx.doi.org/10.1007/BF00006123.

[9] W.A. Wurts and R.M. Durborow, Interactions of pH, Carbon Dioxide, Alkalinity and Hardness in Fish Ponds. SRAC Publication No. 464 December 1992 4p Retrieved April 12, 2015.

[10] D. Chapman, Water Quality Assessment. A Guide to the use of Biota, Sediments and Water in Environmental Monitoring.1stEdn. Cambridge University Press, Cambridge, 1992 Pages 585.

[11] R.C. Engle, Aquaculture Economics and Financing: Management and Analysis. Blackwell Publication. Iowa, USA (Electronic version) 2010.

31

Effect of Soybean Varieties and Nitrogen Fertilizer Rates on Yield, Yield Components and Productivity of Associated Crops Under Maize/Soybean Intercropping at Mechara, Eastern Ethiopia

Wondimu Bekele[1, *], Ketema Belete[2], Tamado Tana[2]

[1]Oromia Agricultural Research Institute, Mechara Agricultural Research Center, West Hararghe Zone, Mechara, Ethiopia
[2]College of Agriculture and Environmental Science, Department of Plant Science, Haramaya University, Dire Dawa, Ethiopia

Email address:
wondubekele@gmail.com (W. Bekele)

Abstract: Due to decreasing land units and decline in soil fertility integrating soybean in to the maize production system is a viable option for increasing productivity and protein source. In view of this, field experiment was conducted during 2012 at Mechara Agricultural Research Center with theobjectives of identifying best compatible combinations of maize with soybean varieties and N rates for maximum yield and yield components of the associated cropsand productivity of intercropping system. Three varieties of soybean (Awasa-95, Cocker-240 and Crowford) were intercropped with early maturing maize variety Melkasa-2 with three rates of nitrogen (32, 64 and 96 kg N ha^{-1}). The experiment waslaid out in factorial arrangement in randomized complete block design in three replications. Highest maize grain yield (2196kg ha^{-1}) was obtained from soybean variety Crowford and 32 kg N ha^{-1} and lowest yield (1352 kg ha^{-1}) was recorded from maize intercropped with soybean variety Awasa-95 at 96 kg N ha^{-1}. The grain yield of intercropped soybean was increased from 586 kg ha^{-1} to 842kg ha^{-1} as the nitrogen rates increased from 32 kg N ha^{-1} to 96 kg N ha^{-1}. The higheist LER (1.10) was obtained from maize intercropped with soybean variety Crowford and lowest LER (1.08) was from maize intercropped with variety Cocker-240 due to main effects of soybean varieties while due to main effects of N, the highest (1.16) and the lowest (1.1) LER were obtained from higher rate of nitrogen (96 kg N ha^{-1}) and lowest rate of nitrogen (32 kg N ha^{-1}), respectively. On the other hand, the highest Gross Monetary Value (17315 Birr ha^{-1}) was recorded from interaction of Cocker-240 at highest rate of nitrogen (96kg N ha^{-1}) which was not significantly different from Awasa-95 at 32 kg N ha^{-1} (15304 birr ha^{-1}) and Crowford at 32 kg N ha^{-1} (15103) while lowest GMV (12362birr ha^{-1}) was obtained from variety Cocker-240 at 32 kg N ha^{-1}. Therefore, variety Awasa-95 at lower rate of nitrogen (32 kg ha^{-1}) could be best in intercropping system to reduce cost of fertilizer and maximize total productivity.

Keywords: Soybean,Intercropping, Land Equivalent Ratio,Gross Monetary Value

1. Introduction

Crop intensification is one of the strategies to increase productivity per unit area of land. For example, intercropping provides potential for the subsistence farmers who operate in low resources (inputs) situation. It is the practice of growing two or more crops simultaneously in the same field. Insurance against the vagaries of weather, disease and pests and higher productivity per unit area are the major reasons for the existence of intercropping. By growing more than one crop at a time in the same field, farmers maximize water use efficiency; maintain soil fertility, and minimize soil erosion, which are the serious drawbacks of monocropping (Francis, 1986).

In Ethiopia, as it is also true in most tropical countries, traditional cropping systems are based on resource poor farmers' subsistence requirements and are not necessarily the most efficient ones (Kidane *et al.*, 2010). Because of this, crop production per unit land area is usually below National average. Therefore, in diversified crop production systems having production constraints, diversified options need to be assessed.

In western Hararghe zone intercropping maize with sorghum, cereal with pulse, maize and Kchat is common. As most of people in Hararghe are based on cereal consumption, protein from pulse is very low. Evaluating the performance of soybean varieties for increasing of soil fertility under intercropping systems could help to maximize yield in the area.

Intercropping cereals and soybean is not anew practice and has been tried in a number of countries for example in Nigeria (Mueneke *et al.*, 2007), Canada (Carruther, *et al.*, 2000) and United States of America. In Africa, soybean is one of the leguminous crops selected for active research for instance, in Zimbabwe the maize yield was enhanced with soybean intercropping through nitrogen transferred from nitrogen fixing soybean to maize during crop development (Mudita *et al.*, 2008). Brophy and Hiechel (1989) reported that the soybean released 10.4% of symbiotically fixed N in to the root zone over its growth period. Martin *et al.* (1991) also reported that the elevated yield and protein level observed in maize and soybean intercrop may be a consequence of Nitrogen transfer from soybean to maize.

Now a day the cost of inorganic fertilizer is increasing and the resource poor farmers are forced to use below recommended rate or null. Therefore, technologies that will reduce N fertilizer input by resource-poor farmers in the area are urgently needed. Nitrogen input through biological N_2 fixation (BNF) by grain legumes can help to maintain soil N reserves as well as substitute for N fertilizer requirement for large crop yields. Different growth habit and maturity period soybean varieties have different nitrogen fixation ability. Since late maturing soybean varieties were able to fix more N2 than early and medium varieties, greater N contribution to any cropping system is expected through their roots, litter and harvest residues. Ogoke *et al.* (2003) reported that a positive N balance by soybean crop was reported due to the effect of increased crop duration and N application. Late maturing soybean varieties are, therefore, able to give higher N benefit compared to early and medium varieties for the improvement of the cropping systems. The objective of this experiment was to identify the appropriate combination of Soybean varieties and nitrogen fertilizer rates on yield, yield components and productivity of Soybean and Maize under intercropping at Mechara.

2. Material and Methods

An on station experiment was conducted for in the 2012/13 cropping season at Mechara Agricultural Research Center (MeARC) west Hararghe Zone, eastern Ethiopia. Three varieties of soybean namely; Awasa-95, Cocker-240 and Crowford and maize variety Melkesa-2 were used for the study. Awassa-95 is relatively early to intermediate maturing variety requiring around 120 days reaching physiological maturity depending on the temperature, altitude and moisture availability of the growing locations. It is suitable for production in intermediate rainfall areas. The areas receiving 500 mm rainfall in growing period is conducive for its production. Crowford is early maturing variety, determinate

growth habit and takes 90-120 days to reach physiological maturity. It grows on soils free from excessive rain fall and at altitude ranges from 1300-1700 m. Cocker-240 is a medium maturity class and indeterminate growth habit with a physiological maturity of 121-150 days. It best grows with altitude ranges 1300-1700 m and temperature 23-25°C (Mandafro *et al.*, 2009).The maize variety Melkasa-2 used in this experiment is an open pollinated variety recommended for moisture stress areas that receive annual rainfall of 600-1000 mm. It is early maturing variety that reaches physiological maturity in 130 days after emergence. It was released in 2004 by Melkasa Agricultural Research Center (MoA, 2011).

The three N levels (32 kg N/ha, 64 kg N/ha and 96 kg N/ha) used in maize/soybean intercropping were from DAP and urea. The sole maize received 46kg N/ha from urea and 18 kg N/ha from DAP. Sole soybean varieties received 18kg N/ha and 46 kg P_2O_5/ha from 100 kg DAP ha^{-1}.The rate used for the sole crops was as recommended for production of each crop.

The intercrop of maize and soybean were in 100% of the sole maize population and 53.3% of soybean population was intercropped as additive series between the two maize rows at the same time. Two seeds per hill of both maize and soybean were planted to ensure germination and good stand of the crops and were thinned to one plant per hill after emergence. The plot size for intercropping was 11.25m^2 (3.75 m width and 3m length). The plot size for sole maize was the same as the intercropped with the row spacing of 0.75 m and 0.25 m. Four rows of soybean were planted in maize rows in between plant spacing of 0.05 m. The plot size for sole soybean was 11.25 m^2 (3.75 m width and 3 m length) containing nine rows, 0.4m and 0.05m row spacing and spacing between plants, respectively. The yield data for experiment were collected from the net plot area of 4.5m^2 (2.25 m x 2 m) both for sole and intercropped. The design of the experiment was randomized complete block design in factorial arrangement in three replications.

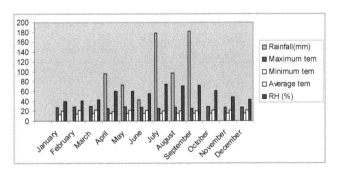

Figure 1. *Weather condition of experimental area during 2012 Source: MeARC weather station, 2012.*

2.1. Soil Condition of Experimental Area

Analysis of soil before planting was done for some physical and chemical properties of soil at Ziway Soil Laboratory of Oromia Agricultural Research Institute (Table 1). The analysis indicated that the soil had low levels of total nitrogen (0.172%) and medium organic matter (2.62%), medium level of available phosphorus (21.3ppm) and high CEC (30.32) as per the criteria developed by Murphy (1968) for Ethiopian soils and Landon (1984) for tropical soils. The pH of the soil was 5.82 showing moderately acidic nature of the soil (Tekalign, 1991). The textural class of the experimental site was silty clay soil.

Table 1. Selected physico-chemical properties of experimental soil.

Soil characteristic	Values
pH (1:2.5 H$_2$O)	5.82
Organic matter (%)	2.62
Total nitrogen (%)	0.172
Available phosphorus (ppm)	21.3
Cation exchange capacity (meq/100g)	30.32

Table 2. Total nitrogen of the experimental plot after harvest in response to the treatments.

Treatments	Total Nitrogen (%)
Maize +Awasa-95+ 32 kg N/ha	0.073
Maize +Awasa-95+ 64 kg N/ha	0.099
Maize +Awasa-95 + 96 kg N/ha	0.114
Maize +Crowford + 32 kg N/ha	0.071
Maize +Crowford + 64 kg N/ha	0.086
Maize +Crowford + 96 kg N/ha	0.099
Maize +Cocker-240 + 32kg N/ha	0.099
Maize +Cocker-240 + 64kg N/ha	0.097
Maize +Cocker-240 + 96kg N/ha	0.085
Sole maize	0.064
Sole cocker	0.099
Sole Crowford	0.085
Sole Awasa-95	0.099
Before planting	0.172

2.2. Post Harvest Soil Analysis

The soil analysis for the samples collected before planting and after harvesting revealed that there was variation in total nitrogen due to variation in cropping practice (Table 2). The soil analysis after harvesting showed that intercropping of maize and soybean resulted in increased total soil nitrogen than sole maize planting. However, all cropping systems reduced total nitrogen compared to total nitrogen of the site before planting (Table 2). The reason for the reduction of total nitrogen could be due to maize and soybean depleted soil nutrients extensively and most of the soil nitrogen was removed through grains and other plant parts of both crops. Other possible losses could be through denitrification, leaching, volatilization and/or their combination. Low soil pH and drought might have affected nodule development and efficiency that ultimately affected the amount of atmospheric nitrogen fixed by soybean.

2.3. Data Collected and Analysis

Data on maize yield components such as number of ears per plant, ear length, thousand kernel weight, grain yield and soybean number of pods per plant,100 seed weight (g), grain yield (kg ha^{-1}) and harvest index (%) were collected. The collected data were analyzed using GenStat Release 13.3 software (Genstat, 2010). Mean separation was carried out using Least Significant Difference (LSD) test at 5% probability level.

3. Results and Discussion

3.1. Maize Yield Components and Yield

Analysis of variance showed that maize stand count at harvest was not significantly affected by main effect of soybean varieties, nitrogen rates and interaction of main effects. Stand count of maize was significantly (p<0.05) affected by cropping system (Table 3). The mean number of stand count of sole cropped maize was higher (21.67/plot) than intercropped maize 20.11/plot (Table 3). The lower stand count in intercropped maize may be due to competition for the same resource with soybean or due to shortage of moisture during early vegetative growth. Similar to this result, Biruk (2007) reported reduction in stand count of intercropped sorghum with common bean varieties.

Table 3. Main effects of the intercropped soybean varieties and nitrogen rates on yield components of maize in maize and soybean intercropping.

Treatments	No. of stand count/plot at harvest	No. of ears per plant	Ear length(cm)	No. of kernels per ear	1000 kernels weight (g)	Harvest index (%)
Soybean varieties						
Awasa-95	20.22	1.044	11.80	396.6	219.1	33.9
Cocker-240	19.78	1.067	12.71	378.2	229.9	37.6
Crowford	20.33	1.133	12.64	398.1	220.8	34.3
LSD (5%)	NS	NS	NS	NS	NS	NS
Nitrogen rates (kgha^{-1})						
32	19.44	1.07	12.24	386.4	226.3	33.8
64	20.67	1.06	12.60	396.0	220.4	37.2
96	20.33	1.12	12.31	390.4	223.8	34.7
LSD (5%)	NS	NS	NS	NS	NS	NS
CV (%)	6.7	11.1	13.8	11.8	8.6	15.4
Cropping system						
Intercropping	20.11b	1.08b	12.39b	391.0	223.3b	35.3
Sole cropping	21.67a	1.26a	14.87a	428.0	294.0a	31.7
LSD (5%)	1.48	0.15	2.19	NS	51.97	NS
CV (%)	5.8	10.8	13.9	10.9	19.1	16.7

Means followed by the same letter(s) in the column are not significantly different at 5% level of significance
NS=not significant

Number of ears per plant, ear length and number of kernels per ear were not significantly affected by main effect of soybean varieties; nitrogen rates and interaction of main effects. However, cropping system had significant effect on number of ears per plant and ear length (Table 3). Sole cropped maize produced significantly more number of ears per plant (1.26) than intercropped 1.08 (Table 3). Similarly, significantly longer ear (14.89cm) was recorded due to sole cropping while shorter ear length (12.39cm) scored due to intercropped maize (Table 3). The reduction in number of ears per plant and ear length in intercropped maize might be due to the reduction in the ear leaf photosynthesis due to competition with soybean that lowers the number of ears per plant. Similar to this result, Wogayehu (2005) and Walelign (2008) reported lower number of ears per plant, ear length and number of kernels per ear of maize from intercropped maize with haricot bean varieties.

Analysis of variance showed that 1000 kernel weight, grain yield and harvest index were not significantly affected by main effect of soybean varieties and nitrogen rates, while grain yield was significantly influenced by interaction of main effects (Table 3 and figure 2). The highest grain yield (2196 kg ha^{-1}) was obtained from combination of maize intercropped with soybean variety Crowford and 32kg N ha^{-1} and the lowest maize grain yield (1352 kg ha^{-1}) was obtained from maize intercropped with soybean variety Awasa-95 at 96 kg N ha^{-1}. Higher grain yield of maize with soybean variety Crowford might be due to good nitrogen fixing abilities related to the higher number of nodules per plant as compared to other varieties, and early maturity of soybean variety Crowford. This might be because intercropping with early maturing legume could lead to increased productivity of the cereal (Rao, 1980).

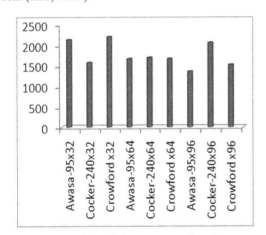

Figure 2. *Interaction effect of the intercropped soybean varieties and nitrogen rate on grain yield (kg ha^{-1}) of maize in maize and soybean intercropping.*

Grain yield of maize was significantly reduced by 31.7% due to intercropping. Similarly, Wandahwa *et al.* (2006) found that intercropping maize and soybean reduced the yield of maize probably because of competition for resources.

Even though the difference was not significant, higher harvest index (35.3%) was recorded from intercropped than sole cropped maize (31.7%) (Table 3). The higher harvest index from intercropped maize could be due to increasing in nitrogen rates in intercropping increased biomass of maize. Similar result was reported by Selamawit (2007) that higher harvest index was from intercropped maize with potato than sole cropped maize.

In this study a severe water stress during early growing period in June (43.3mm) and August (97.7mm) might have contributed for lower yield and yield components of maize.

3.2. Soybean Yield Components and Yield

The analysis of variance showed that number of pod per plant, 100 seed weight and harvest index were significantly (P<0.05) affected by soybean varieties. The highest number of pods per plant (32.22) was obtained from soybean variety Awasa-95 intercropped with maize while the lowest number of pod per plant (26.0) was obtained from variety Crowford (Table 4). Similar to this result, Thole (2007) reported that number of pods per plant of soybean intercropped with maize was significantly reduced by soybean varieties. Number of seeds per pod was not significantly affected by main effect of soybean varieties, nitrogen rates and their interaction and cropping system (Table 4). The number of 100 seed weight was significantly affected by main effect of varieties. The highest 100 seed weight (17.31g) was obtained from soybean variety Crowford intercropped with maize while significantly the lowest 100 seed weight (14.16g) was obtained from soybean variety Awasa-95 (Table 4). The highest 100 seed weight in variety Crowford could be due its larger seed size.

Cropping system significantly (P<0.05) affected 100 seed weight and grain yield. Higher 100 seed weight (15.95 g) was recorded from intercropping (Table 4). This could be the fact that the lower intra species competition between soybean as plant density was lower in intercropping than sole cropped soybean and higher seed weight was recorded from lower plant density (Turk *et al.*, 2003). Similar to this result, Wogayehu (2005) in maize/haricot bean, Biruk (2007), in sorghum/haricot bean and Egbe *et al.* (2010) from sorghum/soybean intercropping study reported higher 100 seed weight of legume components in intercropping than in sole crop.

Grain yield per hectare of soybean was significantly affected by main effect of soybean varieties and nitrogen rates (Table 4). Yield obtained from plot treated with 96 kg N ha^{-1} (842 kg ha^{-1}) was significantly higher than that of 32 kg N ha^{-1} (586 kg ha^{-1}). In this result, the yield of soybean was enhanced by increased level of nitrogen rates. The response of soybean to increase in grain yield might be the soil was deficit for nitrogen required by crop. In agreement with this result, Wandahwa *et al.* (2006) reported that the yield of soybean was increased due to increased nitrogen fertilizer in intercropped maize and soybean.

Cropping system highly significantly (P<0.01) affected the yield of soybean. Sole cropping gave significantly higher grain yield (1754 kg ha^{-1}) than intercropping 703 kg ha^{-1} (Table 4).

Lower grain yield of the intercropped soybean might be due to the competition effect exerted by maize component for limited growth factors in intercropping and lower stand count under intercropping. Pal *et al.* (2001) and Muoneke *et al.* (2007) reported similar yield reductions in soybean intercropped with maize and sorghum and associated the yield depression to interspecific competition and the depressive effect of the cereals.

Here in additive intercropping of maize and soybean, the intercropped soybean grain yield per hectare was reduced by 40% as compared sole cropped (Table 4). Comparably, Huxley and Maingu (1978), in cereals and legumes intercropping system, reported that the grain yield of the legume component declined, on average, by about 52% of the sole crop yield, whereas the cereal yield was reduced by only 11%.

Thus, the general observation in this study showed that yields of soybean component were significantly depressed by maize component in intercropping. This is most likely due to competition for soil nutrient and the reduction in transmitted photosynthetically active radiation to the soybean as a result of shading.

The harvest index of soybean was highly significantly (P<0.01) affected by soybean varieties. The highest harvest index (47%) was from soybean variety Crowford and the lowest harvest index was from Awasa-95 (32.7%) which was significantly not different from variety Cocker-240 (36.7%) (Table 4). The highest harvest index recorded for variety Crowford intercropped with maize might be due to the high grain yield to biomass obtained by the variety as a result of high partitioning of dry matter to the grain. Udealor (2002) and Ano (2005) reported that the differences in harvest index might be due to the inherent varietal characteristics, environmental factors and other cultural practices.

Table 4. Effect of intercropped soybean varieties and nitrogen rates on yield components and yield of soybean in maize and soybean intercropping.

Treatments	No. of pods per plant	No. of seeds per pod	100 seed weight (g)	Grain yield (kgha⁻¹)	Harvest index (%)
Soybean varieties					
Awasa -95	32.22a	2.49	14.16c	679	32.7b
Cocker-240	28.56abc	2.53	16.07ab	755	36.7b
Crowford	26.0c	2.35	17.31a	676	47.0a
LSD (0.05)	4.67	NS	1.25	NS	8.86
Nitrogen rates (kgha⁻¹)					
32	26.33	2.39	15.73	586b	36.1
64	29.67	2.40	15.87	681ab	39.0
96	30.78	2.59	16.23	842a	41.4
LSD (0.05)	NS	NS	NS	212.5	NS
CV (%)	16.2	10.2	7.9	30.2	22.8
Cropping system					
Intercropping	28.9	2.46	15.95a	703b	38.8
Sole cropping	31.8	2.54	14.52b	1754a	33.7
LSD(0.05)	NS	NS	1.21	169.4	NS
CV (%)	19.9	10	9.9	22.4	26.9

Means followed by the same letter(s) within column are not significantly different at 5% level of significance
NS=Not significant

Table 5. Effect of intercropped soybean varieties and nitrogen rates land equivalent ratio (LER) and gross monetary values (GMV) of maize and soybean intercropping.

Treatments	LER			MV		
Soybean varieties	Maize	Soybean	Total	Maize(Birr/Ha)	Soybean(Birr/ha)	GMV(Birr/ha)
Awasa-95	0.70	0.39	1.09	8906	4685	13591
Cocker-240	0.69	0.40	1.08	9213	5208	14421
Crowford	0.69	0.41	1.1	9318	4662	13980
LSD (0.05)	NS	NS	NS	NS	NS	NS
Nitrogen rates (kgha⁻¹)						
32	0.76	0.34	1.1	10210	4046	14256
64	0.65	0.39	1.04	8694	4697	13391
96	0.68	0.48	1.16	8533	5811	14344
LSD (0.05)	NS	NS	NS	NS	NS	NS
CV (%)	17.8	20.7	16.3	19.9	38.4	14.7
Cropping system						
Intercropping	0.69b	0.40b	1.09a	9146b	4850.7b	13996.7
Sole cropping	1.0a	1.0a	1.0a	13387a	12102.6a	-
LSD (5%)	0.17	0.089	NS	2878	1169	-
CV (%)	18.5	20.7	16.0	24	22.4	-

Means followed by the same letter(s) within column are not significantly different at 5% level of significance

3.3. Total Land Productivity and Gross Monetary Evaluation

Analysis of variance showed that partial LER of maize and soybean and total LER were not significantly (P<0.05) affected by the main effects of soybean varieties, N rates and their interaction. The highest total LER due to main effect of variety was obtained from soybean variety Crowford intercropped with maize (1.10) while the lowest value was obtained from Cocker (1.08). With respect to the main effect of N, the highest LER (1.16) was obtained due to application of 96 kg N ha^{-1} and the lowest LER (1.04) was obtained due to 64 kg N ha^{-1} (Table 5).

Productivity was improved in almost all intercrops as depicted by LER values greater than one (Table 5). The total land productivity ranged from 108% in Cocker-240 and maize to 110% Crowford and maize intercrop as compared to sole crops. This indicated that intercropping of maize and soybean was advantageous than sole planting of either maize or soybean. The result also indicated that the intercrops are more advantageous in efficiently utilizing land than the sole cropping of either maize or soybean and it would require 10% more land to get the same yield obtained from the intercropping system. This intercropping system resulted in the highest cumulative total yields than either of maize or soybean.

The LERs, greater than one in this experiment might have resulted from morphological differences of these two species and creating various niches for resources such as sun light, nutrients and moisture. The higher LERs in intercropping than mono-cropping were reported by Adeniyan and Ayola (2006), Bingcheng *et al.* (2008) and Javanmard *et al.* (2009).

Table 6. *Gross monetary value (Birr ha^{-1}) of maize and soybean under intercropping as influenced by interaction of soybean varieties and nitrogen rates.*

Nitrogen rates (kg ha^{-1})			
Soybean varieties	32	64	96
Awasa -95	15304ab	12922b-g	12548b-h
Cocker-240	12362b-i	13584b-e	17315a
Crowford	15103abc	13667bcd	13168b-f
Intercropped mean			13997a
Sole soybean			9146b
Sole maize			13387a
	Soybean varieties x N rates		Cropping system
LSD (0.05)	3565.3		1877
CV (%)	14.7		19.8

Means followed by the same letter(s) within column and row are not significantly different at 5% level of significance

Unlike main effects of soybean varieties and N rates, the analysis of variance showed that the interaction significantly (P<0.05) affected the GMV in the intercropping system. The highest GMV (17,315 ETB ha^{-1}) and the lowest (12,362 ETB ha^{-1}) were obtained from soybean variety Cocker-240 intercropped with maize at 96 kg N ha^{-1} and 32 kg N ha^{-1} respectively (Table 6). This finding was in agreement with the previous studies on maize-soybean intercropping by Raji

(2007), Thole (2007) and Gani (2012) who obtained higher monetary returns from intercropping maize and soybean as compared to sole maize.

4. Conclusion

Due to increasing population, decreasing land units and soil fertility, integrating legumes in to the cereal production system is a viable option in western Hararghe for food security. The statistical analysis revealed that maize yield components and yield were not significantly affected by main effects of varieties and nitrogen except grain yield which was affected by interaction of main effects. The highest maize grain yield (2196kg ha^{-1}) was for soybean variety Crowford at 32 kg N ha^{-1} and the lowest yield (1352 kg ha^{-1}) was recorded from maize intercropped with soybean variety Awasa-95 at 96 kg N ha^{-1}.The main effect of soybean variety significantly affected yield components of soybean such as number of pod per plant, 100 seed weight and harvest index while N and their interaction had no significant effect on yield and yield components of soybean except grain yield which was significantly influenced by main effect of nitrogen. The highest LER due to main effect of soybean varieties (1.10) was recorded from soybean variety Crowford intercropped with maize while the highest LER (1.16) due to main effect of nitrogen rates was recorded from the highest rate of nitrogen (96 kg N ha^{-1}) and the highest GMV (17315 Birr ha^{-1}) was obtained from Cocker-240 and 96 kg N ha^{-1} while the lowest GMV (12362 Birr ha^{-1}) was from Cocker-240 and 32 kg N ha^{-1}. Awasa-95at lowest rate of nitrogen (32 kg N ha^{-1}) which was not significantly different from Cocker-240 at highest rate of nitrogen (96 kg N ha^{-1}) could be better in intercropping system to maximize yield of both crops as well as total productivity.

Ackowledgements

The author thanks Ahmedziyad Abubaker and Wolansa Mokonin for their support in data collection and Oromia Agicultural Research Institute for financial support.

References

[1]	Ano A. O, 2005. Effect of soybean relayed in to yam minisett/maize intercrop on the yield of component crops and soil fertility of yam based system. *Nigeria Journal of. Soil Science*, 15: 20-25.

[2]	BirukTesfaye, 2007. Effects of Planting Density and Varieties of Common bean (*Phaseolusvulgaris* L.) Intercropped with Sorghum (*Sorghum bicolor* L.) on Performance of the Component Crops and Productivity of the System in South Gondar, Ethiopia. M.Sc. Thesis. Haramaya University.

[3]	Brophy, L. S and G. H. Heichel, 1989. Nitrogen Release from Roots of Alfalfa and Soybean in Intercrops. Direct[15]N Labeling Methods.

[4] Carruthers K, Prithiviraj BFQ, Cloutier D, Martin RC, Smith DL, 2000. Intercropping maize with soybean, lupin and forages: yield component responses. *European Journal of Agronomy,* 12: 103-115.

[5] Cotteinie, A, 1980. Soil and Plant Testing as a Base of Fertilizer Recommendations. Soils Bulletins, No. 38, FAO, Rome.

[6] Egbe, O. M, Alibo, S. E and Nwueze, I., 2010. Evaluation of Some Extra-Early- And Early-Maturing Cowpea Varieties for Intercropping With Maize in Southern Guinea Savanna of Nigeria. *Agriculture and Biology Journal of North America.* 1(5) 845-858.

[7] FAO, 2000. Fertilizers and Their Use 4th ed. International Fertilizer Industry Association, FAO, Rome, Italy.

[8] Francis, C. A., 1986. Multiple cropping systems. Vol. 1. Macmillan Publishing Co., New York.

[9] Fukai, S. and B. R. Trenbath, 1993. Presses of Determining Intercrop Productivity and Yield of Component Crops. *Field Crops Research,* 34: 247-271.

[10] Gani, O. K., 2012. Effect of phosphorus fertilizer application on the performance of maize/soybean intercrop in the southern Guinea savanna of Nigeria. *Archives of Agronomy and Soil Science,* 58:2, 189-198.

[11] GenStat, 2015. GenStat Release 15, VSN International Ltd.

[12] Ghosh, P. K., 2004. Growth yield competition and economics of groundnut/cereal fodder intercropping systems in the semi-arid tropics of India. *Field Crops Research,* 88: 227-237.

[13] Huxley, P. A. and Z. Maingu, 1978. Use of a systematic spacing design as an aid to the study of intercropping: Some general considerations. *Experimental Agriculture,* 14: 519-27.

[14] Kidane Georgis, 2010. Inventory of Adaptation Practices and Technologies of Ethiopia. Environment and natural resource working paper 38, FAO, Rome.

[15] Mandefro, N., Anteneh, G., Chimdo, A., and Abebe K., (Eds.), 2009. Improved technologies and resource management for Ethiopian Agriculture. A Training Manual. RCBP, MoARD, Addis Ababa, Ethiopia.

[16] Martin, R. C., Voldeng, H. C. and Smith, D. L. 1990. Intercropping corn and soybean in a cool temperate region: yield, protein and economic benefits. *Field Crops Research.* 23: 295–310.

[17] MoA (Ministery of Agriculture), 2011. Crop Variety Registry, Issue Number 14, Addis Ababa.

[18] Mudita, I.I., Chiduza, S.J. Richardson-Kageler and F.S. Murangu, 2008. Performance of Maize and Soybean Cultivars of Varying Growth Habit in Intercrop in Sub humid Environments' of Zimbabwe. *Journal of Agronomy,* 7(3): 227-237.

[19] Muoneke C.O, Ogwuche M.O, Kalu B.A., 2007. Effect of maize planting density on the performance of maize/soybean intercropping system in a guinea savanna agroecosystem. *Afr. J. Agric. Res.,* 2: 667-677.

[20] Murphy, H.F., 1968. A report on fertility status and other data on some soils of Ethiopia. Expt. Bull. No. 44. College of Agriculture, Haile Selasie I University, Alemaya, Ethiopia. 551p.

[21] Ogoke I.J., Carsky R.J., Togun A.O. & Dashiell K., 2003. Effect of P fertilizer application on N balance of soybean crop in the Guinea savanna of Nigeria. *Agriculture Ecosystem and Environment,* 100: 153-159.

[22] Olsen, S.R., C.V. Cole., F.S. Watanabe and L.A. Dean, 1954. Estimation of Available Phosphorus in Soils by Extraction with Sodium Bicarbonate. USDA Circular, 939: 1-19.

[23] Raji, J. A., 2007. Intercropping soybean and maize in a derived savanna ecology. *African Journal of Biotechnology.* 6 (16): 1885-1887.

[24] Selamawit Getachew., 2007.Effect of Plant Population and Nitrogen Fertilizer on Growth and Yield of Intercropped Potato (*Solanumtuberosum* L.) and Maize (*Zea mays L.)* at Haramaya, Eastern Ethiopia. M.Sc. Thesis, Haramaya University.

[25] Tekelign Tadesse, 1991. Soil, Plant, Water, Fertilizer, Animal Manure and Compost Analysis. Working Document No. 13. International livestock Research center for Africa (ILCA), Addis Ababa.

[26] Thole, A., 2007. Adaptability of soybean varieties to intercropping under leaf stripped and detasseled maize. MSc Thesis. University of Zimbabwe.

[27] Turk M.A., A.M. Tawaha and M.K.J. El-Shatnawi, 2003. Response of Lentil (*Lens culinaris*medic) to plant density, sowing date, phosphorus fertilization and ethephonapplication in the absence of moisture stress. *Journal of Agronomy and Crop Sciences,* 189: 1-6.

[28] Udealor A., 2002. Studies on the growth, yield, organic matter turnover and soil nutrient changes in cassava (*Manihotesculenta* Crantz)/vegetable cowpea (*Vignaunguiculata* L. Walp.) mixtures. Ph. D. Dissertation, University of Nigeria, Nsukka, Nigeria.

[29] Walelign Worku., 2008. Evaluation of haricot bean (*Phaseolus vulgaris* L) Genotypes of diverse growth under sole and intercropping with maize in southern Ethiopia. *Journal of Agronomy,* 7(4): 306-313.

[30] Walkley, A and I.A. Black, 1934. An Examination of Digestion of Degrjareff Method for Determining Soil Organic Matter and Proposed Modification of the Chromic Acid Titration Method. *Soil Science,* 37: 29-38.

[31] Wandahwa P, Tabu IM, Kendagor MK, Rota IA. 2006. Effect of intercropping and fertilizer type on growth and yield of soybeans. *Journal of Agronomy.* 5(1): 69–73.

[32] Willey, R.W., 1979. Intercropping-its importance and research needs. Competition and yield advantages. *Field Crops Research,* 32: 1-10.

[33] Wogayehu Worku., 2005. Evaluation of Common Bean (*Phaseolus vulgaris* L.) Varieties Intercropped with Maize (Zea mays L.) for Double Cropping at Alemaya and Hirna areas, Eastern Ethiopia. M.Sc. Thesis. Haramaya University.

Permissions

List of Contributors

Awal, Mohd Abdul
Environmental Scientist, Ministry of Environment and Forest, Health & Pollution Research Farm, Long Island City, New York, USA

Lechisa Takele and Achalu Chimdi
Wollega University, College of Natural and Computational Science Department of Soil Resource and Watershed Management, P.O. Box 395, Nekemte, Ethiopia

Alemayehu Abebaw
Ambo University, College of Natural and Computational Science Department of Chemistry, PO Box: 19, Ambo, Ethiopia

Yayeh Bitew, Fekremariam Asargew and Oumer Beshir
Adet Agricultural Research Centre, Amhara Agricultural Research Institute, Bahir Dare, Ethiopia

Aïcha Megherbi-Benali, Fawzia Toumi-Benali, Laid Hamel and Mohamed Benyahia
Ecodevelopment Spaces Laboratory, Sciences Environment Department Djilali Liabes University, Sidi Bel-Abbes, Algeria

Zoheir Mehdadi
Vegetal Biodiversity, Conservation and Enhancement Laboratory, Sciences Environment Department, Djilali Liabes University, Sidi Bel-Abbes Algeria

Yohannes Seyoum
Dry land Crop Research Department, Somali Region Pastoral and Agro-pastoral Research, Jijiga, Ethiopia

Firew Mekbib
School of Plant sciences, Haramaya University (HU), Dire Dewa, Ethiopia

Hanan Ibrahim Mudawi
The Environmental, Natural Resources and Desertification Research Institute, National Centre for Research, Ministry of Science and Technology, Khartoum, Sudan

Mohamed Osman Idris
Department of Plant Protection, College of Agriculture, Khartoum University, Khartoum, Sudan

Ermias Assefa, Addis Alemayehu and Teshom Mamo
Southern Agricultural Research Institute, Bonga Agricultural Research Center, Department of Crop Science Research Process, Bonga, Ethiopia

Gemechis Legesse Yadeta
Oromia Agricultural Research Institute, Holeta Bee Research Center (HBRC), Holeta, P. O. Box 22, Ethiopia

Efe Okere, Ebere Samuel Erondu and Nenibarini Zabbey
Department of Fisheries, Faculty of Agriculture, University of Port Harcourt, PMB 5323 Choba, Port Harcourt, Rivers State, Nigeria

Roland Nuhu Issaka and Moro Mohammed Buri
CSIR-Soil Fertility and Plant Nutrition Division, Soil Research Institute, Academy Post Office, Kwadaso, Ghana

Satoshi Nakamura and Satoshi Tobita
Crop Production and Environment Division, Japan International Research Center for Agricultural Sciences Ohwashi, Tsukuba, 305-8686, Japan

Lamia Lajili-Ghezal, Talel Stambouli and Marwa Weslati, Asma Souissi
ESA Mograne, 1121 Mograne, Tunisia

Benjamin Tetteh Anang
Department of Agricultural and Resource Economics, FACS, University for Development Studies, Tamale, Ghana

Joseph Amikuzuno
Department of Climate Change and Food Security, FACS, University for Development Studies, Tamale, Ghana

Tura Bareke Kifle, Kibebew Wakjira Hora and Admassu Addi Merti
Holeta Bee Research Centre, Oromia Agriculture Research Institute, Holeta, Ethiopia

Muluken Mekuyie Fenta
Animal nutritionist, Hawassa University, Wondo Genet College of Forestry and Natural Resources, Wondo Genet, Ethiopia

Halim
Specifications Weed Science, Department of Agrotechnology, Faculty of Agriculture, Halu Oleo University, Southeast Sulawesi, Indonesia

Fransiscus S. Rembon
Specifications Soil Nutrition, Department of Agrotechnology, Faculty of Agriculture, Halu Oleo University, Southeast Sulawesi, Indonesia

Aminuddin Mane Kandari
Specifications Agroclimatology, Department of Agrotechnology, Faculty of Agriculture, Halu Oleo University, Southeast Sulawesi Indonesia

Resman
Specifications Soil Science, Department of Agrotechnology, Faculty of Agriculture, Halu Oleo University, Southeast Sulawesi, Indonesia

Asrul Sani
Specifications Biomathematics, Department of Mathematics, Faculty of Sciences, Halu Oleo University, Southeast Sulawesi, Indonesia

Chuwang Pam Zang
Department of Crop Science, Faculty of Agriculture, University of Abuja, Nigeria

Yohannes Seyoum and Zelalem Fisseha
Dryland Crop Research Department, Somali Region Pastoral and Agro-pastoral Research Institute (SoRPARI), Jijiga, Ethiopia

Firew Mekbib
School of Plant sciences, Haramaya University (HU), Dire-Dewa, Ethiopia

Adefris Teklewold
Crop Research Directorate, Ethiopian Institute of Agriculture Research (EIAR), Addis Ababa, Ethioipia

Belayneh Admassu and Dawit Beyene
Holetta biotech Laboratory, Ethiopian Institute of Agriculture Research (EIAR), Holetta, Ethiopia

I. O. Faboya
Department of Forestry, Ministry of Environment Ekiti State, Ekiti State, Nigeria

S. I. Adebola and O. O. Awotoye
Institute of Ecology and Environmental Studies, Obafemi Awolowo University, Ile-Ife, Nigeria

Vinita Sharma and Alka Dubey
Zoological Survey of India, Northern Regional Centre, Dehradun, Uttarakhand, India

Apostolos Ainalis
Directorate of Coordination and Inspection of Forests, Decentralised Administration Macedonia-Thrace, 46th Agriculture School St, Thessaloniki

Ioannis Meliadis
Lab. of Remote sensing and GIS, Forest Research Institute, N.AG.RE.F., Vassilika, Thessaloniki

Konstantinos Tsiouvaras
Laboratory of Range Management (236), Dept. of Forest and Natural Environment, Aristotle University of Thessaloniki, Thessaloniki

Katerina Ainali
Lab. of Geoinformatics, Rural and Surveying Engineering, National Technical University of Athens, Zografou Campus, Iroon Polytechniou 9, Athens

Dimitrios Platis
Lab. of Ecology, Dept. of Agriculture, Aristotle University of Thessaloniki, Thessaloniki

Panagiotis Platis
Lab. of Range science, Forest Research Institute, N.AG. RE.F., Vassilika, Thessaloniki

Assefa Adane
Chemistry Department of Hawassa College of Teacher Education, Hawassa, Ethiopia

Heluf Gebrekidan and Kibebew Kibret
School of Natural Resources Management and Environmental Sciences of Haramaya University, Dire Dawa, Ethiopia

Ayodele O. J. and Shittu O. S.
Department of Crop, Soil and Environmental Sciences, Ekiti State University, Ado-Ekiti, Nigeria

Azmi Elhag Aydrous
Department of Agricultural Engineering, Faculty of Agriculture, Omdurman Islamic University, Omdurman, Sudan

Abdel Moneim Elamin Mohamed
Department of Agricultural Engineering, Faculty of Agriculture, University of Khartoum, Khartoum, Sudan

Hussein Mohammed Ahmed Abuzied
Department of Landscape and dry land cultivation, Faculty of Agriculture, Omdurman Islamic University, Omdurman, Sudan

Salah Abdel Rahman Salih and Mohamed Abdel Mahmoud Elsheik
Department of Agricultural Engineering, Faculty of Agriculture, Elneelain University, Khartoum, Sudan

Anderson Ndema and Edward Missanjo
Department of Forestry, Malawi College of Forestry and Wildlife, Dedza, Malawi

Khondokar Humayun Kabir and Debashis Roy
Department of Agricultural Extension Education, Bangladesh Agricultural University, Mymensingh, Bangladesh

Wachira P. M.
School of Biological Sciences, University of Nairobi, Nairobi, Kenya

Kimenju J. W.
Department of Plant Science and Crop Protection, University of Nairobi, Nairobi, Kenya

Otipa M.
Kenya Agricultural and Livestock Research Organization, Nairobi, Kenya

Roseline Gusua Caspa
Regional Postgraduate School of Integrated Tropical Forest and Landscape Management (ERAIFT), University of Kinshasa, P.O. Box 15373 Kinshasa, Democratic Republic of Congo

Isaac Roger Tchouamo
Department of Economics and Rural Sociology, University of Dschang, Dschang, Cameroon

Jean-Pierre Mate Mweru
Regional Postgraduate School of Integrated Tropical Forest and Landscape Management (ERAIFT), University of Kinshasa, P.O. Box 15373 Kinshasa, Democratic Republic of Congo

Joseph Mbang Amang
Institute of Agricultural Research for Development (IRAD), P.O. Box 2123 Yaounde, Cameroon

Vasko Nikolov Koprivlenski
Department of Management end Marketing, Faculty of Economics, Agricultural University, Plovdiv, Bulgaria

Maya Dincheva Dimitrova, Ivan Stoyanov Jalnov, Ilian Dimitrov Zheliazkov and Plamen Ivanov Zorovski
Department of Foundation of Agriculture, Faculty of Agronomy, Agricultural University, Plovdiv, Bulgaria

Madete S. K. Pauline, Matofari W. Joseph and Muliro S. Patrick
Department of Dairy and Food Science and Technology, Egerton University, Nakuru, Kenya

Bebe O. Bockline
Department of Animal Sciences, Egerton University, Nakuru, Kenya

Olapade Olufemi Julius and Alimamy Turay
Department of Aquaculture and Fisheries Management - School of Forestry and Horticulture, Njala University, Njala, Sierra Leone

Momoh Rashid Raymond
Department of Extension and Rural Sociology School of Agriculture, Njala University, Njala, Sierra Leone

Wondimu Bekele
Oromia Agricultural Research Institute, Mechara Agricultural Research Center, West Hararghe Zone, Mechara, Ethiopia

Ketema Belete and Tamado Tana

College of Agriculture and Environmental Science, Department of Plant Science, Haramaya University, Dire Dawa, Ethiopia

Index

Printed in the USA
CPSIA information can be obtained
at www.ICGtesting.com
JSHW051437221024
72173JS00006B/1495